Laboratory Measurement of Lipids, Lipoproteins and Apolipoproteins

Edited by

Nader Rifai

G. Russell Warnick

AACC Press

2101 L Street, N.W., Suite 202

Washington, DC 20037-1526

Library of Congress Cataloging-in-Publication Data

Laboratory measurement of lipids, lipoproteins, and apolipoproteins /
edited by Nader Rifai, G. Russell Warnick.
p. cm.
Includes bibliographical references.
ISBN 0-915274-74-4
1. Blood lipids—Analyses. 2. Blood lipoproteins—Analysis.
3. Apolipoproteins—Analysis. I. Rifai, Nader. II. Warnick, G. Russell.
RB46.L33 1994
616.07'561—dc20

94-16803
CIP

Cover design by Jim DeVall
Editorial/production supervision by Sterling Publishing Services

ISBN 0-915274-74-4

Printed in the United States of America

This book is dedicated to all those behind the scenes who have contributed to its publication: the lipid researchers, many of whom labored for years in obscurity until cholesterol entered the limelight; our technical and administrative staff members; and not the least, our wives, Dorothy and Nancy, and our children for their support and the many sacrifices they have made over the years.

Contributors

Joseph D. Artiss, Ph.D.
Associate Professor, Department of Pathology
Wayne State University School of Medicine, and
Technical Director, Chemistry Laboratory
DMC University Laboratories
Detroit, MI

Melissa A. Austin, Ph.D.
Associate Professor, Department of Epidemiology
School of Public Health and Community Medicine
University of Washington
Seattle, WA

Paul Bachorik, Ph.D.
Professor , Department of Pediatrics, and
Associate Professor, Department of Pathology
Lipid Research-Atherosclerosis Unit
The Johns Hopkins University School of Medicine
Baltimore, MD

John D. Belcher, Ph.D.
Assistant Professor, Division of Epidemiology
School of Public Health
University of Minnesota
Minneapolis, MN

John D. Brunzell, M.D.
Professor, Department of Medicine
University of Washington School of Medicine
Seattle, WA

Thomas G. Cole, Ph.D.
Research Associate Professor, Department of Medicine, and
Director, Core Laboratory
Lipid Research Center
Washington University School of Medicine
St. Louis, MO

Gerald R. Cooper, M.D., Ph.D.
Research Medical Officer, Division of Environmental Health Laboratory Sciences
National Center for Environmental Health
Centers for Disease Control and Prevention
Atlanta, GA

Ben Cost, M.D.
TNO Institute for Preventive Health Research
Leiden, The Netherlands

Pierre N.M. Demacker, Ph.D.
Laboratory of General Internal Medicine
University Hospital Nijmegen
Nijmegen, The Netherlands

Marek H. Dominiczak. M.D.
Consultant Biochemist, Western Infirmary Glasgow, and
Hon. Clinical Senior Lecturer, University of Glasgow
Scotland, UK

D. Robert Dufour, M.D.
Associate Professor, Department of Pathology
The George Washington University School of Medicine, and
Adjunct Associate Professor, Department of Pathology
Uniformed Services University of the Health Sciences, and
Chief, Laboratory Service
Veterans Affairs Medical Center
Washington, DC

Donald H. Feldbruegge, M.S.
Administrative Program Specialist, Clinical Chemistry Laboratory
University of Wisconsin Hospital and Clinics
Madison, WI

Gregory F. Grinstead, Ph.D.
Clinical Chemist
Marshfield Laboratories
Marshfield, WI

David J. Hassemer, M.S.
Assistant Director, Outreach Programs
State Laboratory of Hygiene
University of Wisconsin
Madison, WI

Louis M. Havekes, Ph.D.
TNO Institute for Preventive Health Research
Leiden, The Netherlands

L. Omar Henderson, Ph.D.
Supervisory Research Chemist,
Division of Environmental Health Laboratory Sciences
National Center for Environmental Health
Centers for Disease Control and Prevention
Atlanta, GA

John E. Hokanson, M.P.H.
Research Associate, Department of Medicine
University of Washington School of Medicine
Seattle, WA

Paul N. Hopkins, M.D., M.S.P.H.
Associate Professor, Cardiovascular Genetics
Department of Internal Medicine, Cardiology Division
University of Utah School of Medicine
Salt Lake City, UT

Ishwarlal Jialal, M.D. M.R.C. Path., DABCC, FACB, FACN
Associate Professor, Departments of Pathology and Internal Medicine, and
Director, Clinical Chemistry Laboratory
University of Texas Southwestern Medical Center
Dallas, TX

Mary M. Kimberly, Ph.D.
Coordinator, Cholesterol Reference Method Laboratory Network
Division of Environmental Health Laboratory Sciences
National Center for Environmental Health
Centers for Disease Control and Prevention
Atlanta, GA

Sigrid G. Klotzsch, M.S.
Senior Chemist
Schiapparelli Biosystems, Inc.
ENI - Diagnostics Division
Fairfield, NJ

Martin H. Kroll, M.D.
Associate Professor, Department of Pathology, and
Associate Director, Clinical Chemistry
The Johns Hopkins University Hospital
Baltimore, MD

Christine Labeur, Ph.D.
Assistant, Department of Clinical Biochemistry
A.Z. St.-Jan
Brugge, Belgium, and
Innogenetics N.V.
Gent, Belgium

Daniel M. Levine, Ph.D.
Director, Clinical Research Laboratory
The Rogosin Institute
The New York Hospital - Cornell Medical Center
New York, NY

Giuseppe Lippi, M.D.
Post-doctoral Fellow, Northwest Lipid Research Laboratories
University of Washington School of Medicine
Seattle, WA

Paola Lombardi, Ph.D.
TNO Institute for Preventive Health Research
Leiden, The Netherlands

Santica M. Marcovina, Ph.D., D.Sc.
Research Professor, Department of Medicine
Division of Metabolism, Endocrinology and Nutrition
Director, Core Laboratory, Northwest Lipid Research Laboratories
University of Washington School of Medicine
Seattle, WA

Judith R. McNamara, M.T. (ASCP)
Senior Research Assistant, Lipid Metabolism Laboratory
USDA Human Nutrition Research Center on Aging
Tufts University
Boston, MA

Matthew J. McQueen, M.D., Ph.D.
Professor, Department of Pathology
McMaster University
Chief, the Department of Laboratory Medicine
Hamilton Civic Hospitals - Hamilton General Division, and
Hamilton, Ontario, Canada

W. Greg Miller, Ph.D.
Professor, Department of Pathology, and
Director, Clinical Chemistry Laboratory
Medical College of Virginia
Virginia Commonwealth University
Richmond, VA

Christoph Möllers, Dipl. Chem.
Assistant Lecturer, Institut für Klinische Chemie
und Laboratoriumsmedizin
University of Regensburg
Germany

Gary L. Myers, Ph.D.
Chief, Special Activities Branch
Division of Environmental Health Laboratory Sciences
National Center for Environmental Health
Centers for Disease Control and Prevention
Atlanta, GA

James D. Otvos, Ph.D.
Professor, Department of Biochemistry
North Carolina State University
Raleigh, NC

Terry Pry, Ph.D.
Manager, Research and Development Laboratory
Abbott Diagnostics
Abbott Park, IL

Nader Rifai, Ph.D.
Associate Professor, Department of Pathology
Harvard Medical School, and
Director, Clinical Chemistry Laboratory
Children's Hospital
Boston, MA

Maryvonne Rosseneu, D.Sc.
Director, Department of Clinical Biochemistry
A.Z. St.-Jan
Brugge, Belgium

Cristine Scaccini, Ph.D.
Visiting scientist from the National Institute of Nutrition
Rome, Italy

Ernst J. Schaefer, M.D.
Professor, Departments of Medicine and Nutrition, and
Director, Lipid Research Laboratory and
Lipid and Heart Disease Prevention Clinic,
New England Medical Center, School of Medicine, and
Director, Lipid Metabolism Laboratory
USDA Human Nutrition Research Center on Aging
Tufts University
Boston, MA

Gerd Schmitz, Prof. Dr. Med.
Director, Institut für Klinische Chemie und Laboratoriumsmedizin
University of Regensburg
Germany

Evan A. Stein, M.D., Ph.D.
President, Medical Research Laboratories, and
Director, The Christ Hospital Cardiovascular Research Center, and
Clinical Professor, Department of Pathology and Laboratory Medicine
University of Cincinnati Medical Center
Cincinnati, OH

G. Russell Warnick, M.S., M.B.A.
President, Pacific Biometrics, Inc.
Pacific Biometrics Research Foundation
Seattle, WA

Donald A. Wiebe, Ph.D.
Associate Professor, Department of Pathology and
Laboratory Medicine, and
Associate Director, Clinical Chemistry Laboratory
University of Wisconsin Hospital and Clinics
Madison, WI

Lily H. Wu, Ph.D.

Associate Professor, Departments of Pathology and Internal Medicine, and
Director, Department of Cardiovascular Genetics
Biochemistry Laboratory and Reagent Production Laboratory
Associated Regional University Pathologists (ARUP)
University of Utah School of Medicine
Salt Lake City, UT

James T. Wu, Ph.D.

Professor, Department of Pathology, and
Director, Special Chemistry Laboratory and Regional Development Laboratory
Associated Regional University Pathologists (ARUP)
University of Utah School of Medicine
Salt Lake City, UT

Bennie Zak, Ph.D.

Emeritus Professor, Department of Pathology
Wayne State University School of Medicine, and
Consultant, Clinical Chemistry Laboratory
DMC University Laboratories
Detroit, MI

Joachim Ziegenhorn, Ph.D.

Vice President, Quality Assurance
Boehringer Mannheim GmbH
Penzberg, Germany

Contents

Preface

Lipids and lipoproteins are no longer esoteric substances of research interest alone. They have become household words and of great interest to the clinician. Their assessment has taken us far beyond the measurement of serum cholesterol and has allowed much more specific linkage of lipid disorders to genetic and acquired diseases. The lipoproteins of blood plasma carry the cholesterol as one of several lipids. The protein components of lipoproteins are structurally necessary for the synthesis and metabolism of these particles. Major metabolic roles for the individual lipoproteins are now understood; however, additional knowledge is needed in the genetic, biochemical, clinical, and epidemiological spheres.

This book, *Laboratory Measurement of Lipids, Lipoproteins and Apolipoproteins*, provides a broad overview for individuals entering the field of lipoproteins with a basic knowledge of biochemistry. It summarizes the relevant physiological and clinical knowledge and reviews a series of very practical issues that must be understood if accurate quantitation of laboratory parameters is to be achieved. The individual chapters cover modern analytic techniques that are used in thousands of laboratories, such as the measurement of cholesterol, triglycerides, and low- and high-density lipoproteins. In addition, this volume contains detailed reviews of the assays for the individual subclasses of these major lipoprotein species as well as the apolipoproteins and even the products of oxidative damage inflicted on the lipoproteins. Furthermore, new and unusual techniques for assessing the lipoproteins are discussed, including those using their isoelectric, immunochemical, or flotation properties.

Every clinical laboratory involved in lipoprotein measurements will find this book useful. The information herein should enhance the understanding of quality control and improve overall laboratory performance. Laboratories that are dedicated to research on lipoprotein structure and metabolism will find this book to provide a rapid introduction to many of the issues relevant to specific projects.

W. Virgil Brown, M.D.

Introduction

The last decade has brought substantial progress in the measurement and utilization of the lipid analytes in clinical practice. The American Association for Clinical Chemistry (AACC), primarily through its Lipids and Lipoproteins Division, has been an integral part of and has made a major contribution to this progression. Just ten years ago, the Lipid Research Clinics Program reported the findings of its Coronary Primary Prevention Trial, observing that decreasing cholesterol values could reduce the incidence of coronary heart disease. This historic observation was confirmed by several other studies of the effects of drug- and diet-mediated cholesterol lowering. Subsequently, a consensus conference sponsored by the National Heart, Lung and Blood Institute recommended launching a national intervention program. The result was the National Cholesterol Education Program, a beneficial program that has been emulated by other countries. The national and international effort to combat coronary heart disease has generated considerable enthusiasm, which we share. By some estimates, this effort may result in one of the greatest contributions to public health that the field of clinical chemistry has made to date.

The AACC's Lipids and Lipoproteins Division itself grew out of the activities of the Lipid Research Clinics Program, which fostered a highly successful laboratory collaboration among academic scientists who conducted the studies, those in industry who supported the studies, and especially the Centers for Disease Control and Prevention, which had long promoted standardization and method improvement. Division members were involved in the various expert panels of the National Cholesterol Education Program, which established analytical performance goals and guidelines for improving lipid and lipoprotein measurement. Division-sponsored educational workshops and symposia for practicing laboratorians and industry also contributed to the evolution.

The predecessor to this book, *Methods for Clinical Laboratory Measurement of Lipid and Lipoprotein Risk Factors,* published in 1991, had its genesis in the Lipids and Lipoproteins Division and in the atmosphere of method improvement engendered by the National Cholesterol Education Program. The first book, which comprised nine chapters covering the basic lipid analytes, has sold over 2,000 copies and continues to be sought after today. The key chapters on the basic lipid/lipoprotein analytes were authored by ad hoc committees previously convened by the Division to recommend improvements in the methods. The book was published in soft-cover format to keep it affordable and to signal the intent to update the book as measurement technologies evolved.

This second book builds on the first. The basic chapters from the first book have been revised and updated to reflect the considerable progress that has occurred in the intervening three years. In addition, the scope of the initial text has been expanded considerably with the inclusion of nine new chapters covering topics of increasing clinical interest, such as analytes in transition from research to clinical application, the effect of matrix interactions on standardization, and an overview of research methods.

We trust that this new book will be equally valuable to laboratorians. We wish to thank all who have contributed their time, knowledge, and experience to producing the individual chapters. All contributing authors share a common commitment to developing and promoting quality methods that produce clinically useful results.

Nader Rifai
G. Russell Warnick

CHAPTER 1

Apolipoproteins and Lipoproteins in Human Plasma

Marek H. Dominiczak

INTRODUCTION

❖ Lipids play an important role in cell metabolism. Fatty acids and their storage form, triglycerides, are a high-energy metabolic fuel. Cholesterol and phospholipids are essential components of cell membranes. Cholesterol is also a precursor of vitamin D and other steroids. Although cholesterol is a component of virtually every cell, its metabolism takes place mainly in the liver. Cholesterol is delivered to the liver either from the intestine following dietary intake or transported from extrahepatic tissues. It is removed from the liver through incorporation into lipoproteins, by conversion into bile acids, and by secretion into bile.

Since the hydrophobic character of lipids prevents their direct transport in plasma, they are transported between tissues and organs in the form of particles known as lipoproteins. Lipoproteins are not metabolically inert; their metabolism involves enzymes such as lipoprotein lipase (LL), hepatic triglyceride lipase (HTGL), and lecithin:cholesterol acyltransferase (LCAT). In addition, there is an exchange of components between lipoprotein particles facilitated by the cholesteryl ester transfer protein (CETP).

Each lipoprotein contains protein components called apolipoproteins. Apolipoproteins not only channel particle metabolism by binding to specific receptors, but also act as cofactors for enzymes. There is a constant traffic of apolipoproteins between different particles. Lipoprotein receptors present on plasma membranes control the rate of cellular uptake and degradation of lipoprotein particles and, indirectly, the rate of *de novo* cholesterol synthesis. At any time, human plasma contains several lipoproteins which remain in dynamic equilibrium.

This chapter provides an overview of lipoproteins and apolipoproteins and describes their interrelationships. Due to space constraints, only the present status of the field is described. A fascinating account of early work can be found in a recently published review.[1]

THE MEASUREMENT OF LIPOPROTEINS AND APOLIPOPROTEINS

In clinical practice, lipids and lipoproteins are measured to detect and diagnose lipid disorders (dyslipidemias), to assess the associated cardiovascular risk, and to determine subsequent treatment decisions.[2, 3]

Most of the epidemiological evidence on the relationship between plasma lipid concentrations and cardiovascular risk is based on relatively crude assessments of lipid metabolism such as the measurements of total triglyceride (TG) or total plasma cholesterol (TC) concentrations.[4] Although these measurements are satisfactory for population studies, a more accurate diagnosis of lipid/lipoprotein disorders is required to allow the clinician to identify subgroups of patients who, at equivalent concentrations of TC or TG, are at higher (or lower) than average cardiovascular risk.[5] For example, if patients with common hypercholesterolemia are taken as a reference point, at the same TC concentrations, a diagnosis of familial hypercholesterolemia or hyperapobetalipoproteinemia will signify a higher risk, whereas a diagnosis of hyperalphalipoproteinemia will indicate a lower risk.

Because of the multicomponent structure of lipoproteins, routine laboratory methods do not measure the concentration of these particles. Instead, a chosen component is measured and used as a marker of the concentration of the lipoprotein in question. Traditionally, cholesterol measurement has been used for this purpose. The availability of apolipoprotein assays and progress in their standardization is an incentive to use such assays to supplement measurements of TC, TG, and high-density lipoprotein cholesterol (HDL-C). They may be used as markers of lipoprotein concentration in conditions where apolipoprotein (apo) concentrations reflect lipoprotein particle number. For example, apo B-100 is a good marker of very-low-density lipoproteins (VLDL), low-density lipoproteins (LDL), and intermediate-density lipoproteins (IDL) concentrations. The clinical usefulness of apolipoprotein measurements depends on whether they provide an advantage over existing assays in terms of cardiovascular risk assessment[6] (see discussion below). Finally, apolipoproteins may be measured together with lipid variables to diagnose specific lipid disorders such as hyperapobetalipoproteinemia, familial dyslipidemia, or apolipoprotein deficiencies, or to gain insight into the composition of an investigated lipoprotein. The functions of apolipoproteins are summarized in Table 1–1.

THE LIPOPROTEIN PARTICLE

The hydrophobic core of a lipoprotein particle consists of cholesteryl esters and TG. The hydrophilic surface layer contains an amphipathic phospholipid bilayer, non-esterified cholesterol, and apolipoproteins. Lipoprotein particles are charac-

Table 1–1 ❖ The Apolipoproteins

Apolipoprotein	*Main Functions*	*Association with CHD Risk*
Apo AI	Structural for HDL. Ligand for HDL binding. LCAT cofactor.	Yes
Apo AII	Structural for HDL. Ligand for HDL binding. LCAT cofactor. Modulator of LL and HTGL activity (?)	No
Apo AIV	Ligand for HDL binding. LCAT activator.	No
Apo(a)	Structural for Lp(a). Possibly affects fibrinolysis.	Yes
Apo B-48	Structural for chylomicrons.	No
Apo B-100	Structural for VLDL, IDL and LDL. LDL receptor ligand.	Yes
Apo CI	LCAT and LL activator.	No
Apo CII	LCAT and LL activator.	No
Apo CIII	LL inhibitor. Modulator of uptake of triglyceride-rich lipoproteins by LRP.	No
Apo D	Unknown.	No
Apo E	Ligand for LDL receptors and for LRP.	Phenotype, yes
Apo F	Unknown.	No
Apo H	Unknown.	No
Apo J	Membrane protection (?)	No

HDL = High-density lipoprotein, LCAT = Lecithin:cholesterol acyltransferase, LL = Lipoprotein lipase, HTGL = Hepatic triglyceride lipase, VLDL = Very-low-density lipoprotein, LDL = Low-density lipoprotein, IDL = Intermediate-density lipoprotein, LRP = LDL receptor related protein

terized by their size, density, flotation constant, and electrophoretic mobility (see Table 1–2). When separated by ultracentrifugation, the main lipoprotein classes in plasma are chylomicrons, VLDL, IDL, LDL, and HDL.[7] The particle density increases from chylomicrons to HDL, and particle size decreases from chylomicrons to HDL.

Lipoprotein separation may be performed at several levels. Agar electrophoresis separates plasma lipoproteins into pre-beta VLDL, beta LDL, and alpha

Table 1–2 ❖ The Lipoproteins

Lipoprotein	*Density (kg/L)*	*Particle Diameter (nm)*	*Flotation Rate (Sf)*	*Electrophoretic Mobility*
Chylomicrons	< 0.95	80–1200	> 400	Origin
VLDL	0.95–1.006	30–80	60–400	Pre-beta
IDL	1.006–1.019	23–35	20–60	Broad beta
LDL	1.019–1.063	18–25	0–20	Beta
HDL	1.063–1.21	5–12	0–9	Alpha

VLDL = Very-low density lipoprotein, IDL = Intermediate-density lipoprotein, LDL = Low-density lipoprotein, HDL = High-density lipoprotein

HDL fractions.[1] An ultracentrifugal separation method performed according to the Lipid Research Clinics protocol, known as beta-quantification,[8, 9] separates VLDL and combined LDL and HDL fractions. HDL-C is then measured by precipitation and LDL-C is calculated. VLDL, IDL, and LDL (and their subfractions) can be further separated by flotation ultracentrifugation.[7] Discrete subpopulations of LDL[10] and HDL[11] particles can be identified by non-denaturing gradient polyacrylamide gel electrophoresis. Details of these procedures are described elsewhere in this book.

THE METABOLISM OF LIPOPROTEINS

An overview of cholesterol metabolism is given in Figure 1–1. Dietary lipids are transported from the intestine to the liver via the exogenous pathway, and lipids synthesized in the hepatocytes are transported from the liver to peripheral tissues via the endogenous pathway. These two pathways partially overlap at the stage of hydrolysis of TG-rich particles by LL and HTGL. In addition to these pathways, HDL transports lipids from peripheral tissues back to the liver. This is known as reverse cholesterol transport. Thus, the final extrahepatic tissue cholesterol concentration, and probably the degree of its arterial deposition, may be, at least in part, a result of a balance between these processes.

CHYLOMICRONS

The exogenous pathway of lipoprotein metabolism is illustrated in Figure 1–2.

Figure 1–1 ❖ An Overview of Cholesterol Metabolism

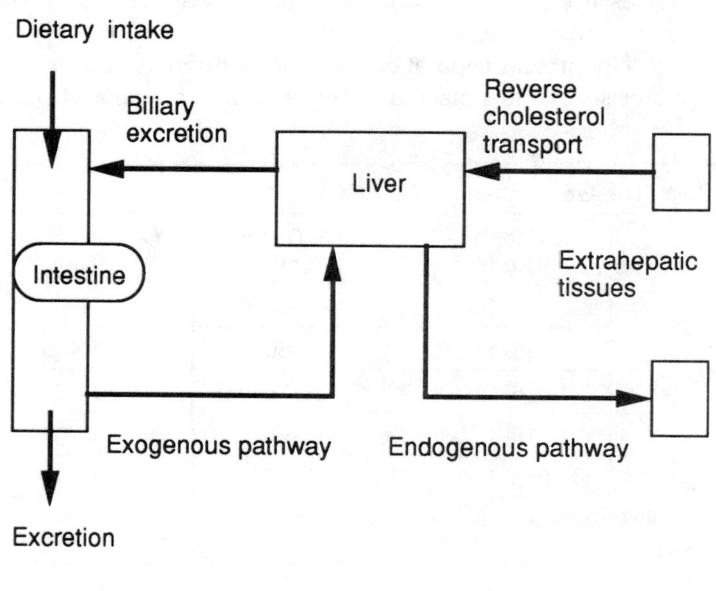

Figure 1-2 ❖ Exogenous Pathway of Lipoprotein Metabolism

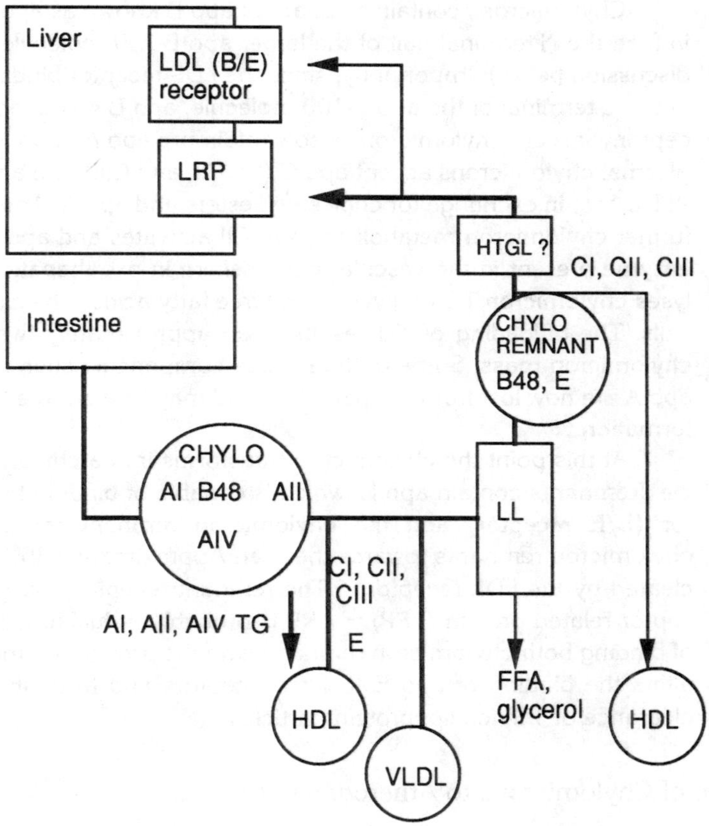

CHYLO = Chylomicrons   LL = Lipoprotein lipase   FFA = Free fatty acids
HTGL = Hepatic triglyceride lipase   LRP = LDL receptor-related protein

The relative size of the lipoprotein particles shown is not to scale.

Dietary cholesterol, mostly as cholesteryl esters, is absorbed in the intestine.[12] Esters are hydrolyzed by cholesterol esterase to release free cholesterol and fatty acids. TG are hydrolyzed by pancreatic and intestinal lipases and absorbed mainly as free fatty acids and monoglycerides. Long-chain fatty acids and cholesterol are then re-synthesized in the intestinal epithelial cells. TG, TC, and apolipoproteins are assembled into large lipoprotein particles known as chylomicrons. Chylomicrons, composed of more than 90% TG, are secreted into the lymph and enter plasma through the thoracic duct. They are normally present in plasma in the postprandial state only, and, if present in high concentration, can add a milky appearance to plasma. The presence of chylomicrons is reflected by an increased plasma TG concentration; if a large amount of chylomicrons is present, there will also be a considerable increase in plasma TC.

Chylomicrons contain a variant of apo B known as apo B-48.[13] Apo B-48 is, in fact, the N-terminal half of the larger apo B-100 molecule present in LDL (see discussion below). Importantly, since the LDL receptor binding domain is present in the C-terminal of the apo B-100 molecule, apo B-48 cannot bind to the LDL receptor. Nascent chylomicrons also contain pro-apo AI, apo AII, and apo AIV.[14] In plasma, chylomicrons accept apo C (CI, CII, and CIII) and apo E from HDL[15] and VLDL[16, 17] in exchange for cholesteryl esters and apo A. Apo C and apo E control further chylomicron metabolism. Apo CII activates and apo CIII inhibits[18] LL, an enzyme present in the vascular endothelium in extrahepatic tissues.[19] LL hydrolyses chylomicron TC to glycerol and free fatty acids, which are then taken up by cells. The offloading of TG results in an approximately twenty-fold decrease in chylomicron mass. Some of the surface components such as phospholipids and apo A are now lost from the particle[20] and may be a source for intravascular HDL formation.[21]

At this point the chylomicron transforms into a chylomicron remnant particle. Remnants contain apo E, which is capable of binding to both the LDL receptor (B/E receptor) and the chylomicron remnant receptor. Apo E targets chylomicron remnants towards the liver. Approximately 35% of chylomicrons are cleared by the LDL receptor.[22] The remnant receptor appears to be the LDL-receptor related protein (LRP).[23] LRP is probably a dual-function receptor capable of binding both chylomicron remnants and the protein α-2-macroglobulin. Apo CI alters the binding of apo E to liver receptors and through this may inhibit the clearance of TG-rich lipoprotein particles.

Relevance of Chylomicrons to Atherogenesis

The postprandial (up to 6–8 h after a meal) increase in chylomicron and TG-rich remnant concentration may exceed the critical level above which deposition of these particles or their components in the arterial wall would occur.[12] Postprandial lipemia may also induce changes in LDL and HDL. However, no excess cardiovascular risk is observed in LL deficiency, where chylomicron concentrations are extremely high.

VLDL

The endogenous pathway of lipoprotein metabolism is illustrated in Figure 1–3. VLDL particles are assembled in the hepatocytes from TG, cholesteryl esters, and apo B-100, and are secreted through the Golgi apparatus. VLDL particles consist of several subpopulations of different sizes which, while smaller than chylomicrons, are also TG-rich. Each particle contains one apo B-100 molecule. Several subpopulations of VLDL particles of varying size exist in plasma. In common with chylomicrons, VLDL also exchanges components with plasma HDL. Cholesteryl esters, apo C (CI, CII, and CIII), and apo E are transferred from HDL to VLDL in exchange for TG.[17] Apo C- and E-enriched VLDL undergo hydrolysis by LL.[19] In

Figure 1–3 ❖ Endogenous Pathway of Lipoprotein Metabolism

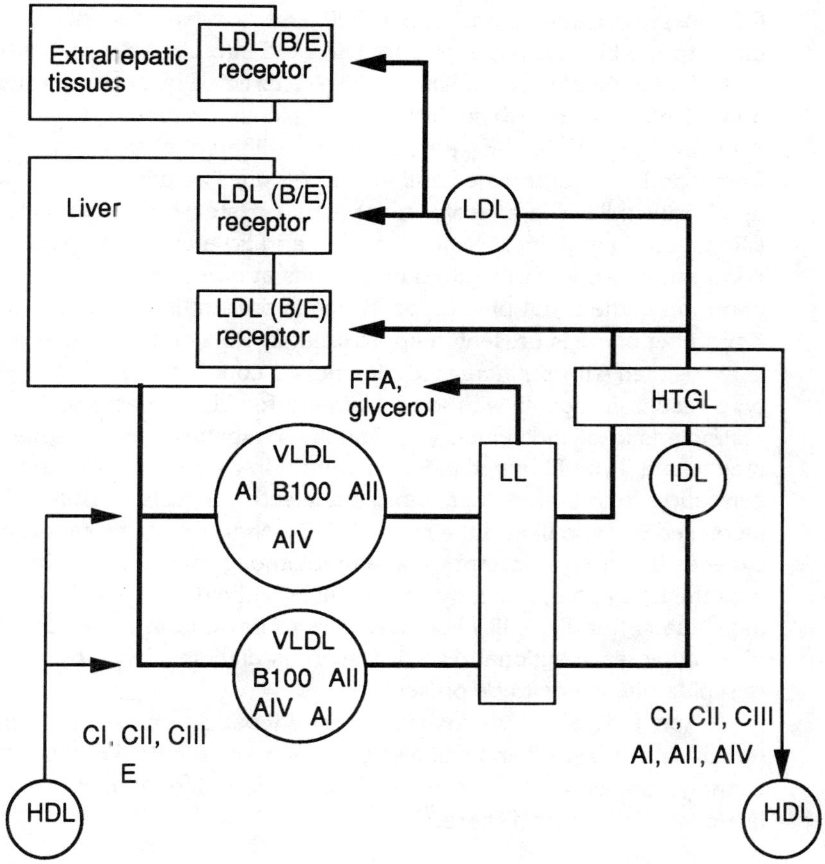

LL = Lipoprotein lipase FFA = Free fatty acids
HTGL = Hepatic triglyceride lipase

The relative size of the lipoprotein particles shown is not to scale.

further analogy to chylomicrons, TG hydrolysis results in the generation of smaller particles known as VLDL remnants. VLDL and VLDL remnants have higher affinity towards the LDL receptor than LDL, and may either be taken up by the liver or remain in plasma and be further hydrolyzed to yield IDL and finally LDL. HTGL is probably involved at this stage,[24] since its deficiency[25] results in the accumulation of VLDL remnants in plasma.

At any stage of the VLDL-LDL cascade, TG-rich particles may be removed from plasma by the liver.[26] Remnants of large VLDL appear to be directly metabolized, whereas remnants of small VLDL enter the IDL-LDL route.[27] In LDL receptor defects, less VLDL is removed directly and more is available for transformation into LDL. This results in an increase in plasma LDL concentration.[26]

IDL AND APO E

IDL possesses apo E and apo B-100 and, similarly to VLDL remnants, either binds to the LDL receptor or is transformed into LDL. Multiple subclasses of IDL exist.[28] The metabolism of IDL particles is directed by apolipoprotein E.[29] Several apo E isoforms exist which differ in their affinity for binding to the LDL receptor.[29] Apo E variants E4 and E2 are products of different alleles at one locus and differ from apo E3 in patterns of amino-acid residues at positions 112 and 158 in the apo E molecule. The E3 isoform contains a cysteine/arginine pattern at these positions, E4 an arginine/arginine pattern, and E2 a cysteine/cysteine pattern. Less common apo E variants with substitutions at other positions also exist. The most common is the E3E3 phenotype, which is present in 60% of the population. The E2E3 phenotype is present in approximately 25% of the population. The E4 allele is associated with an increased plasma TC concentration.[30, 31] The E2E2 phenotype results in apo E with lowest affinity for LDL receptors. This phenotype is characteristic of individuals with familial dysbetalipoproteinemia and is associated with a Type III hyperlipidemic pattern in plasma—that is, increased IDL concentration with broad beta band on electrophoresis.[32] Both TG and TC are increased in Type III and the ratio of VLDL-cholesterol to total plasma TG usually exceeds 0.7. It is important to note that although up to 1% of the population possess the E2E2 phenotype, overt Type III hyperlipidemia occurs in only 1 in 10,000 individuals. For Type III hyperlipidemia to develop in individuals with the E2E2 phenotype, an additional disorder, such as diabetes, hypothyroidism, or genetic dyslipidemia, needs to be present.

Apo E is also synthesized in extrahepatic tissues: the brain, kidney, adrenals, and spleen,[33] and it is also present in the amyloid plaque.[34, 35] Most interestingly, an ε4 allele has recently been found to occur more frequently in late-onset Alzheimer disease.[36, 37]

Relevance of IDL to Atherogenesis

IDL particles are atherogenic. β-VLDL can be directly taken up by macrophages and form foam cells.[38] Patients with Type III hyperlipidemia are prone to coronary heart disease (CHD). Diabetic individuals also have an increased IDL concentration,[39] and this may contribute to the increased incidence of CHD in the diabetic population.[40]

LDL AND APO B-100

IDL continue to lose TG, apo C, and apo E, probably through the interaction with HDL. This results in their transformation into LDL. LDL is the main carrier of cholesterol in plasma. LDL contain very little TG and free cholesterol, with cholesteryl esters being its major component. Apo B is virtually the only apolipoprotein present in LDL. Each LDL particle contains one molecule of apo B. Between 75% and 80% of LDL is metabolized in the liver.

LDL is metabolized mainly through a receptor pathway.[41] The LDL receptor binds apo B-100 and apo E. A detailed description of the LDL receptor is beyond the scope of this chapter; however, several excellent reviews have been published.[42-44] The LDL receptor undergoes regulation which controls the rate of LDL metabolism and, through this, its plasma concentration. The rate of receptor synthesis–and thus LDL clearance–is inversely related to the amount of intracellular free cholesterol. In normolipidemic individuals, about half of LDL is metabolized through the LDL receptor route.[45] LDL receptor synthesis is stimulated by thyroid hormone[46] and estrogens.[47] Mutations of the LDL receptor affect both LDL binding and receptor processing.[42-44]

In addition to changes in its concentration, plasma LDL may vary in composition. The ongoing transfer of cholesteryl esters from HDL to VLDL by CETP may indirectly affect LDL composition. The cholesterol content of LDL is increased when there is high CETP activity[48, 49] or high LCAT activity.[50] Both of these conditions result in increased cholesterol esterification, transfer of esters from HDL to TG-rich particles, and enrichment of LDL in cholesterol. It is also possible that cholesterol-poor LDL may be selectively cleared from plasma.[51] The enrichment of LDL in cholesteryl esters is reflected in an increase in the LDL-C/apo B ratio.

Since 90% of plasma apo B-100 is present in LDL,[52] the measurement of apo B reflects LDL concentration. An increased apo B concentration signals an increased number of LDL particles. It has been suggested that apo B measurement is the best discriminator for the identification of men with early-onset CHD.[53] It is also a predictor of risk of myocardial infarction[54] and of angiographically defined CHD.[55]

LDL Subclass Phenotypes

LDL particles in plasma are heterogenous. One scheme defines four LDL subclasses[56] which can be identified within the LDL density band. LDL_1 is the largest in size and most buoyant fraction, whereas LDL_3 and LDL_4 are small and dense. The predominance of LDL_1 and LDL_2 in plasma, which occurs in 75% of subjects, is known as pattern A, whereas the predominance of LDL_3 is called pattern B.[10] The LDL subclass phenotype is influenced by an allele with an autosomal-dominant inheritance. Individuals with pattern B are also characterized by higher plasma TG, higher VLDL concentration and VLDL mass, increased IDL concentrations, and lower HDL concentrations. Apo B concentration is higher and apo AI lower in pattern B. Changes associated with pattern B are known as atherogenic lipoprotein phenotype (ALP).[57, 58]

Disorders of LDL metabolism may result from defects in the LDL receptor, as in familial hypercholesterolemia (FH);[59] from defects in apo B, as in familial defective apo B-100 (FDB), where glutamine has been substituted by arginine at position 3500 of apo B-100;[60, 61] and from increased apo B synthesis or secretion, as in familial combined hyperlipidemia (FCH).[62, 63] In addition, hyperapobetalipoproteinemia, also due to overproduction of apo B-100, may occur at relatively low plasma TC concentrations[64-66] and is associated with increased cardiovascular risk. Diagnosis of hyperapobetalipoproteinemia is a major clinical reason for routine measurement of apo B.

Relevance of LDL to Atherogenesis

The atherogenic potential of LDL stems from its ability to accumulate in the extracellular matrix and in cells resident in the arterial intima. While the LDL receptor-mediated cellular uptake occurs in a controlled way due to receptor regulation, the LDL uptake mediated by the macrophage scavenger receptor is uncontrolled and results in cellular cholesterol overload, which is manifested by generation of foam cells.[67] Native LDL is not readily taken up by the macrophage; modification of LDL is required before this can take place. The most important known modification is LDL oxidation.[68] Nonenzymatic glycation of LDL may also be clinically significant.[69] There is still relatively little evidence that LDL oxidation occurs in plasma, although plasma auto-antibodies to oxidized LDL have recently been identified.[70] The bulk of LDL oxidation seems to occur extravascularly after LDL has penetrated the endothelial barrier. Oxidized LDL is present in the arteriosclerotic plaque.[71]

CHD is more prevalent in patients with smaller LDL particles of higher density.[72] Pattern B is also prevalent in diabetic and nondiabetic patients with hypertriglyceridemia; in diabetes, the LDL subfraction pattern is not related to LDL glycation.[73] It has been hypothesized that small, dense LDL more easily penetrate the endothelial barrier. This would facilitate their deposition in the arterial intima. In addition, it appears that small, dense LDL are more susceptible to oxidation than LDL_1.[74]

HDL

HDL is the smallest and most dense lipoprotein particle. Proteins account for 50% of the HDL mass. There are two main HDL subclasses, HDL_2 and HDL_3. HDL_2 is larger and richer in lipids than HDL_3. HDL_3 is smaller and denser.[75] At least seven HDL subfractions have been identified by gradient polyacrylamide gel electrophoresis.[11] Apo A (AI and AII) are the main apolipoproteins of HDL, but HDL also contains apo C and apo E. Approximately 60% of HDL particles contain apo AI and AII; the rest contain only apo AI.[76]

The role of HDL appears to be the transport of cholesterol from peripheral cells to the liver, known as reverse cholesterol transport.[75] Reverse cholesterol transport is illustrated in Figure 1–4. The mechanisms of reverse transport have not been completely elucidated. The liver and intestine secrete nascent, discoid HDL containing apo A and apo E. Some apo-C-containing discoid forms are generated from VLDL and chylomicrons during TG hydrolysis. Free cholesterol released from cells is taken up by nascent HDL and converted into cholesteryl ester by LCAT.[50] This converts nascent HDL into a spherical form. Apo AI and apo AII are cofactors of LCAT. Also, LCAT is inhibited by cholesteryl esters. The newly formed cholesteryl esters may either be transported to the liver or exported to TG-rich lipoprotein particles. Transfer of cholesteryl esters to VLDL and chylomicrons occurs through the action of CETP[77] in exchange for TG. CETP activity tends to be high in conditions associated with low HDL,[78, 79] in dyslipidemia,[80] and in dia-

Figure 1-4 ❖ *Reverse Cholesterol Transport*

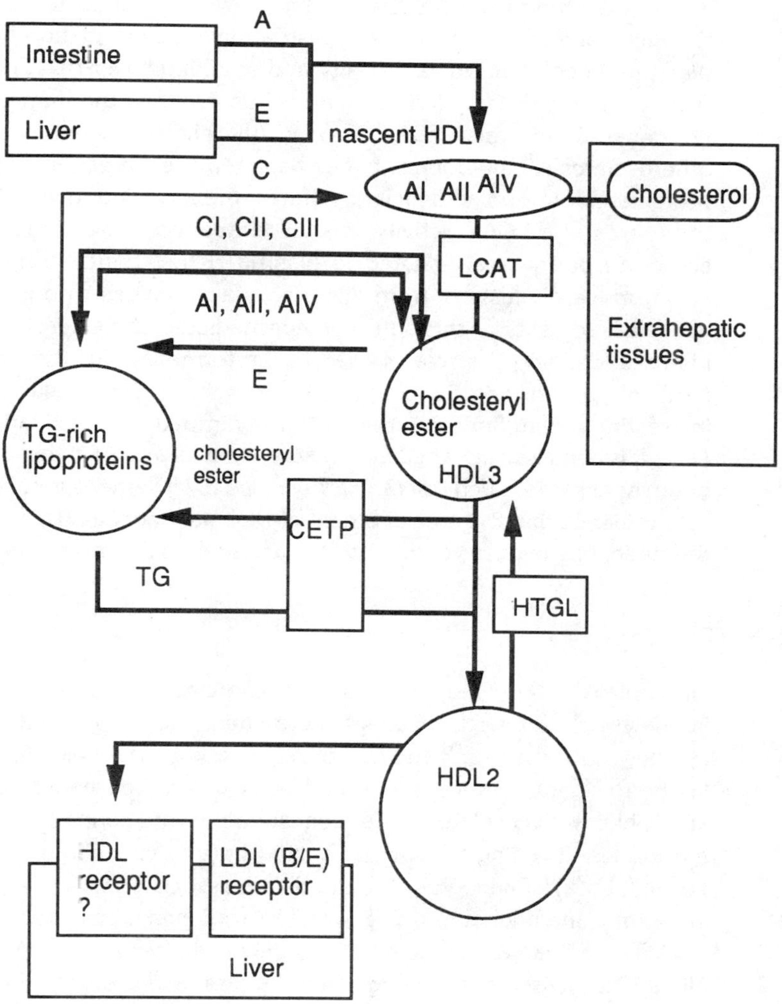

LCAT = Lecithin:cholesterol acyltransferase TG = Triglycerides
HTGL = Hepatic triglyceride lipase CETP = Cholesteryl ester transfer protein

The relative size of the lipoprotein particles shown is not to scale.

betes mellitus.[81] This process results in the enrichment of HDL in TG and an increase in HDL size.[80] The liver HDL uptake is probably mediated by a specific receptor(s) capable of binding apo A,[82] but due to its apo E content, some HDL may be taken up by either the LDL receptor or the remnant receptor. Cholesteryl esters transferred to VLDL will eventually be found in LDL.

Relevance of HDL to Atherogenesis

There is a strong negative relationship between plasma HDL-C concentration and the incidence of CHD.[83] There is also an inverse correlation between HDL-C and plasma TG concentration. Obesity and smoking lower HDL. Low HDL-C has also been observed in Type II diabetes mellitus. On the other hand, exercise and ethanol consumption raise HDL.[84, 85] Apo AII deficiency does not lead to premature atherosclerosis,[86] but longevity has been reported in patients with hyperalphalipoproteinemia. In apo AI deficiency states, the association with CHD was observed to be greater if LCAT activity was also impaired.[5] The aortic deposition of lipids seems to be inversely related to plasma HDL-C concentration.[75] HDL particles can remove cholesterol from lipid-rich macrophages through either scavenging the released lipids or through receptor-mediated cholesterol efflux.[87] The mechanisms underlying the role of HDL in the formation (or the prevention of formation) of atherosclerotic plaques have not yet been elucidated.[76] The most interesting recent finding is that HDL may impede LDL oxidation.[88]

It is controversial whether measurements of apo AI in the assessment of cardiovascular risk add much discriminatory value to HDL measurement.[6, 54] There is also little evidence that the measurement of HDL subfractions (HDL_2 and HDL_3) is more discriminating in terms of cardiovascular risk than the measurement of total HDL.

LIPOPROTEIN(a)

Lipoprotein(a) [Lp(a)] appears to link lipid metabolism with vascular thrombosis and fibrinolysis.[89] On electrophoresis, Lp(a) migrates as pre-beta lipoprotein. Lp(a) is synthesized in the liver[90] and comprises an apo B-100 molecule linked by a disulfide bridge to apo(a). Apo(a) has a molecular weight between 300,000 and 800,000[91] and is highly glycosylated. Lp(a) concentrations are controlled by a series of autosomal alleles at a single locus designated $Lp(a)^F$, $Lp(a)^B$, $Lp(a)^{S1}$, $Lp(a)^{S2}$, $Lp(a)^{S3}$, $Lp(a)^{S4}$, $Lp(a)^0$. Phenotypes S3 and S4 are associated with low Lp(a) concentrations in plasma, and phenotypes B, S1, and S2 with high concentrations.[89]

There is an antigenic similarity between plasminogen and apo(a).[92] Apo(a) has a high degree of homology with the protease domain and kringles 4 and 5 of plasminogen.[93] There are 37 copies of plasminogen kringle 4 in the apo(a) molecule. In spite of this, apo(a) is resistant to cleavage by plasminogen activators because of one amino-acid substitution within the protease domain. Apo(a) has, however, some proteolytic activity against artificial substrates.[94, 95]

Lp(a) inhibits clot lysis in plasma. By competing with plasminogen for its binding sites on the cell surface, it decreases the plasminogen activation by tissue plasminogen activator.[96, 97] Lp(a) also binds to fibrin.[98] Lp(a) binds to the LDL receptor, although with lesser affinity than apo B.[99, 100]

Relevance of Lp(a) to Atherogenesis

Lp(a) concentration is higher in patients with CHD and has been shown to be an independent risk factor for CHD.[101–103] Lp(a) concentrations seem to be important in the assessment of cardiovascular risk in women of all ages and in younger men.[101] Lp(a) is increased in individuals who have a parental history of premature CHD,[104, 105] and an association between Lp(a) concentration and cerebrovascular disease has been observed.[106] There is evidence that Lp(a) is deposited in the arterial intima and in the atherosclerotic plaques. An increased fatty streak formation was observed in a transgenic mouse line expressing human apo(a).[107] The ability of Lp(a) to deposit in the extracellular matrix, together with its potential interference with fibrinolytic mechanism, suggest a direct role in atherogenesis.

RELATIONSHIPS BETWEEN DISORDERS OF LIPOPROTEIN AND APOLIPOPROTEIN METABOLISM

The relationship between plasma TC and cardiovascular risk was established in large-scale prospective epidemiological studies.[2-4] In parallel, in recent years there has been enormous progress in the characterization of discrete phenotypically and genetically defined lipid disorders and of the associated cardiovascular risk. This led to increasing knowledge of interrelationships between existing lipid disorders and between lipid disorders and other conditions.

The presence of familial hypercholesterolemia and atherogenic apo(a) phenotypes are associated with increased risk of CHD.[104, 108] As mentioned above, the expression of Type III hyperlipidemia in individuals with E_2E_2 phenotype is dependent on the background presence of other disorders.[32] The atherogenic lipid phenotype probably predisposes to the expression of the gene(s) responsible for lipid and apo B increases in familial combined hyperlipidemia.[109] Lipid abnormalities are linked to hyperinsulinemia and insulin resistance. The association of hyperinsulinemia and hypertension is well established. Syndrome X, described by Reaven, consists of glucose intolerance, hypertension, increased plasma TG concentration, and low HDL-C[110] and carries an increased risk of CHD. LDL phenotype B is also associated with insulin resistance.[111] The interactions described above emphasize the importance of pattern recognition in the comprehensive assessment of cardiovascular risk.

FUNCTIONS OF APOLIPOPROTEINS OTHER THAN LIPID TRANSPORT AND METABOLISM

There are several lipoproteins for which a lipid transport/metabolism function has not been defined.[103, 112]

Recently characterized apo J is associated with the HDL fraction and can be involved in inhibition of complement-mediated cell lysis and in sperm maturation. It is hypothesized that due to its capacity to interact with hydrophobic molecules, apo J has a membrane-protective role.[112] Apart from plasma and urine, apo J has been detected in cerebrospinal fluid, breast milk, and seminal fluid.

Similar to apo J and apo E (discussed earlier), apo D is expressed in a variety of extrahepatic tissues such as spleen, adrenal, lungs, brain, and testes.[113] Finally, apo H, a 43–54 kDa protein present in all main lipoprotein fractions, has been reported to affect platelet response to ADP.[112] The functions of apo F and apo G remain unknown.

CONCLUSIONS

It is now well established that lipids participate in the formation of atherosclerotic plaque. Lipid abnormalities are a major determinant of cardiovascular risk. Cardiovascular risk factors such as dyslipidemia, smoking, hypertension, and glucose intolerance are additive in terms of cardiovascular mortality.[2, 3] Although many relationships and risk determinants have been identified, the assessment of cardiovascular risk of an individual patient is still far from perfect. Further refinement of the laboratory diagnosis of disorders associated with cardiovascular risk is required.

Although it is now clear that apolipoproteins perform functions other than lipid transport and metabolism, their full biological significance remains to be elucidated. ❖

Acknowledgments: This work was partially supported by grants from the Chest Heart and Stroke Association (Scotland), the Kidney Research Fund, and the Scottish Hospitals Endowment Research Trust (SHERT). The author is grateful to Ms. S. Cathcart, Dr. E. Kilpatrick, and Dr. S. Kilpatrick for their helpful comments, and Ms. J. Gardiner for her excellent secretarial assistance.

REFERENCES

1. Fredrickson DS. Phenotyping. On reaching base camp (1950-1975). [Review]. Circulation 1993;87(Suppl III):III-1–III-15.
2. European Atherosclerosis Society. International task force for prevention of coronary heart disease. Prevention of coronary heart disease: scientific background and new clinical guidelines. Nutr Metab Cardiovasc Dis 1992;2:113–56.
3. The Expert Panel. Summary of the second report of the National Cholesterol Education Panel (NCEP) Expert Panel on Detection, Evaluation and Treatment of High Blood Cholesterol in Adults (Adult Treatment Panel II). JAMA 1993;269: 3015–23.
4. Rosenfeld L. Atherosclerosis and the cholesterol connection: evolution of a clinical application. [Review]. Clin Chem 1989;35:521–31.
5. Schaefer EJ, Genest Jr JJ, Ordovas JM, Salem DN, Wilson WF. Familial lipoprotein disorders and premature coronary artery disease. [Review]. Curr Opin Lipidol 1993;4:288–98.
6. Bhatnagar D, Durrington PN. Clinical value of apolipoprotein measurement. [Review]. Ann Clin Biochem 1991;28:427–37.
7. Havel RJ, Eder HA, Bragdon J. Distribution and chemical composition of ultracentrifugally separated lipoproteins in human serum. J Clin Invest 1955;34: 1345–53.
8. Manual of laboratory operations. Lipid Research Clinic Program. Revised. Washington DC: US Government Printing Office, 1982:63–77.
9. Cathcart S, Dominiczak MH. The measurement of lipoprotein subfractions in plasma using a tabletop ultracentrifuge. Ann Clin Biochem 1990;27:459–64.

10. Kraus RM. Low-density lipoprotein subclasses and risk of coronary artery disease. [Review]. Curr Opin Lipidol 1991;2:248–52.
11. Blanche PJ, Gong EL, Forte TM, Nichols AV. Characterisation of human high-density lipoproteins by gradient gel electrophoresis. Biochim Biophys Acta 1981; 665:408–19.
12. Patsch JR. Postprandial lipaemia. [Review]. In: Shepherd J, ed. Lipoprotein metabolism. Baillieres Clin Endocrinol Metab 1987;1:551–80.
13. Powell LM, Wallis SC, Pease RJ, Edwards YH, Knott TJ, Scott J. A novel form of tissue specific RNA processing produces apolipoprotein B48 in intestine. Cell 1987;50:831–40.
14. Beisiegel U, Utermann G. An apolipoprotein homolog of rat apolipoprotein AIV in human plasma: isolation and partial characterisation. Eur J Biochem 1979;93: 601–8.
15. Havel RJ, Kane JP, Kashyap ML. Interchange of apoproteins between chylomicrons and high density lipoproteins during alimentary lipemia in man. J Clin Invest 1973;52:32–8.
16. Blum CB. Dynamics of apolipoprotein E metabolism in humans. J Lipid Res 1982;23:1308-16.
17. Huff MW, Breckenridge WC, Strong WLP, Wolfe BM. Metabolism of apolipoproteins C-II, C-III and B in hypertriglyceridemic men: changes after heparin-induced lipolysis. Arteriosclerosis 1988;8:471–9.
18. Ginsberg HN, Le N-A, Goldberg IJ, Gibson JC, Rubinstein A, Wang-Iverson P et al. Apolipoprotein B metabolism in subjects with deficiency of apolipoproteins CIII and AI. J Clin Invest 1986;78:1287–95.
19. Olivecrona T, Bengtsson-Olivecrona G. Lipoprotein lipase and hepatic lipase. Curr Opin Lipidol 1990;1:220–30.
20. Schaefer EJ, Jenkins LL, Brewer HB. Human chylomicron apoprotein catabolism. Biochem Biophys Res Commun 1978;80:405–12.
21. Tall AR, Small DM. Plasma high-density lipoproteins. [Review]. N Engl J Med 1978;299:1232–6.
22. Thuren T, Wilcox RW, Sisson P, Waite M. Hepatic lipase hydrolysis of lipid monolayers. Regulation by apolipoproteins. J Biol Chem 1991;266:4853–61.
23. Beisiegel U, Weber W, Ihrke G, Herz J, Stanley KK. The LDL-receptor-related protein, LRP, is an apolipoprotein E-binding protein. Nature 1989;341:162–4.
24. Havel RJ. The formation of LDL: mechanisms and regulation. [Review]. J Lipid Res 1984;25:1570–6.
25. Clay HA, Hopkins GJ, Ehnholm C, Barter PJ. The rabbit as an animal model of hepatic lipase deficiency. Biochim Biophys Acta 1989;1002:173–81.
26. Grundy SM. Multifactorial etiology of hypercholesterolemia. [Review]. Arterioscler Thromb 1991;11:1619–35.
27. Packard CJ, Boag DE, Clegg RJ, Bedford DK, Shepherd J. Effects of 1, 2-cyclohexanedione modification on the metabolism of VLDL apoprotein B: potential role of receptors in IDL catabolism. J Lipid Res 1985;26:1058–67.
28. Musliner TA, Giotas C, Krauss RM. Presence of multiple subpopulations of lipoproteins of intermediate density in normal subjects. Arteriosclerosis 1986;6:79–87.
29. Mahley RW, Innerarity TL, Rall JSC, Weisgraber KH, Taylor JM. Apolipoprotein E: genetic variants provide insights into its structure and function. [Review]. Curr Opin Lipidol 1990;1:87–95.
30. Utermann G. Apolipoprotein E polymorphisms in health and disease. [Review]. Am Heart J 1987;113:433–40.
31. Enholm C, Lukka M, Kaussi T, Nikkila E, Utermann G. Apoprotein E polymorphism in the Finnish population: gene frequencies and relation to lipoprotein concentrations. J Lipid Res 1986;27:227–35.

32. Brewer HB, Zech LA, Gregg RE, Schwartz D, Schaeffer EJ. Type III hyperlipoproteinemia: diagnosis, molecular defects, pathology and treatment.[Review]. Ann Int Med 1983;98:623–40.
33. Blue ML, Williams DL, Zucker S, Khan SA, Blum CB. Apolipoprotein E synthesis in human kidney, adrenal gland and liver. Proc Natl Acad Sci USA 1983;80:283–7.
34. Strittmatter WJ, Saunders AM, Schmechel, Pericakvance M, Enghild J, Salvesen GS, et al. Apolipoprotein E: high-avidity binding to β-amyloid and increased frequency of type 4 allele in late onset familial Alzheimer disease. Proc Natl Acad Sci USA 1993;90:1977–81.
35. Strittmatter WJ, Weisgraber KH, Huang DY, Dong LM, Salvesen GS, Pericakvance M, et al. Binding of human apolipoprotein E to synthetic amyloid β peptide: isoform-specific effects and implications for late onset Alzheimer disease. Proc Natl Acad Sci USA 1993;90:8098–102.
36. Saunders AM, Schmader K, Breitner JCS, Benson MD. Brown WT, Goldfarb L, et al. Apolipoprotein ε4 allele distributions in late-onset Alzheimer's disease and in other amyloid-forming diseases. Lancet 1993;342:710–11.
37. Scott J. Apolipoprotein E and Alzheimer's disease. [Editorial]. Lancet 1993;342:696.
38. Koo C, Wernette-Hammond ME, Innerarity TL. Uptake of canine β-very low density lipoproteins by mouse peritoneal macrophages is mediated by a low density lipoprotein receptor. J Biol Chem 1986;261:11194–201.
39. Winocour PH, Durrington PN, Ishola M, Anderson DC. Lipoprotein abnormalities in insulin-dependent diabetes mellitus. Lancet 1986;1:1176–8.
40. American Diabetes Association. Consensus statement. Role of cardiovascular risk factors in prevention and treatment of macrovascular disease in diabetes. Diabetes Care 1989;12:573–9.
41. Brown MS, Goldstein JL. A receptor-mediated pathway for cholesterol homeostasis. Science 1986;232:34–47.
42. Bradley WA, Gianturco SH. Lipoprotein receptors. [Review]. Curr Opin Lipidol 1990;1:216–21.
43. Brown MS, Goldstein JL. Lipoprotein receptors in the liver. Control signals for plasma cholesterol traffic. [Review]. J Clin Invest 1983;72:743–7.
44. Havel RJ, Hamilton RL. Hepatocytic lipoprotein receptors and intracellular lipoprotein catabolism. [Review]. Hepatology 1988;8:1689–704.
45. Shepherd J, Packard CJ. Lipid transport through the plasma: the metabolic basis of hyperlipidaemia. [Review]. In: Shepherd J, ed. Lipoprotein metabolism. Baillieres Clin Endocrinol Metab 1987;1:495–514.
46. Thompson GR, Soutar AK, Spengel FA, Jadhav A, Gavigan SJ, Myant NB. Defects of receptor-mediated low density lipoprotein catabolism in homozygous familial hypercholesterolemia and hypothyroidism in vivo. Proc Natl Acad Sci USA 1981;78:2591–5.
47. Windler EE, Kovanen PT, Chao YS, Brown MS, Havel RJ, Goldstein JL. The estradiol-stimulated lipoprotein receptor of rat liver: a binding site that membrane mediates the uptake of rat lipoproteins containing apoprotein B and E. J Biol Chem 1980;255:10464–71.
48. Hesler CB, Swenson TL, Tall AR. Purification and characterization of a human plasma cholesterol ester transfer protein. J Biol Chem 1987;262:2275–82.
49. Inazu A, Brown ML, Hesler CB, Agellon LB, Koizumi J, Takata K et al. Increased high-density lipoprotein levels caused by a common cholesteryl-ester transfer protein gene mutation. N Engl J Med 1990;323:1234–8.
50. Glomset JA. The plasma lecithin: cholesterol acyltransferase reaction. [Review]. J Lipid Res 1968;9:155–67.

51. Vega GL, Grundy SM. Kinetic heterogeneity of low density lipoproteins in primary hypertriglyceridemia. Arteriosclerosis 1986;6:395–406.
52. Durrington PH, Bolton CM, Hartog M. Serum and lipoprotein apolipoprotein B levels in normal subjects and patients with hyperlipoproteinaemia. Clin Chim Acta 1978;82:151–60.
53. Kwiterowich PO, Coresh HH, Bachorik PS, Derby CA, Pearson TA. Comparison of plasma levels of apolipoproteins B and A-I and other risk factors in men and women with premature coronary artery disease. Am J Cardiol 1992;69:1015–21.
54. Stampfer MJ, Sacks FM, Salvini S, Willett WC, Hennekens CH. A prospective study of cholesterol, apolipoproteins and the risk of myocardial infarction. N Engl J Med 1991;325:373–81.
55. Schmidt SB, Wasserman AG, Muesing RA, Schlesselman SE, La Rosa JC, Roos AM. Lipoprotein and apolipoprotein levels in angiographically defined coronary atherosclerosis. Am J Cardiol 1985;55:1459–62.
56. Krauss RM, Burke DJ. Identification of multiple subclasses of plasma low density lipoproteins in normal humans. J Lipid Res 1982;23:97–104.
57. Austin MA, King M-C, Vranizan KM, Krauss RM. Atherogenic lipoprotein phenotype: a proposed genetic marker for coronary heart disease risk. Circulation 1990;82:495–506.
58. Nishina PM, Johnson JP, Naggert KJ, Krauss RM. Linkage of atherogenic lipoprotein phenotype to the low density lipoprotein receptor locus on the short arm chromosome 19. Proc Natl Acad Sci USA 1992;89:708–12.
59. Bild DE, Williams RR, Brewer HB, Herd JA, Pearson TA, Stein E. Identification and management of heterozygous familial hypercholesterolemia: summary and recommendations from an NHLBI workshop. Am J Cardiol 1993;72:1D–5D.
60. Innerarity TL, Mahley TL, Weisgraber KH, Bersot TP, Krauss RM, Vega GL et al. Familial defective apolipoprotein B-100: a mutation of apolipoprotein B that causes hypercholesterolemia. J Lipid Res 1990;31:1337–49.
61. Tybjaerg-Hansen A, Gallager A, Vincent J, Houlston R, Talmud P, Seed AM et al. Familial defective apolipoprotein B-100: detection in the United Kingdom and Scandinavia, and clinical characteristics of ten cases. Atherosclerosis 1990;80: 235–42.
62. Goldstein JL, Schrott HG, Hazzard WR, Bierman EL, Motulsky AG. Hyperlipidaemia in coronary heart disease. II. Genetic analysis of lipid levels in 176 families and delineation of a new inherited disorder, combined hyperlipidaemia. J Clin Invest 1973;52:1544–68.
63. Kissebah AH, Alfarsi S, Evans DC. Low density lipoprotein metabolism in familial combined hyperlipidemia: mechanism of the multiple lipoprotein phenotypic expression. Arteriosclerosis 1984;4:614–24.
64. Teng B, Sniderman AD, Soutar AK, Thompson GR. Metabolic basis of hyperapobetalipoproteinemia: turnover of apolipoprotein B in low density lipoproteins and its precursors and subfractions compared with normal and familial hypercholesterolemia. J Clin Invest 1986;77:663–72.
65. Sniderman A, Shapiro S, Marpole D, Skinner B, Teng B, Kwiterovich PO. Association of coronary atherosclerosis with hyperapobetalipoproteinemia (increased protein but normal cholesterol levels in human plasma low density (beta) lipoprotein). Proc Natl Acad Sci USA 1980;77:604–8.
66. Teng B, Thompson GR, Sniderman A, Forte TM, Krauss RM, Kwiterovich PO Jr. Composition and distribution of low density lipoprotein fractions in hyperapobetalipoproteinemia, normolipidemia and familial hypercholesterolemia. Proc Natl Acad Sci USA 1983;80;6662–6.
67. Badimon JJ, Fuster V, Chesebro JH, Badimon L. Coronary atherosclerosis. A multifactorial disease. [Review]. Circulation 1993;87 (suppl II): II-3–II-16.

68. Steinberg D, Parthasarathy S, Carew TE, Khoo JC, Witztum JL. Beyond cholesterol: modifications of low-density lipoprotein that increase its atherogenicity. [Review]. N Engl J Med 1989;320:915–22.
69. Witztum JL, Mahoney EM, Branks MJ, Fisher M, Elam R, Steinberg D. Nonenzymatic glucosylation of low-density lipoprotein alters its biologic activity. Diabetes 1982;31:283–91.
70. Salonen JT, Yla-Hertuala S, Yamamoto R, et al. Autoantibodies against oxidised LDL and progression of carotid atherosclerosis. Lancet 1992;339:883–7.
71. Yla-Hertuala S, Palinski W, Rosenfeld ME, et al. Evidence for the presence of oxidatively modified low density lipoprotein in atherosclerotic lesions of rabbit and man. J Clin Invest 1989;84:1086–95.
72. Austin MA, Breslow JL, Hennekens CH, Buring JE, Willett WC, Krauss RM. Low-density lipoprotein subclass patterns and risk of myocardial infarction. JAMA 1988;260:1917–21.
73. Stewart JM, Kilpatrick ES, Small M, Dominiczak MH. Low density lipoprotein particle size in type 2 diabetic patients and age matched controls. Ann Clin Biochem 1994 (in press).
74. De Graaf J, Hak-Lemmers HLM, Hectors MPC, Demacker PNM, Hendriks JCM, Stalenhoef AFH. Enhanced susceptibility to in vitro oxidation of the dense low density lipoprotein subfraction in healthy subjects. Arteriosclerosis 1991;11:298–306.
75. Miller NE. High-density lipoprotein: a major risk factor for coronary atherosclerosis. [Review]. In: Shepherd J, ed. Lipoprotein metabolism. Baillieres Clin Endocrinol Metab 1987;1:603–22.
76. Cheung M, Albers JJ. Distribution of high density lipoprotein particles with different apoprotein composition: particles with AI and AII and particles with AI but no AII. J Lipid Res 1982;23:747–53.
77. Cholesteryl ester transfer protein. [Editorial]. Lancet 1991;338:666–7.
78. Ruhling K, Zane-Langhenning R, Till U, Thielmann K. Enhanced net mass transfer of HDL cholesteryl ester to apo B-containing lipoproteins in patients with peripheral vascular disease. Clin Chem Acta 1989;184:289–96.
79. Yamashita S, Hui DY, Wetterau JR, Sprecher DL, Harmony JAK, Sakai N et al. Characterisation of plasma lipoproteins in patients heterozygous for human plasma CETP deficiency: plasma CETP regulates high-density lipoprotein concentrations and composition. Metabolism 1991;40:756–63.
80. Bagdade JD, Ritter MC, Subbaiah PV. Accelerated cholesteryl ester transfer in plasma of patients with hypercholesterolemia. J Clin Invest 1991;87:1259–65.
81. Bagdade JD, Ritter MC, Subbaiah PV. Accelerated cholesteryl ester transfer in patients with insulin-dependent diabetes mellitus. Eur J Clin Invest 1991;21:161–7.
82. Tozuka M, Fidge N. Purification and characterization of two high-density lipoprotein-binding proteins from rat and human liver. Biochem J 1989;261:239–44.
83. Gordon JG, Rifkind BM. High-density lipoprotein: the clinical implications of recent studies. [Review]. N Eng J Med 1989;321:1311–6.
84. Heiss G, Johnson NJ, Reiland S, Davis CE, Tyroler HA. The epidemiology of plasma high-density lipoprotein cholesterol levels: The Lipid Research Clinics Program Prevalence Study: summary. Circulation 1980;62(Suppl IV):IV-116–IV-136.
85. Davis CE, Gordon D, LaRosa J, Wood PDS, Halperin M. Correlations of plasma high-density lipoprotein cholesterol levels with other plasma lipid and lipoprotein concentrations: The Lipids Research Clinics Program Prevalence Study. Circulation 1980;62(Suppl IV):IV-24–IV-30.
86. Schaefer EJ. Clinical biochemical and genetic features in familial disorders of high density lipoprotein. [Review]. Arteriosclerosis 1984;4:303–22.

87. Schmitz G, Williamson E. High-density lipoprotein metabolism, reverse cholesterol transport and membrane protection. [Review]. Curr Opin Lipidol 1991:2: 177–89.
88. Parthasarathy S, Barnett J, Fong LG. High density lipoprotein inhibits the oxidative modification of low density lipoprotein. Biochem Biophys Acta 1990; 1044:275–83.
89. Loscalzo J. Lipoprotein(a): a unique risk factor for atherothrombotic disease. [Review]. Arteriosclerosis 1990;10:672–9.
90. Bersot TP, Innerarity TL, Pitas RE, Rall SC, Weisgraber LJ, Mahley RW. Fat feeding in humans induces lipoproteins of density less than 1.006 that are enriched in apoprotein(a) and that cause lipid accumulation in macrophages. J Clin Invest 1986;77:622–30.
91. Utermann G, Menzel HJ, Kraft HG, Duba HC, Kemmler HG, Seitz C. Lp(a) glycoprotein phenotypes. J Clin Invest 1987;80:458–65.
92. Karadi I, Kostner GM, Gries A, Nimpf J, Romics L, Malle E. Lipoprotein(a) and plasminogen are immunochemically related. Biochim Biophys Acta 1988;960: 91–7.
93. McLean JW, Tomlinson JE, Kuang W-J, Eaton DL, Chen EY, Fless GM et al. cDNA sequence of human apolipoprotein(a) is homologous to plasminogen. Nature 1987;330:132–7.
94. Jurgens G, Marth E, Kostner GM, Holasek A. Investigation of the Lp(a) lipoprotein aggregating and enzymatic activity associated with the Lp(a) polypeptide. Artery 1977;3:13–26.
95. Salonen E-M, Jauhiainen M, Zardi L, Vaheri A, Ehnholm C. Lipoprotein(a) binds to fibronectin and has serum proteinase activity capable of cleaving it. EMBO J 1989;8:4035–40.
96. Miles LA, Fless GM, Levin EG, Scanu AM, Plow EF. A potential basis for the thrombotic risks associated with lipoprotein(a). Nature 1989;339:301–3.
97. Hajjar KA, Gavish D, Breslow JL, Nachmann RL. Lipoprotein(a) modulation of endothelial surface fibrinolysis and its potential role in atherosclerosis. Nature 1989;339:303–5.
98. Loscalzo J, Weinfeld M, Fless G, Scanu AM. Lipoprotein(a), fibrin binding and plasminogen activation. Arteriosclerosis 1990;10:240–5.
99. Floren C-H, Albers JJ, Bierman EL. Uptake of Lp(a) by cultured fibroblasts. Biochem Biophys Res Commun 1981;102:636–9.
100. Krempler F, Kostner GM, Rascher A, Haslauer F, Bolzano K, Sandhofer F. Studies on the role of specific cell surface receptors in the removal of lipoprotein(a) in man. J Clin Invest 1983;71:1431–41.
101. Rhoads GG, Dahlen G, Berg K, Morton NE, Dannenberg AL. Lp(a) lipoprotein as a risk factor for myocardial infarction. JAMA 1986;256:2540–4.
102. Scanu AM. Update on lipoprotein (a). [Review]. Curr Opin Lipidol 1991;2:253–8.
103. Bhatnagar D, Durrington PN. Does measurement of apolipoproteins add to the clinical diagnosis and management of dyslipidemias? [Review]. Curr Opin Lipidol 1993;4:299–304.
104. Berg K, Dahlen G, Borrenson A-L. Lp(a) phenotypes, other lipoprotein parameters and family history of heart disease in middle-aged males. Clin Genet 1977; 16:347–52.
105. Durrington PN, Hunt L, Ishola M, Arrol S, Bhatnagar D. Apolipoproteins (a), AI and B and parental history in men with early onset ischaemic heart disease. Lancet 1988;1:1070–3.
106. Zenker G, Koltringer P, Bone G, Niederkorn K, Pfeiffer K, Jurgens G. Lipoprotein (a) as a strong indicator for cerebrovascular disease. Stroke 1986;17:942–5.
107. Wade DP. Lipoprotein (a). [Review]. Curr Opin Lipidol 1993;4:244–9.

108. Utermann G, Hoppichler F, Dieplinger H, Seed M, Thompson G, Boerwinkle E. Defects in the low density lipoprotein receptor gene affect lipoprotein (a) levels: multiplicative interaction of two gene loci associated with premature atherosclerosis. Proc Natl Acad Sci USA 1980;86:4171–4.
109. Austin MA, Brunzell JD, Fitch WL, Krauss RM. Inheritance of low density lipoprotein subclass patterns in familial combined hyperlipidaemia. Arteriosclerosis 1990;10:520–30.
110. Reaven GM. Role of insulin resistance in human disease. [Review]. Diabetes 1988;37:1595–607.
111. Selby JV, Austin MA, Newman B, Zhang D, Quesenberry CP Jr, Mayer EJ et al. LDL subclass phenotypes and the insulin resistance syndrome in women. Circulation 1993;83:381–7.
112. Jordan-Starck TC, Witte DP, Aronow BJ, Harmony JAK. Apolipoprotein J: a membrane policeman? [Review]. Curr Opin Lipidol 1992;3:75–85.
113. Provost PR, Villeneuve L, Weech PK, Milne RW, Marcel YL, Rassart E. Localization of the major sites of apolipoprotein D gene transcription by in situ hybridization. J Lipid Res 1991;32:1959–70.

CHAPTER 2

Overview of the Diagnosis and Treatment of Lipid Disorders

Ernst J. Schaefer and Judith R. McNamara

INTRODUCTION

❖ Current guidelines published by the National Cholesterol Education Program (NCEP) Adult Treatment Panel focus on dietary treatment of all adults over age 20 if their serum low-density lipoprotein cholesterol (LDL-C) values are ≥ 160 mg/dL (4.1 mmol/L), or ≥ 130 mg/dL (3.4 mmol/L) in the presence of two or more coronary heart disease (CHD) risk factors, or ≥ 100 mg/dL (2.6 mmol/L) in the presence of CHD.

CHD risk factors include: male ≥ 45 y, female ≥ 55 y or with premature menopause and not on hormonal replacement therapy, family history of CHD, cigarette smoking, hypertension, high-density lipoprotein cholesterol (HDL-C) < 35 mg/dL (0.91 mmol/L), or diabetes. A risk factor is subtracted if the HDL-C value is ≥ 60 mg/dL (1.6 mmol/L).[1, 2] After an adequate trial of a diet restricted in total fat (≤ 30% of calories), saturated fat (< 7%) and cholesterol (< 200 mg/d), and after ruling out secondary causes of hypercholesterolemia, persons with LDL cholesterol values ≥ 190 mg/dL (4.9 mmol/L), or ≥ 160 mg/dL (4.1 mmol/L) in the presence of two or more CHD risk factors, or ≥ 130 mg/dL (3.4 mmol/L) in the presence of CHD, are candidates for pharmacologic therapy with agents such as anion exchange resins, niacin, hydroxymetlyglutaryl-CoA (HMG CoA) reductase inhibitors, or a combination of these agents. Guidelines for the entire population and for children and adolescents have also been formulated which focus on dietary modification in the general population, to include < 30% fat, < 10% saturated fat, < 300 mg cholesterol/d, and use of diet therapy and in some cases resins in high-risk children.[3, 4] The rationale for all of these guidelines is that increased LDL-C concentrations have clearly been shown to be associated with premature CHD in prospective studies, and that lowering LDL-C with either diet therapy or a combination of diet and drug treatment has been shown to reduce CHD risk prospectively.

LIPOPROTEIN METABOLISM

Plasma cholesterol and triglycerides (TG), along with phospholipids and proteins, are carried on lipoprotein particles. As shown in the overview of lipoprotein metabolism presented in Figure 2–1,[5] dietary fats are packaged by the intestine into large, TG-rich chylomicrons (Step 1), which rapidly lose much of their TG after entry into plasma due to lipolysis (Step 4). Apolipoprotein (apo) B-48 is the important structural protein of chylomicrons and contains the initial 48% of the apo B-100 sequence. During lipolysis, chylomicron remnants are formed which contain apo E and cholesterol obtained from other lipoproteins. These remnants are rapidly taken up by the liver via an apo E receptor-mediated process (Step 5).

The liver synthesizes very-low-density lipoproteins (VLDL) (Step 2), which are the major TG-carrying lipoproteins in serum or plasma obtained from fasting subjects. Much of VLDL TG and phospholipids are rapidly removed via lipolysis, and in the process VLDL can be converted to low-density-lipoproteins (LDL) (Step 6) or catabolized directly (Step 7). LDL generally are the major cholesterol-carrying lipoproteins in the serum or plasma and are mainly catabolized in the liver and other tissues by an LDL receptor-mediated process (Steps 8 and 9). Apo B-100 is the major structural protein of LDL and VLDL. When present in excessive amounts, LDL, as well as chylomicron and VLDL remnants, can be deposited in the arterial wall, modified, and taken up by macrophages, causing atherosclerosis.[6] For additional information on this process, refer to Chapter 17.

High-density lipoproteins (HDL) are synthesized in both the liver and the intestine (Step 3). They can pick up lipids (phospholipids, TG) and proteins (apo AI, apo E, C apolipoproteins) from TG-rich lipoproteins (chylomicrons and VLDL) (Steps 4 and 6). The major protein of HDL is apo AI. HDL accept cholesterol from tissues (Step 10) and deliver it to the liver directly (Step 12) or transfer it to other lipoproteins, such as chylomicron or VLDL remnants, which are then taken up by the liver. In the liver, cholesterol can either be excreted directly into bile, converted to bile acids, or re-utilized in lipoprotein production.

Another lipoprotein not shown in Figure 2–1 is lipoprotein(a) or Lp(a). This lipoprotein consists of an LDL particle with apo(a) attached to it by a disulfide bond. Apo(a) is a unique glycoprotein with protein domains, known as kringles, which have homology with similar protein domains in plasminogen. Apo(a) contains multiple and variable numbers of kringle 4-like domains and one kringle 5-like domain. Variability in Lp(a) concentrations relates to apo(a) molecular weight, number of kringle 4-like domains, and differences in hepatic apo(a) secretion rates.[7–11] Increased concentrations may interfere with fibrinolysis or clot lysis.[12–14] Like LDL, Lp(a) can be deposited directly in the arterial wall, promoting atherosclerosis.[8] Increased Lp(a) concentrations have been associated with premature CHD in most studies.[15–31]

Figure 2–1 ❖ *An Overview of Lipoprotein Metabolism*

Open arrows signify synthetic pathways, thin closed arrows signify transfer pathways, thick closed arrows signify catabolic pathways, dotted arrows signify minor pathways, and solid boxes or symbols signify receptors. Fats absorbed in the intestine are packaged into large TG-rich particles known as chylomicrons (Step 1). These lipoproteins undergo lipolysis (removal of TG) to form chylomicron remnants (Step 4), which are taken up by the liver via an apo E receptor (Step 5). The liver can also secrete TG-rich lipoproteins known as VLDL (Step 2). Following lipolysis, these particles can be converted to LDL (Step 6) or taken up by the liver via an apo E receptor (Step 7). LDL are catabolized mainly by the liver (Step 8) or other tissues (Step 9) via LDL receptors that recognize both apo B-100 and apo E but not apo B-48. If LDL are modified, they can also be taken up by scavenger receptors on macrophages (Step 11). HDL are synthesized by both the liver and the intestine (Step 3). HDL pick up lipid and protein constituents from chylomicrons and VLDL as these particles undergo lipolysis (Steps 4 and 6). HDL pick up free cholesterol from peripheral tissues (Step 10) and macrophages (Step 11) and are catabolized mainly in the liver (Step 12).

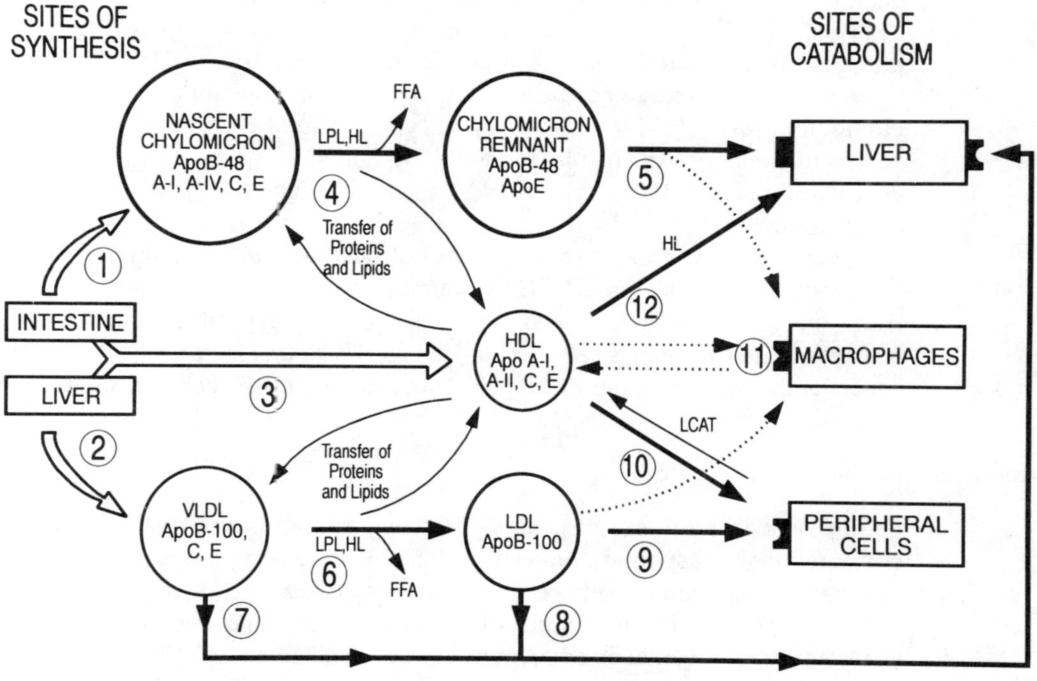

LPL = lipoprotein lipase HL = hepatic lipase
LCAT = lecithin:cholesterol acyltransferase
Apo = apolipoprotein FFA = free fatty acids
HDL = high-density lipoprotein LDL = low-density lipoprotein
VLDL = very-low-density lipoprotein

RATIONALE FOR TREATMENT OF LIPID DISORDERS

Plasma LDL-C values are increased by diets high in saturated fat and cholesterol mainly because of decreased LDL receptor-mediated catabolism. Animals that consume such diets develop high LDL-C and atherosclerosis. Such diets can also increase HDL-C concentrations; this change may be compensatory. Human populations on diets high in saturated fat and cholesterol have increased LDL-C concentrations and have significantly greater incidence of CHD due to atherosclerosis than populations on diets low in saturated fat and cholesterol.[1-3] In U.S. society, increased LDL-C concentrations as well as decreased HDL-C values are independent risk factors for premature CHD.[1-3, 32-34] Women have higher HDL-C values than men and have a lower risk of CHD.

Prospective studies indicate that dietary treatment or combined dietary and drug therapy which decreases LDL-C can reduce subsequent CHD morbidity and mortality.[28, 29, 35-45] Moreover, studies indicate that aggressive lipid modification in CHD patients to get LDL-C concentrations to < 100 mg/dL (2.6 mmol/L) can result in stabilization of existing coronary atherosclerosis, as well as some regression of this process.[38, 46-56]

The NCEP guidelines focus on LDL-C values because LDL-C has clearly been shown to be an independent risk factor for CHD and because lowering of CHD risk has been documented to occur with LDL-C lowering by diet and/or drugs in patients with increased LDL-C concentrations and/or CHD. Because of these studies, consensus has been reached on use of drugs for lowering LDL-C,[57] but not for lowering TG or raising HDL-C.[58] Decreased HDL-C has been shown to be an independent CHD risk factor, and HDL-C increases have been shown to reduce CHD risk within both the Lipid Research Clinics Study and the Helsinki Heart Study.[29, 30, 42, 59, 60]

Increased TG concentrations have not been shown to be an independent CHD risk factor.[33] Increased TG concentrations are correlated with decreased HDL-C values. It should be noted, however, that no prospective study has as yet been completed in patients selected for decreased HDL-C or increased TG concentrations. For these reasons, the NCEP guidelines focus on LDL-C concentrations.

EVALUATION OF THE PATIENT

Serum or plasma total cholesterol (TC) and HDL-C concentrations could be measured in the fasting or non-fasting state for screening purposes.[2] Plasma values are approximately 3% lower than serum values and should be adjusted upward if EDTA is the anticoagulant used. For screening programs, compact analyzers using finger-stick specimens are now available for TC, TG, and HDL-C.

Lipid values have been classified as listed in Table 2–1 for adults over age 20. According to guidelines issued by the NCEP Adult Treatment Panel II, patients without CHD who have a desirable TC value of < 200 mg/dL (5.2 mmol/L) and an HDL-C value of ≥ 35 mg/dL (0.9 mmol/L) should have their values checked again within 5 y.[2] In addition, education on the recommended general population diet guidelines (≤ 30% fat, < 10% saturated fat, and < 300 mg of cho-

Table 2–1 ❖ Lipid Abnormalities in Adults*

	Desirable	*Borderline High Risk for CHD*	*High Risk for CHD*	*High Risk for Pancreatitis*
Cholesterol	< 200 (5.2)	200–239 (5.2–6.2)	≥ 240 (6.2)	—
LDL Cholesterol	< 130 (3.4)	130–159 (3.4–4.1)	≥ 160 (4.1)	—
HDL Cholesterol	> 60 (1.6)	—	< 35 (0.9)	—
Triglyceride	< 200 (2.3)	200–400 (2.3–4.5)	> 400 (4.5)	> 1000 (11.3)
TC/HDL	< 5.0	5.0–6.0	> 6.0	

*All values in mg/dL; values in parentheses are in mmol/L. Elevated triglycerides have been shown to be a risk factor for pancreatitis (> 1000 mg/dL or 11.3 mmol/L), but have not been clearly shown to be an independent risk factor for CHD.

lesterol/d), exercise, and risk factor reduction should be provided to all patients. If the patient has TC values in the borderline range of 200–239 mg/dL (5.2–6.2 mmol/L), information about other CHD risk factors should be obtained.[2] These risk factors are listed in Table 2–2.

In the absence of CHD (prior myocardial infarction, angina), if a patient has a TC value in the borderline range, a normal HDL-C value, and less than two CHD risk factors, information on diet and CHD risk reduction should be provided and the patient's values checked within the next 1–2 y. If the patient has CHD, a TC value ≥ 240 mg/dL (6.2 mmol/L), a borderline TC value in the presence of two or more CHD risk factors, or an HDL-C value < 35 mg/dL (0.9 mmol/L), it is recommended that a fasting (at least 9 h) lipid profile (TC, TG, HDL-C) be measured. Laboratories should maintain a coefficient of variation (CV) of ≤ 3% and a bias of

Table 2–2 ❖ NCEP ATP II Major Risk Factors for CHD in Addition to Elevated LDL Cholesterol*

- Male ≥ 45 years, Female ≥ 55 years or with premature menopause without estrogen replacement therapy
- Family history of premature CHD (definite myocardial infarction or sudden death before age 55 in a male first-degree relative or before age 65 in a female first-degree relative)
- Current cigarette smoking
- Hypertension (≥ 140/90 mmHg or taking antihypertensive medication)
- HDL cholesterol < 35 mg/dL (0.9 mmol/L)
- Diabetes mellitus

*A risk factor should be subtracted if the HDL-C concentration is ≥ 60 mg/dL (1.6 mmol/L).

≤ 3% for their TC assay as compared to the National Reference System for Cholesterol (NRS/CHOL).[61] The formula commonly used to calculate LDL-C is:[62]

$$\text{LDL-C} = \text{TC} - \text{HDL-C} - \text{TG}/5$$

when values are in mg/dL, or TG/2.22 when values are in mmol/L.

It has been documented that this formula cannot be used for values obtained on serum or plasma from non-fasting individuals or those whose TG values are > 400 mg/dL (4.5 mmol/L). Moreover, there is considerable variability in calculated LDL-C concentrations when TG values are 200–400 mg/dL (2.3–4.5 mmol/L) as compared to values obtained by ultracentrifugation.[63] Our data indicate that about 5% of the population have fasting TG values > 400 mg/dL (4.5 mmol/L), and that an additional 14% have calculated LDL-C values that are not placed in the appropriate NCEP categories as compared to values obtained by ultracentrifugation.[63]

The Lipoprotein Working Group of the NCEP has recommended development of more direct methods of measuring LDL-C. Such a direct LDL-C assay has been developed and is now available from both Sigma Diagnostics (St. Louis, MO) and the Genzyme Corporation (Cambridge, MA). The principle of the assay is to immunoprecipitate chylomicrons, VLDL, and HDL from serum or plasma, leaving LDL behind, and then measure the cholesterol content of the LDL by automated enzymatic techniques. This assay has been evaluated by Pacific Biometrics (Seattle, WA) and found to correlate very highly with LDL-C measured following ultracentrifugation (r = 0.96) based on analysis of 74 serum samples obtained from subjects with TG concentrations of 26–936 mg/dL (0.29–10.58 mmol/L).[64] Within-run precision is reflected by a CV ranging from 1.1 to 2.8% and a between-run CV ranging from 3.0 to 3.4%.[64] Moreover, it was documented that this assay gave accurate and reliable results on non-fasting as well as fasting serum. Similar results have been obtained in the laboratory of Dr. Thomas Cole of Washington University School of Medicine, as well as in our laboratory. For additional information on this assay, refer to Chapter 7.

An LDL-C concentration of 160 mg/dL (4.1 mmol/L) represents approximately the 75th percentile for middle-aged Americans. LDL-C values have been classified by NCEP as shown in Table 2–1.[1, 2] Since values are subject to considerable biologic variation, it is important to confirm the presence of abnormalities by repeat determinations. Hospitalization or acute illness can also affect lipid values, and therefore lipid determinations should generally be carried out when subjects are in the free-living state. Getting LDL-C and HDL-C at the time of screening may be more cost-effective. For additional information on biological variations, refer to Chapter 3.

Although an increased TG concentration has not been clearly shown to be an independent risk factor for CHD, TG concentrations are inversely associated with low concentrations of HDL-C, which has been shown to be a significant risk factor for CHD. Moreover, markedly increased TG concentrations are associated with pancreatitis.

Common secondary causes of lipid abnormalities are shown in Table 2-3.[1,2] These secondary causes should be screened for, and treated if possible, prior to

Table 2-3 ❖ Secondary Causes of Lipid Abnormalities*	
●	Obesity
●	Diet
●	Lack of Exercise
●	Cigarette Smoking
●	Alcohol Intake
●	Diabetes Mellitus
●	Hypothyroidism
●	Obstructive Liver Disease
●	Renal Insufficiency
●	Nephrotic Syndrome
●	Medications (hormones, beta blockers, diuretics)

*Screen by measuring glucose, TSH, creatinine, urinary protein, transaminases, and alkaline phosphatase.

initiating drug therapy for lipid disorders. Some forms of dyslipidemia are genetic and therefore can be identified by family studies.

In some studies, low apo AI values and decreased LDL particle size as well as increased apo B concentrations have been reported to be superior to LDL-C and HDL-C concentrations as markers for CHD.[65-70] However, these have all been case-control studies. In our view, apo AI and apo B concentrations are less affected by dietary modification and the use of beta blockers than are TG, LDL-C, and HDL-C concentrations and LDL particle size.[21, 71] Most CHD patients have modified their diet and are often on beta-adrenergic blocking agents, and these are confounders in such case-control studies.[22] Supporting this view is our finding that LDL size is not an independent CHD risk factor after controlling for the effects of beta blockers and LDL-C and HDL-C concentrations.[71] Moreover, of six prospective studies assessing the utility of apo AI and apo B versus LDL-C and HDL-C concentrations as CHD markers, none has documented that apolipoprotein values added information about CHD risk.[26,72-76] In addition, adequate standardization for apo AI and apo B is not available at the present time, and gels for running LDL sizing are not commercially available. For these reasons, it is our view that apo AI, apo B, and LDL size measurements are not likely to have clinical utility and cannot be recommended for CHD risk assessment at this time. However, more data from large-scale prospective studies are required before a definitive conclusion can be reached in this regard.

In contrast, an increased Lp(a) concentration is likely to be a significant independent CHD risk factor. This concept is supported by many case-control studies and four of six prospective studies.[15-30] Problems associated with Lp(a) measurement include lack of assay standardization and considerable heterogeneity within the protein. When calibrated to total mass of the particle, an Lp(a) value of > 40 mg/dL is clearly above the 90th percentile, but many authorities feel that a value > 30 mg/dL is a high-risk value. Our data indicate that familial Lp(a) excess (values above the 90th percentile) is a highly heritable disorder found in 15–20% of kindreds with premature CHD.[23] Niacin administration has been shown to lower Lp(a) concentrations, as well as to reduce CHD morbidity and mortality in unselected men with CHD.[41, 44, 77] For these reasons, it is our view that Lp(a) concentrations should be measured in patients with CHD, and patients whose values are increased [Lp(a) > 30 mg/dL or Lp(a) cholesterol > 10 mg/dL] should be treated with niacin if they can tolerate it. The cholesterol content of Lp(a) can be assessed, and this method may aid in the standardization process.[78]

The NCEP has not issued recommendations with regard to Lp(a). Moreover, two prospective studies have been negative with regard to the value of Lp(a) as a CHD marker,[27, 28] and no prospective trial has been carried out in patients with increased Lp(a) concentrations in the presence or absence of CHD. Definitive recommendations with regard to Lp(a) await such studies.

FAMILIAL LIPOPROTEIN DISORDERS

An overview of familial hypercholesterolemic states is provided in Table 2–4. By far the most common of these disorders is familial combined hyperlipidemia, in which affected kindred members may have increased LDL-C alone (above the 90th percentile of age- and gender-adjusted norms), increased TG alone, or increases of both parameters. Both abnormalities must be present in the family to make the diagnosis.[23, 79, 80] These patients also often have decreased HDL-C values due to enhanced degradation of HDL protein (Step 12 in Figure 2–1).[23] They have also been shown to have overproduction of VLDL apo B-100, but not TG (Step 2).[81, 82] Approximately 15% of patients with premature CHD have this disorder.[23] Sporadic or polygenic hypercholesterolemia is also quite common and is observed in approximately 2% of kindreds with premature CHD.[23, 79] Treatment consists of diet and, if necessary, niacin, HMG CoA reductase inhibitors, gemfibrozil, or combinations of these medications with anion exchange resins.[83]

A much more rare form of combined increases of TC and TG is known as familial dysbetalipoproteinemia (Type III hyperlipoproteinemia), in which affected subjects have accumulations of chylomicron remnants and VLDL in the fasting state. These patients usually are homozygous for a mutation in the apo E protein (apo E 2/2 phenotype), or in rare cases have apo E deficiency, resulting in defective hepatic clearance of chylomicron remnants and VLDL particles as well as increased VLDL production (Steps 2, 5, and 6).[84–88] They may also have tubo-eruptive and palmar xanthomas. Diagnosis requires documentation of a VLDL-C/TG ratio > 0.3 after measurement of VLDL-C following ultracentrifugation and documentation of the apo E 2/2 phenotype by isoelectric focusing, or of the genotype by DNA analysis. Treatment consists of diet, niacin, gemfibrozil, or HMG CoA reductase inhibitors. Patients with both familial combined hyperlipidemia and familial dysbetalipoproteinemia often have obesity, glucose intolerance, and hyperuricemia.[86]

Table 2–4 ❖ Familial Hypercholesterolemic States

	Familial Combined Hyperlipidemia	*Familial Hypercholesterolemia*	*Familial Dysbetalipoproteinemia*
Physical findings:	Arcus senilis	Arcus senilis, tendinous xanthomas	Arcus senilis, tubo-eruptive and planar xanthomas
Associated findings:	Obesity, glucose intolerance, hyper-uricemia, HDL deficiency	—	Obesity, glucose intolerance, hyperuricemia
Mode of inheritance:	Autosomal dominant	Autosomal co-dominant	Autosomal recessive
Defect:	Overproduction of hepatic VLDL apo B-100	Defective LDL receptor or defective apo B-100	Defective apo E
Estimated population frequency:	1:50	1:500	1:5000
Estimated frequency in CHD patients:	15%	1–3%	0.5%
CHD risk:	Moderate	High	Low
Treatment:	Diet, niacin, resin, statins, gemfibrozil	Diet, niacin and resin, statins and resin, or triple therapy	Diet, niacin, gemfibrozil, statins

Isolated increases of LDL-C are also found in patients with a disorder known as familial hypercholesterolemia, often associated with tendinous xanthomas. These patients generally have marked hypercholesterolemia (> 350 mg/dL [9.0 mmol/L]) with normal TG values, and have defects at the LDL receptor locus or abnormalities of the apo B protein (apo B-3500 mutation).[79, 80, 88–90] The major metabolic abnormality in these individuals is an impaired ability to catabolize LDL (Steps 8 and 9).[91] Approximately 0.1% of the population and 1% of patients with premature CHD have this disorder due to mutations at the LDL receptor locus.[23, 79] Apo B mutations are even less common. Treatment generally consists of diet and a combination of medications (resins and niacin, resins and HMG CoA reductase inhibitors).[92–94]

Isolated deficiency of HDL-C (below the 10th percentile of normal) can be genetic in nature, and is then known as familial hypoalphalipoproteinemia. This disorder is found in approximately 5% of patients with premature CHD.[23, 95–97] In our own recent studies, patients in families with HDL deficiency and premature CHD had decreases in all lipoproteins containing apo AI (Lp AI, Lp AI/AII) and increases in lipoproteins containing apo B as well as in those containing both apo B

Table 2–5 ❖ Familial Hypertriglyceridemic States

		Severe Hypertriglyceridemia	
	Familial Hypertriglyceridemia	*Early Onset*	*Adult Onset*
Physical findings:	None	Lipemia retinalis, eruptive xanthomas	Lipemia retinalis, eruptive xanthomas
Associated findings:	Obesity, glucose intolerance, hyperuricemia, HDL deficiency	HDL deficiency, pancreatitis	Obesity, glucose intolerance, hyperuricemia, HDL deficiency, pancreatitis, paresthesias, emotional lability
Mode of inheritance:	Autosomal dominant	Autosomal recessive	Autosomal recessive
Defect:	Overproduction of hepatic VLDL triglyceride	Lipoprotein lipase deficiency, apo C-II deficiency	Overproduction of VLDL triglyceride, delayed catabolism of chylomicrons and VLDL
Estimated population frequency:	1:50	< 1:10,000	1:1000
Estimated frequency in CHD patients:	15%	—	1:500
CHD risk:	Low	—	Low
Treatment:	Diet, niacin, gemfibrozil	Diet	Diet, gemfibrozil, fish oil

and apo E (LpB:E).[95] Treatment of HDL deficiency consists of diet, weight reduction if indicated, and an exercise program.[58] In patients with CHD, treatment with niacin, gemfibrozil, or HMG CoA reductase inhibitors may raise their HDL-C concentrations and optimize their LDL-C values. Efforts should be made to optimize the LDL-C values in these patients according to NCEP guidelines.[1, 2]

An overview of familial hypertriglyceridemic states is provided in Table 2–5. By far the most common of these disorders is familial hypertriglyceridemia, an autosomal-dominant disorder in which obesity, glucose intolerance, hyperuricemia, and HDL deficiency are often present.[23, 79] The disorder is associated with overproduction of hepatic VLDL TG but not VLDL apo B-100 (Step 2).[81, 83] Some patients may have defects in VLDL clearance as well (Steps 6 and 7). CHD risk appears to be increased in those kindreds in whom HDL deficiency is also present. Approximately 15% of patients with premature CHD appear to have this disorder.[23] HDL concentrations are low in these subjects because of enhanced degradation (Step 12). Treatment with diet, exercise, and abstinence from the use of alcohol and exogenous estrogens is recommended. In patients with CHD, lipid values can be optimized by use of HMG CoA reductase inhibitors, niacin, or gemfibrozil.

Severe hypertriglyceridemia, wherein TG values are above 1000 mg/dL (11.2 mmol/L), is often observed in middle-aged or elderly individuals who are obese and have glucose intolerance and hyperuricemia. These subjects usually have familial hypertriglyceridemia or familial combined hyperlipidemia which is exacerbated by other factors, such as obesity and diabetes mellitus. These patients generally also have HDL deficiency and may develop lipemia retinalis and eruptive xanthomas. They are at increased risk for developing pancreatitis due to TG deposition in the pancreas and may have paresthesias and emotional lability. These patients often have delayed chylomicron and VLDL clearance and excess VLDL production (Steps 2, 4, 6, and 7). Treatment consists of a calorie-restricted Step 2 diet. In patients with diabetes mellitus, it is crucial to control blood glucose concentrations as much as possible. Medications that are effective in lowering TG to < 1000 mg/dL (11.2 mmol/L) in these patients, thereby reducing their risk of pancreatitis, include gemfibrozil and/or fish-oil capsules (6–10 capsules/d).

Patients who present with severe hypertriglyceridemia in childhood or early adulthood and who are not obese often have a deficiency of the enzyme lipoprotein lipase or its activator protein, apo CII, resulting in markedly impaired removal of TG, which affects chylomicron and VLDL catabolism (Steps 4, 6, and 7). These patients are at increased risk for recurrent pancreatitis, and it is important to restrict their dietary fat to less than 20% of calories. Niacin and/or gemfibrozil are generally ineffective in these patients; however, fish-oil capsules may be helpful in certain patients to keep their TG concentrations below 1000 mg/dL (11.2 mmol/L) and minimize their risk of pancreatitis.

TREATMENT GUIDELINES FOR INCREASED LDL-C

The NCEP Adult Treatment Panel has developed guidelines for the diagnosis and treatment of individuals over age 20 who have increased TC concentrations associated with increased LDL-C concentrations.[1, 2] The NCEP's decision values for LDL-C are given in Table 2–6. Concentrations of LDL-C requiring the initiation of diet and drug therapy as well as the goals of therapy are dependent upon the presence or absence of CHD or CHD risk factors (see Table 2–6) and the presence of secondary causes of lipid abnormalities (see Table 2–3).

Diet Therapy

Diet therapy is the cornerstone of the treatment of lipid disorders.[2] According to our estimates, 50% of the saturated fat and 70% of the cholesterol in the U.S. diet come from hamburgers, cheeseburgers, meat loaf, butter, whole milk products, cheeses, beef dishes, hot dogs, ham, luncheon meats, pork, doughnuts, cookies, and cakes. Hence, such foods should be restricted. Instead, foods such as poultry (white meat without skin), fish, skimmed milk, non-fat or low-fat yogurt, and low-fat cheeses are recommended. The use of fruits, vegetables, and grains is encouraged.

Table 2-6 ❖ NCEP ATP II LDL Cholesterol Decision Values			
	Initiate Diet Therapy	*Initiate Drug Therapy After Diet Therapy*	*Goal of Therapy*
< 2 CHD risk factors	≥ 160 mg/dL (4.1 mmol/L)	≥ 190 mg/dL (4.9 mmol/L)	< 160 mg/dL) (4.1 mmol/L)
2 or more CHD risk factors	≥ 130 mg/dL (3.4 mmol/L)	≥ 160 mg/dL (4.1 mmol/L)	< 130 mg/dL (3.4 mmol/L)
CHD	≥ 100 mg/dL (2.6 mmol/L)	≥ 130 mg/dL (3.4 mmol/L)	< 100 mg/dL (2.6 mmol/L)

Oils rich in saturated fat, such as coconut oil, palm oil, and palm kernel oil, should be restricted. Oils that can be used are unsaturated vegetable oils containing polyunsaturated and mono-unsaturated fatty acids, such as canola, soybean, olive, and corn oil. However, such oils should only be used in moderation because they are pure fat and very calorie rich.

Excellent patient dietary pamphlets are available from the American Heart Association and the National Cholesterol Education Program. NCEP guidelines on dietary therapy are given in Table 2–7. The benefits of the NCEP's Step 1 and Step 2 diets in terms of lowering LDL-C are not striking unless patients restrict total fat and calories, increase exercise, and lose excess weight.[94] Patients who are unable to get an adequate response from diet modification after being given pamphlets and counseling can be referred to a registered dietitian for instructions on the Step 2 diet. In most cases, diet therapy should be tried for at least 6 m prior to initiating drug therapy.

Drug Therapy

An overview of available lipid-lowering medications is provided in Table 2–8. These medications can be divided into two general classes: those that are effec-

Table 2–7 ❖ NCEP ATP II Dietary Therapy*			
Nutrient	*Average U.S. Diet*	*Step 1 Diet*	*Step 2 Diet*
Total Fat	36%	≤ 30%	≤ 30%
Saturated Fat	15%	< 10%	< 7%
Polyunsaturated Fat	6%	< 10%	< 10%
Mono-unsaturated Fat	15%	< 15%	< 15%
Cholesterol (mg/d)	400–500	< 300	< 200
Total Calories	To achieve and maintain desired weight		

*% indicates percentage of total calories.

Table 2–8 ❖ Lipid Medications

	Resins	*Niacin*	*HMG CoA Reductase Inhibitors**	*Gemfibrozil*
Patient acceptance and compliance:	Often poor	Often poor	Generally excellent	Generally excellent
Side effects:	Constipation, bloating, decreased absorption of certain medicines	Flushing, itching, gastritis, hepatotoxicity, hyperuricemia, hyperglycemia	Hepatotoxicty, myositis, GI side effects	Myositis, hpeatotoxicity, GI side effects, coumadin interaction
Usual dose:	8–10 g po BID	1 g po BID or TID with food	10, 20, or 40 mg po QD	600 mg po BID
LDL reduction:	10–20%	10–20%	25–35%	10%
Triglyceride reduction:	**	40%	20%	35%
HDL increase:	5%	15–20%	5–15%	5–20%
CHD risk reduction documented:	Yes 19% 7 y	Yes 20% 5 y	Yes 50% 2 y	Yes 34% 5 y
Long-term safety documented:	Yes	Yes	Yes	Yes

*Current HMG CoA reductase inhibitors include lovastatin, pravastatin, simvastatin, and fluvastatin.
**May increase triglycerides.

tive in decreasing LDL-C by more than 15% and those that are effective in decreasing TG concentrations by more than 15%. Currently, three classes of agents meet the LDL-lowering criteria: anion exchange resins (cholestyramine and colestipol), niacin, and HMG CoA reductase inhibitors (lovastatin, pravastatin, simvastatin, fluvastatin). Of these three types of drugs, patient acceptance and compliance with resins and niacin is often poor, while with HMG CoA reductase inhibitors it is generally excellent. Moreover, long-term safety in adults has now been documented with all three types of agents, and CHD risk reduction has also been reported. Because of efficacy and compliance, HMG CoA reductase inhibitors are the drugs of choice for LDL-C lowering in older patients (over age 50) and in patients with CHD. However, more long-term, large-scale study data on the HMG CoA reductase inhibitors are urgently awaited. Therefore, niacin and resins remain the drugs of choice in younger patients without CHD.

There are currently three agents that decrease TG concentrations by more than 15%: niacin, gemfibrozil, and lovastatin. All of these agents generally also lower LDL concentrations and increase HDL concentrations. Both niacin and gemfibrozil have been shown to lower CHD risk prospectively.

For younger patients with increased LDL-C only, the drugs of choice are niacin or anion exchange resins, while for CHD patients or older subjects (over age 50), HMG CoA reductase inhibitors should be utilized. The combination of a reductase inhibitor and an anion exchange resin is very effective, as is the combination of resins and niacin.

It should be noted that in postmenopausal women who have had a hysterectomy, estrogen replacement is quite effective in lowering LDL-C and raising HDL-C, but should not be used in the case of hypertriglyceridemia. Use of estrogens has been associated with a significant reduction in CHD mortality in postmenopausal women. However, these were observational studies and not randomized placebo-controlled studies.

For patients with increases in both LDL-C and TG, the drugs of choice are niacin and HMG CoA reductase inhibitors.

For patients with hypertriglyceridemia only and normal LDL-C concentrations, there are as yet no clear guidelines for the use of medication. If the patient's fasting TG concentration is in excess of 1000 mg/dL (11.3 mmol/L) while on a restricted diet, use of a medication to reduce the risk of pancreatitis is recommended. However, prior to taking this step, the physician should make sure that these patients are not taking estrogens, thiazides, or beta blockers, are not using alcohol, and do not have uncontrolled diabetes mellitus. Calorie and fat restriction and exercise are important in these patients. The drug of choice for these patients is generally gemfibrozil because most such patients have glucose intolerance. In the absence of glucose intolerance, niacin can be tried. In patients for whom these agents are not effective, or if additional TG reduction is needed, fish-oil capsules (1 g) at a dose of 3–5 capsules twice daily are effective in lowering TG.

In patients with moderate hypertriglyceridemia, especially in the presence of HDL-C deficiency, lifestyle changes including weight reduction and an exercise program are very helpful, as are cessation of the use of beta blockers. If patients have established CHD, the use of HMG CoA reductase inhibitors to optimize their LDL-C concentrations should be considered. The goal of therapy in CHD patients is to get their LDL-C concentrations to $<$ 100 mg/dL (2.59 mmol/L). In patients with moderate hypertriglyceridemia and/or HDL deficiency in the absence of CHD, only lifestyle modification (diet and exercise) and following the guidelines for LDL-C lowering can currently be recommended.

Properties of Lipid Medications

Cholestyramine and colestipol are anion exchange resins which bind bile acids, increase conversion of liver cholesterol to bile acids, and up-regulate LDL receptors in liver, increasing LDL catabolism (Step 8 and 9 of Figure 2–1) and decreasing plasma LDL-C by about 20%. Side effects include bloating and constipation, increased TG, and interference with the absorption of digoxin, tetracycline, d-thyroxine, phenylbutazone, and coumadin (these drugs should be given 1 h before or 4 h after resin). Cholestyramine (4 g scoops) or colestipol (5 g scoops) can be started at 1 scoop twice per day and gradually increased to 2 scoops twice per day (the scoops are half the price of the packets) or two scoops three times daily. Constipation may require treatment. Cholestryamine use has been shown to lower LDL-C by 12% and to reduce CHD risk prospectively by 19% over a 7-y period in middle-aged asymptomatic hypercholesterolemic men.

Niacin decreases VLDL (Step 2) and LDL production (Step 6) and raises HDL-C values (by 20%). Niacin should be started at 100 mg po BID with meals and gradually increased to 1 g po BID or TID with meals. Side effects include flushing, gastric irritation and elevations of uric acid, glucose, and liver enzymes in some patients. Niacin should not be used in patients with liver disease, a history of ulcers, or in diabetic patients not on insulin. Long-acting niacin causes less flushing but more gastrointestinal side effects than regular niacin. One aspirin daily will minimize flushing and can be used initially. Niacin should be discontinued if liver enzymes increase to over twice the upper normal limit. Niacin has been shown to decrease TC concentrations by 10% and reduce the recurrence of myocardial infarction by 20% over a 5-y period of administration in men with CHD. The use of niacin was also associated with an 11% reduction in all causes of mortality 10 y after cessation of niacin. Recently, niacin in combination with clofibrate has been shown to reduce mortality in CHD patients as compared to usual care.

Gemfibrozil is given at a dose of 600 mg po BID and is generally well tolerated. The drug is very effective in lowering TG and VLDL (by 35%) by decreasing production and enhancing breakdown of VLDL (Steps 2 and 6). The drug usually lowers LDL by 10–15% and increases HDL by 5–15%. Rarely, patients may experience gastrointestinal symptoms, muscle cramps (increased CPK, 1%), or intermittent indigestion. The drug should not be used in patients with renal insufficiency, and it is also known to potentiate the action of coumadin. The drug may raise LDL-C concentrations in some hypertriglyceridemic patients. Gemfibrozil has been shown to lower LDL-C by 9%, increase HDL-C by 10% and reduce CHD prospectively by 34% over a 5-y period in middle-aged asymptomatic hypercholesterolemic men.

HMG CoA reductase inhibitors are usually started at 10 mg/d and can be increased to a maximal dose of 40 mg/d po. These drugs are generally well tolerated but may occasionally cause liver enzyme increases, CPK increases with myalgias and myositis, and gastrointestinal upset. Possible differences in the side-effect profile of these agents is currently being explored. The drugs inhibit HMG CoA reductase, the rate-limiting enzyme in cholesterol biosynthesis, causing up-regulation of LDL receptors, enhancing LDL catabolism (Steps 8 and 9), and decreasing plasma LDL-C by 30–45%. They may also decrease VLDL and LDL production (Steps 2 and 6). These agents decrease plasma TC and TG and moderately increase HDL-C.[53] At 40 mg/d for lovastatin, pravastatin, and simvastatin, LDL-C reductions of 27%, 34%, and 38%, respectively, have been reported. CHD risk reduction and benefit in atherosclerosis has been reported for lovastatin and pravastatin. Safety and efficacy in CHD risk reduction have recently been established. However, the results of several large-scale studies with these agents are urgently awaited. In our view, these agents are the drugs of choice for LDL lowering in CHD patients and patients over age 50.

Probucol is an anti-oxidant at a dose of 500 mg po BID, and is a second-line drug in lowering LDL-C by 10–15%, which can be used in familial hypercholesterolemia for increasing non-receptor mediated LDL catabolism (Steps 8 and 9). The drug also lowers HDL-C by 15–25% by decreasing its production (Step 3). It may cause gastrointestinal side effects. Long-term safety and efficacy in CHD risk reduction have not been established. Recent results from a Swedish study indicate no significant benefit in CHD risk reduction or improvement in femoral atherosclerosis. Therefore, probucol cannot be recommended as a lipid-lowering agent.

Drug Combinations

HMG CoA reductase inhibitors and resins, as well as niacin and resin in combination, are very effective in lowering LDL-C concentrations. The combination of gemfibrozil and HMG CoA reductase inhibitors can be used, but myositis incidence may be increased; therefore, this combination should be used with caution.

GOALS OF THERAPY

The NCEP's goals of therapy are to get the LDL-C values to < 100 mg/dL (2.6 mmol/L) in CHD patients, to < 130 mg/dL (3.4 mmol/L) in patients with two or more CHD risk factors, and to < 160 mg/dL (4.1 mmol/L) in all patients. In our view, in CHD patients it may also be prudent to lower TG to < 200 mg/dL (2.25 mmol/L), to lower Lp(a) concentrations to < 30 mg/dL, to increase HDL-C concentrations to > 35 mg/dL (0.9 mmol/L), and to try to lower the TC/HDL-C ratio to < 5.0. In patients with severe hypertriglyceridemia and pancreatitis, the goal of therapy is to reduce TG concentrations to < 500 mg/dL (3.65 mmol/L).

CONCLUSION

This chapter is designed to provide laboratory personnel with an overview of the diagnosis and treatment of lipoprotein disorders and NCEP guidelines for therapy. Current diagnostic tests that are recommended include serum TC and HDL-C for screening in the fasting or non-fasting state, and serum TC, TG, HDL-C, and calculated LDL-C in the fasting state. Routine measurements of apolipoproteins AI and B are as yet not recommended because of lack of standardization of available assays and of prospective data documenting that these assays are superior to standard lipid measurements in CHD risk prediction. A direct assay of LDL-C concentrations is now available and should be an important contribution to the diagnosis and management of lipid disorders, especially since it allows clinicians to assess LDL-C with greater accuracy, and in patients in the non-fasting state or with hypertriglyceridemia. Lp(a) cholesterol may also be an important addition to routine lipid assessment in the future. ❖

REFERENCES

1. The Expert Panel. Report of the National Cholesterol Education Program Expert Panel on Detection, Evaluation, and Treatment of High Blood Cholesterol in Adults. Arch Intern Med 1988;148:36–69.
2. The Expert Panel. Summary of the second report of the National Cholesterol Education Program (NCEP) Expert Panel on Detection, Evaluation, and Treatment of High Blood Cholesterol in Adults (Adult Treatment Panel II). JAMA 1993;269: 3015–23.
3. Expert Panel. Blood cholesterol levels in children and adolescents. National Institutes of Health Publication No. 91-2732, 1-119. Washington DC: US Government Printing Office, 1990.
4. Expert Panel. Population strategies for blood cholesterol reduction. National Institutes of Health Publication No. 90-3046, 1-39. Washington, DC: U.S. Government Printing Office, 1991.
5. Schaefer EJ, Levy RI. The pathogenesis and management of lipoprotein disorders. N Eng J Med 1985;312:1300–10.
6. Steinberg D, Parthasarathy S, Carew TE, Khoo JC, Witztum, JL. Beyond cholesterol: modifications of low-density lipoprotein that increase its atherogenicity. New Engl J Med 1989;320: 915–24.
7. Berg K. A new serum type system in man: the Lp system. Acta Pathol Microbiol Scand. 1963;59:369–82.
8. McLean JW, Tomlinson JE, Kuang WJ, et al. cDNA sequence of human apolipoprotein (a) is homologous to plasminogen. Nature 1987;330:132–7.
9. Utermann G. The mysteries of lipoprotein(a). Science 1989;246:904–10.
10. Lackner C, Boerwinkle E, Leffert CC, Rahmig T, Hobbs HH. Molecular basis of apolipoprotein(a) isoform heterogeneity as revealed by pulsed-field gel electrophoresis. J Clin Invest 1991;87:2153–61.
11. Boerwinkle E, Leffert CC, Lin J, Lackner C, Chiesa G, Hobbs HH. Apolipoprotein(a) gene accounts for greater than 90% of the variation in plasma lipoprotein(a) concentrations. J Clin Invest 1992;90:52–60.
12. Loscalzo J, Weinfeld M, Fless GM, Scanu AM. Lipoprotein (a), fibrin binding, and plasminogen activation. Arteriosclerosis 1990;10:240–5.
13. Miles LA, Fless GM, Levin EG, Scanu AM, Plow EF. A potential basis of the thrombotic risks associated with lipoprotein (a). Nature 1989;339:301–3.
14. Hajjar KA, Gavish D, Breslow JL, Nachman RL. Lipoprotein (a) modulation of endothelial cell surface fibrinolysis and its potential role in atherosclerosis. Nature 1989;339:303–5.
15. Dahlen GH, Guyton JR, Altar M, Farmer JA, Kautz JA, Gotto AM. Association of levels of lipoprotein (a), plasma lipids, and other lipoproteins with coronary artery disease documented by angiography. Circulation 1986;74:758–65.
16. Armstrong VW, Cremer P, Eberle E, et al. The association between serum Lp(a) concentrations and angiographically assessed coronary atherosclerosis. Atherosclerosis 1986;62:249–57.
17. Zenker G, Költringer P, Bone G, Niederkorn K, Pfeiffer K, Jurgens G. Lipoprotein(a) as a strong indicator for cerebrovascular disease. Stroke 1986;17:942–5.
18. Murai A, Miyahara T, Fujimoto N, Matsudo M, Kameyama M. Lp(a) lipoprotein as a risk factor for coronary heart disease and cerebral infarction. Atherosclerosis 1986;59:199–204.
19. Hoefler G, Harnoncourt F, Paschke E, Mitrl W, Pfeiffer KH, Kostner GM. Lipoprotein Lp(a): a risk factor for myocardial infarction. Arteriosclerosis 1988;8: 398–401.

20. Sandkamp M, Funke H, Schulte H, Kohler E, Assmann G. Lipoprotein (a) is an independent risk factor for myocardial infarction at a young age. Clin Chem 1990; 36:20–3.
21. Genest J Jr, Jenner JL, McNamara JR, et al. Prevalence of lipoprotein (a) [Lp(a)] excess in coronary artery disease. Am J Cardiol 1991;67:1039–45.
22. Genest JJ Jr, McNamara JR, Ordovas JM, et al. Lipoprotein cholesterol, apolipoprotein A-I and B and lipoprotein(a) abnormalities in men with premature coronary artery disease. J Am Coll Cardiol 1992;19:792–802.
23. Genest JJ Jr, Martin-Munley SS, McNamara JR, et al. Familial lipoprotein disorders in patients with premature coronary artery disease. Circulation 1992;85: 2025–33.
24. Jenner JL, Ordovas JM, Lamon-Fava S, et al. Effects of age, sex, and menopausal status on plasma lipoprotein (a) levels. The Framingham Offspring Study. Circulation 1993;87:1135–41.
25. Rosengren A, Wihelmsen L, Eriksson E, Risberg B, Wedel H. Lipoprotein (a) and coronary heart disease: a prospective case-control study in the general population sample of middle aged men. Br Med J 1990;301:1248–51.
26. Sigurdsson G, Baldursdottir A, Sigvalderson H, Agnarsson G, Thorgeirsson G, Sigfusson N. Predictive value of apolipoproteins in a prospective survey of coronary artery disease in men. Am J Cardiol 1992;69:1251–4.
27. Jauhiainen M, Koskinen P, Ehnholm C, et al. Lipoprotein (a) and coronary heart disease risk: a nested case-control study of the Helsinki Heart Study participants. Atherosclerosis 1991;89:59–67.
28. Ridker PM, Hennekens CH, Stampfer MJ. A prospective study of lipoprotein (a) and the risk of myocardial infarction. JAMA 1993;270:2195–9.
29. The Lipid Research Clinics Program. The Lipid Research Clinics Coronary Primary Prevention Trial. I. Reduction in incidence of coronary heart disease. JAMA 1984; 251:351–64.
30. The Lipid Research Clinics Program. The Lipid Research Clinics Coronary Primary Prevention Trial. II. The relationship of reduction in incidence of coronary heart disease to cholesterol lowering. JAMA 1984;251:365–74.
31. Schaefer EJ, Lamon-Fava S, Jenner JL, et al. Lipoprotein (a) levels and risk of coronary heart disease in men. The Lipid Research Clinics Coronary Primary Prevention Trial. JAMA 1994;271:999–1003.
32. Kannel WB, Castelli WP, Gordon T. Cholesterol in the prediction of atherosclerotic disease: new perspectives based on the Framingham Study. Ann Int Med 1979; 90: 85–91.
33. Anderson KM, Wilson PWF, Odell PM, Kannel WB. An updated coronary risk profile. A statement for health professionals. AHA medical/scientific statement science advisory. Circulation 1991; 83:356–62.
34. Stamler J, Wentworth D, Neaton JD. Is the relationship between serum cholesterol and risk of premature death from coronary heart disease continuous and graded? Findings in 356,222 primary screenees of the Multiple Risk Factor Intervention Trial (MRFIT). JAMA 1986;256:2823–8.
35. Dayton S, Pearce ML, Hashimoto S, Dixon WJ, Tomiyasu U. A controlled clinical trial of a diet high in unsaturated fat in preventing complications of atherosclerosis. Circulation 1969;40 (Suppl II):11–63.
36. Hjermann I, Holme I, Byre KV, Leren P. Effect of diet and smoking intervention on the incidence of coronary heart disease. Lancet 1981;2:1303–10.
37. Holme I, Hjermann I, Helgelend A, Leren P. The Oslo Study: diet and anti-smoking advice: additional results from a 5 year primary prevention trial in middle aged men. Prev Med 1985;14:279–92.

38. Miettinen M, Karvonen MJ, Turpeiner O, Elosuo R, Paavilainen F. Effect of cholesterol lowering diet on mortality from coronary heart disease and other causes. A twelve year clinical trial in men and women. Lancet 1972;2(782):835–8.
39. Ornish D, Brown SK, Scherwitz LW, et al. Can life-style changes reverse coronary heart disease? Lancet 1990;326:129–33.
40. Carlson LA, Rosenhamer G. Reduction of mortality in the Stockholm Ischemic Heart Disease Study by combined treatment with clofibrate and nicotinic acid. Acta Med Scand 1988;223:405–18.
41. Frick MH, Elo O, Haapa K, et al. Helsinki Heart Study: primary prevention trial with gemfibrozil in middle-aged men with dyslipidemia. N Engl J Med 1987;317: 1237–45.
42. Manninen V, Elo O, Frick MH, et al. Lipid alterations and decline in the incidence of coronary heart disease in the Helsinki Heart Study. JAMA 1988;260:641–51.
43. Manninen V, Tenkanen L, Koskinen P, et al. Joint effects of serum triglyceride and LDL cholesterol and HDL cholesterol concentrations on coronary heart disease risk in the Helsinki Heart Study: implication for treatment. Circulation 1992; 85:37–45.
44. Canner PL, Berge KG, Wenger NK, et al. Fifteen-year mortality in Coronary Drug Project patients: long-term benefit with niacin. J Am Coll Cardiol 1986;8:1245–55.
45. Leren P. The effect of plasma cholesterol lowering diet in male survivors of myocardial infarction. Acta Med Scand 1966;466(Suppl):92–116.
46. Alderman E, Haskell WL, Fain JM, et al. Beneficial angiographic and clinical response to multifactor modification in the Stanford Coronary Risk Intervention Project (SCRIP). [Abstract] Circulation 1991;84:140.
47. Blankenhorn DH, Nessim SA, Johnson RL, Sanmarcio ME, Azen SP, Cashin-Hemphill L. Beneficial effects of combined colestipol-niacin therapy on coronary atherosclerosis and coronary venous bypass grafts. JAMA 1987;257:3233–40.
48. Blankenhorn DH, Azen SP, Kramsch DM, et al. Coronary angiographic changes with lovastatin therapy. The monitored atherosclerosis regression study (MARS). Ann Intern Med 1993;1119: 969–76.
49. Brensike JF, Levy RI, Kelsey SF, et al. Effects of therapy with cholestyramine on progression of coronary atherosclerosis: results of the NHLBI Type II Coronary Intervention Study. Circulation 1984;69:313–24.
50. Brown BG, Zhao XQ, Sacco DE, Albers JJ. Lipid lowering and plaque regression: new insights into prevention of plaque disruption and clinical events in coronary disease. Circulation 1993;87:1781–91.
51. Brown BG, Albers JJ, Fisher LD, et al. Regression of coronary artery disease as a result of intensive lipid-lowering therapy in men with high levels of apolipoprotein B. N Eng J Med 1990;323:1289–98.
52. Buchwald H, Varco RL, Matts JP, Long JM, Fitch LL, Campbell GS. Effect of partial ileal bypass surgery on mortality and morbidity from coronary heart disease in patients with hypercholesterolemia: report of the Program on Surgical Control of the Hyperlipidemias (POSCH). N Eng J Med 1990;323:946–55.
53. Fuster V, Badimon L, Badimon JJ, Chesbro JH. The pathogenesis of coronary artery disease and the acute coronary symptoms. N Eng J Med 1992;326:242-56, 310–18.
54. Kane JP, Malloy MJ, Ports TA, Phillips NR, Diehl JC, Havel RJ. Regression of coronary atherosclerosis during treatment of familial hypercholesterolemia with combined drug regimens. JAMA 1990;264:3007–12.
55. Watts GF, Lewis B, Brunt JNH, et al. Effects on coronary artery disease of lipid lowering diet: a diet plus cholestyramine in the St. Thomas Atherosclerosis Regression Study (STARS). Lancet 1992;339:563–9.

56. Schuler G, Hambrecht R, Schlierf G, et al. Regular exercise and low fat diet: effects on progression of coronary artery disease. Circulation 1992;86:1–11.
57. NIH Consensus Conference. Lowering blood cholesterol to prevent heart disease. JAMA 1985;253:2080–6.
58. NIH Consensus Conference. Triglyceride, HDL cholesterol and coronary heart disease. JAMA 1993;269:505–10.
59. Gordon DJ, Knoke J, Probstfeld JL, Superko R, Tyroler HA. High density lipoprotein cholesterol and coronary heart disease in hypercholesterolemic men: the Lipid Research Clinics Coronary Primary Prevention Trial. Circulation 1986;74: 1217–25.
60. Gordon DJ, Rifkind BM. High-density lipoprotein: the clinical implications of recent studies. N Eng J Med 1989;321:1311–16.
61. Laboratory Standardization Panel. Recommendations for improving cholesterol measurements. National Institutes of Health Publication No. 90-2964, 1-64. Washington DC: US Government Printing Office, 1990.
62. Friedewald WT, Levy RI, Fredrickson DS. Estimation of the concentration of low density lipoproteins cholesterol without use of the preparative ultracentrifuge. Clin Chem 1972;18: 499–502.
63. McNamara JR, Cohn JS, Wilson PWF, Schaefer EJ. Calculated values for low density lipoprotein cholesterol in the assessment of lipid abnormalities and coronary disease risk. Clin Chem 1990;36:36–42.
64. Leary ET, Tjersland G, Warnick GR: Evaluation of the Genzyme immunoseparation reagent for direct quantitation of LDL cholesterol. Clin Chem 1993;39:1124–5.
65. Avogaro P, Bittolo Bon G, Cazzolato G, Quinci GB. Are apolipoproteins better discriminators than lipids for atherosclerosis? Lancet 1979;1:901–3.
66. Sniderman A, Shapiro S, Marpole D, Skinner B, Teng B, Kwiterovich PO Jr. Association of coronary atherosclerosis with hyperapobetalipoproteinemia (increased protein but normal cholesterol levels in human plasma low density lipoproteins). Proc Natl Acad Sci USA. 1980;77:604–8.
67. Whayne TF, Alaupovic P, Curry MD, Lee ET, Anderson PS, Snecter E. Plasma apolipoprotein B and VLDL-, LDL-, and HDL-cholesterol as risk factors in the development of coronary artery disease in male patients examined by angiography. Atherosclerosis 1981;39:411–24.
68. Kwiterovich PO, Jr., Bachorik PS, Smith HH, et al. Hyperapobetalipoproteinaemia in two families with xanthomas and phytosterolaemia. Lancet 1981;1:466–9.
69. Maciejko JJ, Holmes DR, Kottke BA, Zinsmeister AR, Dinh DM, Mao SJT. Apolipoprotein A-I as a marker of angiographically assessed coronary artery disease. N Engl J Med 1983;309:385–9.
70. Austin MA, Breslow JL, Hennekens CH, Buring JE, Willett WC, Krauss RM. Low density lipoprotein subclass patterns and risk of myocardial infarction. JAMA 1988;260:1917–21.
71. Campos H, Genest JJ, Blijlevens E, et al. Low density lipoprotein particle size and coronary artery disease. Arterioscler Thromb 1992;12:187–95.
72. Ishikawa T, Fidge N, Thelle DS, Forde DH, Miller NE. The Tromso Heart Study: serum apolipoprotein A-I concentration in relation to future coronary heart disease. Eur J Clin Invest 1978;8:179–82.
73. Salonen JT, Salonen R, Penttila I, et al. Serum fatty acids, apolipoproteins, selenium and vitamin antioxidants and the risk of death from coronary artery disease. Am J Cardiol 1985;56:226–31.
74. Cremer P, Muche R. The Gottingen Risk, Incidence and Prevalence Study (GRIPS): recommendations for the prevention of coronary heart disease. Therapeutische Umschau 1990;47:482–91.

75. Stampfer MJ, Sacks FM, Salvini S, Willett WC, Hennekens CH. A prospective study of cholesterol, apolipoproteins, and the risk of myocardial infarction. N Engl J Med 1991;325:373–81.
76. Coleman MP, Key TJ, Wang DY, et al. A prospective study of obesity, lipids, apolipoproteins, and ischemic heart disease in women. Atherosclerosis 1992;92: 177–85.
77. Carlson LA, Hamsten A, Asplund A. Pronounced lowering of serum lipoprotein Lp(a) in hyperlipidemic subjects treated with nicotinic acid. J Intern Med 1989; 226:271–6.
78. Seman LJ, Jenner JL, McNamara JR, Schaefer EJ. Quantitation of plasma lipoprotein(a) by cholesterol assay of lectin bound lipoprotein(a). Clin Chem 1994; 40:400–3.
79. Goldstein JL, Schrott HG, Hazzard WR, Bierman EL, Motulsky AG. Hyperlipidemia in coronary heart disease. II. Genetic analysis of lipid levels in 176 families and delineation of a new inherited disorder, combined hyperlipidemia. J Clin Invest 1973;52:1544–68.
80. Goldstein JL, Brown MS. Familial hypercholesterolemia. In: Stanbury JB, Wyngaarden JB, et al, eds. The metabolic basis of inherited disease, 5th ed. New York: McGraw Hill, 1983:672-712.
81. Janus ED, Nicoll AM, Turner PR, Magill P, Lewis B. Kinetic bases of the primary hyperlipidemias:studies of apolipoprotein B turnover in genetically defined subjects. Eur J Clin Invest 1980;10:161–72.
82. Chait A, Albers JJ, Brunzell JD. Very low density lipoprotein overproduction in genetic forms of hypertriglyceridemia. Eur J Clin Invest 1980;10:17–22.
83. Schaefer EJ. Hyperlipoproteinemia. In: Rakel RE (ed). Conn's current therapy. Philadelphia: W.B. Saunders, 1991:515-24.
84. Rall SC Jr, Weisgraber KH, Innerarity TL, Mahley RW. Structural basis for receptor binding heterogeneity of apolipoprotein E from type III hyperlipoproteinemic subjects. Proc Natl Acad Sci USA 1982;79:4696–700.
85. Schaefer EJ, Gregg RE, Ghiselli G, et al. Familial apolipoprotein E deficiency. J Clin Invest 1986;78:1206–19.
86. Schaefer EJ. Dietary and drug treatment. In: Brewer HB Jr., moderator. Type III hyperlipoproteinemia: diagnosis, molecular defects, pathology and treatment. Ann Int Med 1983;98:623–40.
87. Ordovas JM, Litwack-Klein LE, Schaefer MM, Wilson PWF, Schaefer EJ. Apolipoprotein E isoform phenotyping methodology and population frequency with identification and apo E1 and apo E5 isoforms. J Lipid Res 1987;28:371–80.
88. Fredrickson DS, Levy RI, Lees RS. Fat transport in lipoproteins: an integrated approach to mechanisms and disorders. N Engl J Med 1967;276:34–44, 94–103, 148–56, 215–25, 273–81.
89. Brown MS, Goldstein JL. The LDL receptor concept: clinical and therapeutic implications. In: Stokes J III, Mancini M, eds. Hypercholesterolemia: clinical and therapeutic implications. Atherosclerosis Reviews 1987, Vol 18. New York: Raven Press. 1988:85-94.
90. Innerarity TL, Weisgraber KH, Arnold KS, et al. Familial defective apolipoprotein B-100: low density lipoproteins with abnormal receptor binding. Proc Natl Acad Sci 1987;84:6919–25.
91. Langer T, Strober W, Levy RI. The metabolism of low density lipoprotein in familial type II hyperlipoproteinemia. J Clin Invest 1972;51:1528–36.
92. Kane JP, Malloy MJ, Tun P, et al. Normalization of low density lipoprotein levels in heterozygous familial hypercholesterolemia with a combined drug regimen. New Engl J Med 1981;304:251–8.

93. Mabuchi H, Sakai T, Sakai Y. Reduction of serum cholesterol in heterozygous patients with familial hypercholesterolemia: additive effects of compactin and cholestyramine. New Eng J Med 1983;308:609–19.
94. Hunninghake DB, Stein FA, Dujorne CA, et al. The efficacy of intensive dietary therapy alone or combined with lovastatin in outpatients with hypercholesterolemia. New Eng J Med 1993;328:1213–19.
95. Genest JJ Jr, Bard JM, Fruchart JC, Ordovas JM, Schaefer EJ. Familial hypoalphalipoproteinemia in premature coronary artery disease. Arterioscler Thromb 1993;13:1728–37.
96. Vergani C, Bettale A. Familial hypoalphalipoproteinemia. Clin Chim Acta 1981; 114:45–52.
97. Third JLHC, Montag J, Flynn M, Freidel J, Laskarzewski P, Glueck CJ. Primary and familial hypoalphalipoproteinemia. Metabolism 1984;33:136–46.

CHAPTER 3

Preanalytical Variations in Lipid, Lipoprotein, and Apolipoprotein Testing

Nader Rifai, D. Robert Dufour, and Gerald R. Cooper

INTRODUCTION

❖ Accurate determination of serum lipids is dependent on control of both analytical and preanalytical factors. Preanalytical variation in subjects results from differences in lifestyle, altered lipid metabolism due to disease, the source of the specimen, and conditions of sample collection. Variation can arise from biological, behavioral, and clinical factors, as well as variability in specimen collection and handling (see Figure 3–1). This chapter discusses the factors that contribute to preanalytical variations and addresses the importance of standardization of patient preparation for lipid testing.

BIOLOGICAL VARIATIONS

Intra-individual Variation

As for most laboratory tests, intra-individual variation in lipid values is generally smaller than person-to-person or inter-individual variation. An assessment of biological components of variation of total cholesterol (TC), high-density lipoprotein cholesterol (HDL-C), and apolipoproteins (apo) AI and B showed an index of individuality much less than 0.6. (The index of individuality is calculated by dividing the intra-individual coefficient of variation [CV] by the inter-individual CV.) This finding indicates that the use of conventional population-based reference intervals in an interpretation is of little value and may be misleading.[1] Population-based reference ranges are not useful unless the index of individuality is greater than 1.4.[2]

Figure 3–1 ❖ *Components of Preanalytical Variation*

BIOLOGICAL

Intra-individual
Age
Gender
Race

BEHAVIORAL

Diet
Obesity
Cigarette Smoking
Alcohol Intake
Caffeine Intake
Exercise
Stress

CLINICAL

Disease-Induced*

Endocrine & Metabolic
Hypothyroidism
Hypopituitarism
Diabetes Mellitus
Acute Intermittent Porphyria
Pregnancy

Renal
Nephrotic Syndrome
Chronic Renal Failure

Hepatic
Congenital Biliary Atresia

Storage Disease
Gaucher Disease
Glycogen Storage Disease
Tay-Sachs Disease

Acute & Transient
Burns
Infections
Myocardial Infarction

Others
Anorexia Nervosa
Systemic Lupus Erythematosus

Drug-Induced*

Antihypertensives
Thiazides
Chlorothalidone
Beta-blockers

Immunosuppressives
Cyclosporine
FK506
Prednisolone

Sex Steroids
Estrogen
Progestin

SAMPLE COLLECTION & HANDLING

Fasting Status
Anticoagulants & Preservatives
Capillary vs. Venous
Hemoconcentration
Specimen Storage

*Limited list

Therefore, health-based cutpoints from large population studies were derived by experts for the classification of hyperlipoproteinemia.

With recent improvements in analytical performance of lipid assays, it has become easier to define the day-to-day biological variations in lipid concentrations. Knowledge of such variations is important in that it provides both laboratorians and clinicians with a framework for comparing results to evaluate the efficacy of therapy. With improvement in methods, a major fraction of the variation in lipid measurements may be due to intra-individual variability rather than analytical variability. One recent study attributed from 69% (for low-density lipoprotein cholesterol [LDL-C]) to 96% (for very-low-density lipoprotein cholesterol [VLDL-C] and triglyceride [TG]) of total variance to within-person variation.[3]

Serum TC is the most stable lipid analyte; day-to-day biological variation averages 6.1%,[4, 5] although some individuals may vary by up to 11%.[6] (The percentages for biological variations expressed in this chapter represent the CV, which is mathematically derived by dividing the standard deviation by the mean.) Time of day has little effect on TC concentration, as within-day variation is generally less than 3%.[6, 7] TC tends to be higher in winter than in summer by an average of 2.5%.[8] In women, TC concentrations may fluctuate during the menstrual cycle, averaging 10–20% lower in the late luteal and menstrual phases.[9, 10] For the average woman, TC concentrations increase to about 14 mg/dL (0.36 mmol/L) around ovulation and fall to a nadir during menstruation.[11] During pregnancy, TC gradually rises and has increased 50–80% by the time of delivery; thus, lipid testing of pregnant women should be deferred.[12, 13]

TG concentrations show marked intra-individual variation. Even excluding the known marked postprandial fluctuation, fasting TG concentrations differ by an average of 23% over one or more months, and in some persons may fluctuate as much as 40% around the mean value.[4, 6] During the day, TG concentrations typically rise from a nadir at 3:00 AM to reach peak values in mid-afternoon, then progressively fall throughout the evening; average diurnal variation is approximately 30%.[14] Because of this marked random variation, it is difficult to assess seasonal, menstrual, and other types of rhythmic fluctuations of TG.

HDL-C concentrations show an average intra-individual variation of 7% over one month to a year, but seldom differ by more than 12%. Seasonal variation in HDL-C is significant and is slightly less than that of TC and LDL-C.[8]

When LDL-C concentration is derived from measurements of TC, HDL-C, and TG, biological variability is relatively high, at an average of 9.5%.[4] As with TC, LDL-C concentration is approximately 2.5% higher in winter than in summer,[6, 8] and in women is lower in the late luteal phase and during menstruation.[10, 11]

Data from apolipoprotein measurements show similar variability. Apo AI biological variation averages 7–8%; there are apparently no changes during the menstrual cycle.[10, 15] The intra-individual variation for both HDL-C and apo AI is greater in smokers than in non-smokers.[16] Apo B has an average within-day biological variation of 6.5%, with a day-to-day variation of 8–10%.[15] As with TC, apo B increases markedly during pregnancy.

Lipoprotein(a) [Lp(a)] variations are less well known. A recent study reported an average biological variation for Lp(a) of 8.6%.[17] Apo AI concentrations tend to be highest in the evening and fall to a nadir at 6 AM.[18] It is not clear whether these differences are governed by sleep-wake or light-dark cycles. There is no significant diurnal variation in apo B or Lp(a) values.[18, 19]

Cooper et al. have proposed that the total TC intra-individual variation should be less than 5% to reliably classify the risk of coronary heart disease (CHD).[5, 6] Intra-individual variation (CV_I) is due to a combination of biological variation (CV_B), which includes collection variables (discussed below), diurnal variation, and other inherent changes in the person, as well as analytical imprecision (CV_A), which is dependent on the instrumentation used. The total variability can be expressed mathematically as

$$CV_I^2 = \left[\frac{CV_B^2}{n \times K} + \frac{CV_A^2}{K}\right]^{1/2}$$

where K is the number of specimens and n is the number of replicate laboratory determinations per specimen. If the total variation is more than 5%, then it will be necessary to obtain more than one sample to reduce the variation to the acceptable 5% level. Both analytical and biological variation must be reduced to allow estimates of risk of atherosclerosis to be made on the fewest number of specimens possible.

Table 3–1 shows the total individual variation obtained by varying the number of samples, the analytical variation, and by altering the biological variation from the average TC CV_B of 6.5%[6] through rigorous patient preparation. Although the table indicates the need for utilization of at least two specimens to reach this goal, which is also the recommendation of the National Cholesterol Education Program (NCEP),[20] single specimens may suffice for classification of patients whose TC values are far from the cutpoints. In a recent study, single samples resulted in accurate classification of risk category for patients with TC concentrations below 185 mg/dL (4.79 mmol/L), between 215 and 225 mg/dL (5.57–5.83 mmol/L), or over 255 mg/dL (6.60 mmol/L) when using a method with analytical variation of 1.3%.[21] Because of greater measurement imprecision (CV = 4.4%), LDL-C values could be correctly classified as to risk category with a single specimen for concentrations below 116 mg/dL (3.00 mmol/L) or above 174 mg/dL (4.51 mmol/L).[21] Because of the new NCEP risk categories in patients with a history of CHD, the lower value would no longer be reliable. Thus, two measurements will continue to be needed for LDL-C, except for patients with the highest results.

Effects of Age, Sex, and Race on the Lipoprotein Profile

Numerous demographic factors within the population are correlated with variation in lipid and lipoprotein concentrations. In newborns, TC, TG, and most lipoprotein and apolipoprotein concentrations rise rapidly from the low values in cord blood to 80% of adult values by 4 d of age.[22] Lp(a) rises more slowly, continuing to increase gradually through at least the first 6 months of life.[23, 24] In children, lipid concentrations of both boys and girls remain stable until just before puberty, when a transient fall in TC, TG, HDL-C, and LDL-C occurs.[22] At puberty, lipid val-

Table 3-1 ❖ Effect of Multiple Specimens on Total Individual Variation (CV_I% of TC [200 mg/dL or 5.17 mmol/L])

Samples per person = K	CV_A = 5%		CV_A = 3%		CV_A = 2%		CV_A = 1%	
	n = 1	*n = 2*	*n = 1*	*n = 2*	*n = 1*	*n = 2*	*n = 1*	*n = 2*
CV_B 6.5%, Number of Replicate Measurements to Reach CV_I 5%								
1	8.2	7.4	7.2	6.8	6.8	6.7	6.6	6.5
2	5.8	5.2	5.1	4.8	4.8	4.7	4.7	4.6
3	4.7	4.3	4.1	3.9	3.9	3.8	3.8	3.8
4	4.1	3.7	3.6	3.4	3.4	3.3	3.3	3.3
CV_B 5.0%, Number of Replicate Measurements to Reach CV_I 5%								
1	7.1	6.1	5.8	5.4	5.4	5.2	5.1	5.0
2	5.0	4.3	4.1	3.8	3.8	3.7	3.6	3.6
3	4.1	3.5	3.4	3.1	3.1	3.0	2.9	2.9
4	3.5	3.1	2.9	2.7	2.7	2.6	2.5	2.5

CV_A = Analytical coefficient of variation.
CV_B = Biological coefficient of variation.
The dotted line indicates the number of specimens required to meet the NCEP guideline of <5% total variability in cholesterol measurement. If biological variation is at the average value of 6.5%, this goal can be achieved with duplicate samples if the analytical CV is <2%, but requires 3 or 4 specimens if analytical variation is greater. If biological variation were only 5%, which may be achieved with careful patient preparation, duplicate measurements on a single specimen would be close to the guideline with a single specimen for methods with analytical CV <2%.

ues begin to diverge; in boys, TC decreases slightly due to a significant fall in HDL-C (and apo AI).[25, 26] After puberty, HDL-C and apo AI remain stable, at least until age 55.[26] In girls, HDL-C and apo AI increase gradually from menarche to the time of menopause.[25, 27] Lp(a) is similar in men and women of all ages, although it is slightly higher in young and adolescent girls.[28]

In black children, HDL-C, apo AI, Lp(a), and VLDL-C concentrations are noticeably higher than those of white children by about age 9, while LDL-C and apo B concentrations are lower.[29, 30] In adulthood, these sex- and race-related differences persist.[31] Little data exists on other minority groups in the population. In two studies, there were no differences in most lipid measurements between Hispanic and non-Hispanic whites.[32, 33] In contrast, Lp(a) is significantly lower in Hispanic whites, intermediate in Asians and non-Hispanic whites, and highest in blacks.[34-36] Similar plasma TC and TG changes were observed with respect to age and sex in the Lipid Research Clinics prevalence study of 60,502 participants in a survey of 11 separate and well-defined North American populations.[37]

In all individuals, TC concentrations gradually increase with age, although the magnitude of increase is not as great in more recent studies as it was 20 or more years ago.[38] In the Second National Health and Nutrition Examination Survey, mean TC concentration increased from approximately 180 mg/dL (4.66 mmol/L) at age 20 to approximately 230 mg/dL (5.96 mmol/L) in men and 250 mg/dL (6.48 mmol/L) in women by age 65.[39] This increase is due to increasing LDL-C concentration and is accompanied by an increase in apo B values. The TG concentration also increases with age.[25, 40] Lp(a) appears to rise slightly with age, particularly in women after menopause.[41]

BEHAVIORAL VARIATIONS

The major behavioral factors that affect the lipid, lipoprotein, and apolipoprotein serum concentrations are diet, obesity, cigarette smoking, alcohol and caffeine intake, exercise, and stress. Since these lifestyle elements are controllable, it is imperative that subjects maintain their usual behavior for several days before blood specimens for lipid testing are obtained.

Diet

The effect of diet on lipid and lipoprotein concentrations is well established. The extent of this effect, however, varies among individuals. It has been shown that an increase in dietary intake of cholesterol can cause serum TC concentration to rise over 5% in only 30% of the studied population.[6] Various fatty acids appear to have different effects on the lipoprotein profile. In general, diets rich in mono- and polyunsaturated fatty acids cause serum TC, LDL-C, apo B, and TG to decrease, while diets rich in saturated fat, mainly palmitic acid, cause serum TC and LDL-C to increase.[42, 43] However, stearic acid, which is also a saturated fatty acid, does not appear to increase LDL-C.[43]

Omega-3 fatty acids, known as fish oil, have been shown to consistently decrease TG and VLDL in both normal and hypertriglyceridemic subjects, possibly by inhibiting the hepatic synthesis of VLDL apo B and VLDL TG.[44] In contrast, the documented effects of fish oil on TC, LDL-C, and HDL-C have been inconsistent.

The Cholesterol Lowering Atherosclerosis Study observed that when total and saturated fat intakes are being reduced, protein and carbohydrate are preferred substitutes for fat calories rather than mono- or polyunsaturated fats.[45] Furthermore, the reduction in fat intake was reflected in a significant decrease in TC concentrations.[45]

The effect of dietary intake of fiber on serum lipids remains controversial.[46] However, water-soluble fibers were shown to reduce TC.[47, 48]

Vegetarians are known to have a healthier lipid profile than non-vegetarians. One study has demonstrated that LDL-C is 37% lower and HDL-C is 12% higher in strict vegetarians than in non-vegetarians.[49]

Changes in eating habits have been shown to significantly alter lipid and lipoprotein profiles within a relatively short period of time. When the low-fat and high-fiber diet of a group of Tarahumara Indians was substituted with an "affluent" diet for a period of 5 weeks, increases in TC and LDL-C (up to 39%) and TG concentrations (19%) as well as body weight (7%) were seen.[50]

Since diet modification can affect lipid concentrations, subjects should maintain customary dietary intake and eating habits before blood samples are collected for lipid and lipoprotein studies. When subjects have changed their dietary habits significantly, the NCEP guidelines suggest a waiting period of 3–6 months before performing any lipid testing.[20]

Obesity

Obesity has been shown to be associated with increased risk of CHD. In addition, obese individuals have higher TG, TC, and LDL-C and lower HDL-C compared to non-obese controls.[6] After weight loss, an obese individual will experience a decrease in TG of about 40%, a decrease in TC and LDL-C of about 10%, and an increase in HDL-C of about 10%.[51] In a study of identical twins, obesity was independently associated with TC, LDL-C, HDL-C, and TG, as well as systolic and diastolic blood pressure and glucose tolerance.[52]

Current evidence suggests that body fat distribution has a stronger association with various morbidities, such as hyperlipidemia, hypertension, diabetes, and CHD, than just body weight.[53] The intra-abdominal fat accumulation (waist-to-hip ratio) in obese women was associated with higher TG and VLDL-C and lower HDL-C, independent of total obesity.[54]

Repeated weight gain and loss in obese individuals would precipitate significant variations in serum lipid and lipoprotein concentration. *Therefore, subjects should avoid any alterations in lifestyle that can cause weight changes at the time when blood specimens are collected.*

Cigarette Smoking

Cigarette smoking is an established independent risk factor for CHD. Smoking-associated physiological changes that probably contribute to the high risk of heart disease include increased clotting factors and carboxyhemoglobin concentrations, increased blood viscosity, and altered lipoprotein profile.[55] The strong association of smoking to white blood cell count, hematocrit, and heart rate suggests that smoking-induced risk is more associated with occlusive thrombosis than with atherosclerosis.[56] In addition, oxidized LDL produced after exposure to cigarette smoke was shown to cause the accumulation of cholesterol ester in macrophages *in vitro*.[57] Therefore, oxidized LDL could also contribute to foam cell formation and atherosclerosis in smokers.

Compared to non-smokers, cigarette smokers have significantly higher serum TG, VLDL-C, and LDL-C and lower apo AI and HDL-C (mainly the HDL_3-C fraction).[58, 59] Furthermore, the relationship between smoking and HDL-C was shown to be dose-dependent in both men and women.[60] Therefore, the extent of smoking will effect the degree of alteration in the lipoprotein profile.

Subjects should not change their smoking pattern at the time when blood specimens are collected.

Alcohol Intake

Several epidemiological studies have demonstrated a relationship between alcohol consumption, lipoprotein profile changes, and cardiovascular mortality.[61] The alcohol-induced alterations in the lipoprotein pattern depend on the amount of alcohol consumed, individual susceptibility, genetic variables, and dietary factors. Therefore, these changes differ among moderate and heavy drinkers. Moderate alcohol drinkers (1.2 oz/d) have increased concentrations of HDL-C (mainly the HDL_3-C fraction), apo AI, and apo AII compared to non-drinkers.[62, 63] However, moderate alcohol consumption in patients with primary hypertriglyceridemia usually results in a profound increase in TG.[64] With increased alcohol intake, LDL-C and HDL_2-C are reduced and TG is increased.[63] In chronic alcoholic men with normal liver function and structure, TG is usually in the normal range, HDL-C is increased, and LDL-C is decreased.[65]

Lp(a) concentration appears to be influenced by alcohol differently from the other lipids. Lp(a) concentrations decrease by about 33% initially then return to baseline after 6 w of alcohol intake.[66]

Alcohol consumption has been shown to have a U-shaped relationship to heart disease.[6] The actual mechanism of that link and the molecular basis of the alcohol-induced lipid changes remain unclear.

Subjects should maintain their usual pattern of alcohol intake prior to blood specimen collection.

Caffeine Intake

Several epidemiological studies have investigated the link between coffee consumption, cardiovascular morbidity, and altered lipid concentrations. These correlations have been inconsistent and have been largely dismissed as being due to confounding variables. Intake of coffee in Norway has been associated with a trend toward an atherogenic diet and unhealthy lifestyle.[67] Recently it was suggested that the inconsistency in these findings may result from the coffee brewing methods used. The strongest evidence for a direct association between coffee consumption and increased TC and LDL-C values comes from Scandinavia, where coffee is normally brewed by boiling.[68] Consumption of filtered coffee was reported to have a substantially lesser effect on cholesterol values than consumption of coffee brewed by boiling. The serum concentrations of apo AI, apo AII, apo B, HDL-C, and VLDL-C appear to be unaffected by consumption of coffee, even when brewed by boiling.[69] However, in another study, boiled Scandinavian-style coffee was found to contain more lipid material than drip-filtered coffee, and this was reported to raise serum LDL-C and VLDL-C concentrations.[70]

Replacing regular coffee with decaffeinated coffee did not alter TG, TC, or HDL-C in healthy men and women.[71] However, a randomized clinical trial found that consumption of filtered regular coffee led to a statistically significant increase in the plasma concentration of TC, which is due to increases in both LDL-C and HDL-C concentrations.[72]

Subjects should maintain their usual daily intake of caffeine during the days prior to lipid testing, but abstain from drinking coffee and cream 12 h prior to collecting blood specimens.

Exercise

Several epidemiological investigations have reported an association between a sedentary pattern of living and an increased risk of developing the clinical manifestations of CHD.[73] A lower level of physical fitness is associated with higher serum concentrations of LDL-C and TG and higher risk of death from cardiovascular disease.[74] Exercise exerts a preventative effect on the progression of coronary atherosclerosis, possibly by altering the lipoprotein profile. In hyperlipidemic subjects, additional favorable alteration in the lipoprotein profile was seen when physical exercise accompanied dietary treatment.[75, 76]

Strenuous exercise causes a decrease in serum TG, LDL-C, and apo B and an increase in HDL-C (mainly the HDL_2-C fraction) and apo AI.[6, 77] Furthermore, the extent of these changes is dependent on the intensity and type of exercise training program.[77] Acute exercise causes a significant rise in HDL-C which is due to an increase in HDL_3-C.[78] Moderate and regular physical exercise also appears to have a favorable effect on serum lipids. Adults who walk 2.5 to 4 h each week have lower TC and higher HDL-C than those who do not walk regularly.[79] In previously sedentary women who walked briskly an average of 155 min each week for a period of 1 y, TC decreased by 6.5% and HDL-C increased by 27%.[80]

Subjects should maintain their usual level of exercise on days before lipid testing, but avoid any strenuous exercise 24 h prior to blood specimen collection.[6]

Stress

A high level of stress as well as a Type A personality have been linked to increased risk of CHD. Furthermore, several studies have demonstrated an increase in TC concentration during stressful situations,[81] which is possibly due to dietary changes that accompany stress.[82] The stressors employed in these trials were diverse and included both chronic stressors and acute laboratory challenges. Some evidence suggests that stress affects the lipid concentrations of males and females differently.[83] In addition, elective hospitalization was shown to cause HDL-C and apo AI to decrease by approximately 10%.[84]

The exact mechanism of the interplay of stress with CHD and lipid changes remains unclear. Therefore, it is not advisable to perform lipid testing for cardiac risk assessment in a stressful period such as during academic examinations or hospitalization, even for elective procedures.

To minimize the effect of acute stress on lipid testing, subjects who report to the laboratory should relax for at least 15 min prior to blood sample collection.

CLINICAL VARIATIONS

An individual's lipoprotein profile is markedly altered in acute infectious and metabolic diseases. In addition, drug-induced hyperlipidemia has been reported in several patient populations. To obtain the subject's usual lipid and lipoprotein concentrations, testing should preferably be performed in the absence of any secondary dyslipoproteinemia. Otherwise, the disorder or the medication causing the disruption in lipid values should be noted on the patient's analytical report.

Disease-Induced Secondary Lipid Alterations

Several endocrine, metabolic, renal, hepatic, and storage diseases precipitate secondary hyper- or hypolipidemia. Hypothyroidism and diabetes mellitus are perhaps the most common disorders to cause secondary hyperlipidemia. Serum TC and LDL-C are increased in 30% of patients with hypothyroidism because of increased production and decreased removal of LDL.[85] Mild and severe hypertriglyceridemia are seen in patients with Type I and Type II diabetes mellitus.[6, 86] The CARDIA Study found that higher concentrations of insulin are associated positively with unfavorable values of TC and LDL-C, TG, apo B, and blood pressure, and negatively with HDL-C, HDL_2-C, HDL_3-C, and apo AI.[87] Insulin treatment of non-insulin-dependent diabetics usually results in a significant change toward normalization of TG, HDL-C, and LDL-C concentrations.[88] Lp(a) appears to be increased in both Type I and Type II patients.[89] Evidence suggests that a direct correlation exists between diabetic control and Lp(a) concentrations.[89, 90]

The interplay of insulin with the lipoprotein lipase enzyme is thought to be responsible for the hyperlipidemic status in the diabetic population.[91] Patients with nephrotic syndrome or chronic renal failure who are undergoing hemodialysis have significantly higher serum TC and LDL-C, TG, apo B, and Lp(a), and lower concentrations of apo AI compared to controls.[92–94] Furthermore, increases in Lp(a) concentrations were documented in patients with pronounced proteinuria of various origins.[95] The exact nature of these abnormalities remains uncertain. However, it has been recently suggested that lipid alterations in patients with nephrotic syndrome are attributable to reversible increases in lipoprotein production.[96] Obstructive biliary tract disease causes the production of the abnormal lipoprotein "lipoprotein X." Since this cholesterol-rich particle is poorly removed from circulation, TC concentrations in these patients are markedly increased (over 1000 mg/dL [25.90 mmol/L]).[97]

Acute myocardial infarction is associated with variable decreases in TC, LDL-C, apo AI, and apo B and an increase in Lp(a).[98–100] The extent of the decreases is usually dependent on the original lipid values. For example, little change would be seen in TC when the original concentration is below 200 mg/dL (5.17 mmol/L). Lipid concentrations remain stable within 24 h after the infarction and then decrease gradually to a lower plateau at which they remain for 6–8 weeks.[6] Changes in serum lipids similar to those encountered after a myocardial infarction were reported after a cerebral infarction.[101]

Blood specimens should be obtained within 24 h after a myocardial or cerebral infarction or 3 months following the infarct in order to reflect accurately the person's usual lipid values.

Independently of the infectious agent, the cause of the illness, and the clinical condition of the patient, infections and inflammations cause TG and Lp(a) to increase and TC and HDL-C to decrease.[102, 103] Various forms of dyslipoproteinemia are produced by numerous other disorders, such as acute intermittent porphyria, glycogen storage disease, Gaucher disease, Tay-Sachs disease, rheumatoid arthritis, anorexia nervosa, and systemic lupus erythematosus. The treatment of the primary disorder will usually result in the normalization of the lipoprotein profile. Therefore, *serum lipid assessment should only be conducted after a complete recovery from illness.*

Drug-Induced Secondary Lipid Alterations

Medications that alter lipoprotein metabolism and serum concentrations include diuretics, some beta-blockers, sex steroids, glucocorticoids, and cyclosporine, among others. Thiazides and chlorothalidone, the most commonly used diuretics in the treatment of mild essential hypertension, cause increases in serum TC, LDL-C, VLDL-C, TG, and apo B concentrations, and decreases in HDL-C and apo AI concentrations compared to controls.[104] Other diuretics are also known to disrupt the lipoprotein profile.[105] In addition, propranolol, a non-cardioselective beta-blocker which is also used in the treatment of hypertension, significantly increases serum TG and decreases HDL-C (mainly the HDL_2-C fraction).[106]

Various investigators have reported conflicting findings concerning the effect of oral contraceptives on the lipid and lipoprotein concentrations. This discrepancy can be largely attributed to the variations in the hormonal content of the medications used in these studies, mainly the differences in the estrogen to progestin ratio.[107, 108] Oral contraceptives with a high progestin content cause an increase in serum TC and LDL-C and a decrease in HDL-C.[109] Opposite changes can be expected in women taking oral contraceptives with a high estrogen content and in post-menopausal women receiving an estrogen supplement.[110] Estrogen treatment, both in women and in men suffering from prostatic carcinoma, appears to influence Lp(a) concentration.[111]

Immunosuppressive agents are also reported to alter lipid metabolism. Prednisolone increases serum TC, LDL-C, HDL-C, VLDL-C, TG, apo AI, and apo B, possibly by increasing lipoprotein production.[112] Cyclosporine markedly increases TC, LDL-C, and apo B, and decreases Lp(a).[113, 114] However, hypertriglyceridemia is the most profound lipid change after prednisolone and azathioprine therapy, and hypercholesterolemia is the more common finding after cyclosporine therapy in transplant patients.[115] In contrast, FK 506 appears to decrease TC concentration.[116] The actual mechanisms for the distinct effect of the two immunosuppressive agents, FK 506 and cyclosporine, on the lipoprotein values are currently unknown.

Because of the effect of various pharmacological agents on lipoprotein metabolism, medications should be suspended, if possible, for several days or weeks, depending on the drug and its effects, before lipids studies are conducted. Otherwise, any medication that is known to alter lipid values should be noted on the laboratory report.

Pregnancy

Physiologic and endocrine systems are greatly affected during pregnancy, altering the concentration of several biochemical parameters. Serum TC, LDL-C, TG, apo AI, apo AII, apo B, and Lp(a) increase significantly during pregnancy, mainly in the second and third trimesters.[6, 13, 117] The concentrations of these analytes usually return to normal within 10 w postpartum, unless the woman is breast feeding.[118] The increased mobilization of lipids probably reflects the rise in maternal metabolic demand as a result of pregnancy. Therefore, the assessment of the lipoprotein profile during pregnancy does not reflect the usual pattern of the subject.

Lipid studies should be performed at least 3 m postpartum, or 3 m following cessation of lactation.

VARIATION DUE TO SPECIMEN COLLECTION AND HANDLING

Fasting versus Non-fasting

The effect of food ingestion on lipid measurements has long been established. A controlled dietary study showed that ingestion of a typical fat-containing meal caused a significant increase in TG that persisted for at least 9 h.[7] VLDL-C also increases, while LDL-C falls significantly; both changes persist for at least 9 h. The increased VLDL-C and decreased LDL-C are greater for calculated values than for measured values, but are significant for both regardless of the methodology used. For calculated LDL-C, the differences seen are the result of not only inflated VLDL-C, but also of lipoprotein compositional changes. Most, but not all, subjects in studies using precipitation methods experience a slight but significant transient fall in HDL-C, while in ultracentrifugation studies subjects experience a post-prandial increase in HDL-C; the cause of this discrepancy is not yet known.

In the second report of the NCEP Adult Treatment Panel, it is recommended that initial screening for lipid disorders include both TC and HDL-C measurement if the patient is not fasting.[20] If this recommendation is to be followed, it is imperative that the laboratory verify that its method for HDL-C is not artifactually altered in the non-fasting state. TC, Lp(a), apo AI, and apo AII are unchanged after food ingestion.[7, 119] Some studies have reported an increase in apo B after meals,[18] while others report no change; the latter finding is consistent with constancy of the sum of VLDL-C and LDL-C in the post-prandial state.[7, 119] *Therefore, except for measurements of TC, Lp(a), apo AI, and apo AII, lipid studies should be done only after a fast of at least 12 h.*

Anticoagulants and Preservatives

The type of specimen collected is an important consideration in interpreting lipid values. Most laboratories use serum or EDTA plasma. At 4° C, TC and TG concentrations are stable for at least 4 d in either serum or plasma.[6] EDTA has a theoretical advantage in that metal-ion induced oxidation and enzymatic cleavage of lipoproteins is retarded. Because of its osmotic effect, EDTA causes an artifactual fall in lipoprotein concentrations. In the past, most reports stated that this causes a reduction of 3%;[6] however, a recent study reports that the average difference between plasma and serum using tubes with disodium EDTA is now 4.7%, because the EDTA concentration has been made 50% higher in blood collection tubes.[120] Since EDTA concentration is dependent on volume of blood added, the actual decrease in TC in unfilled tubes may be as much as 10%.[121] *It would be prudent to insist on fully filled evacuated tubes to minimize the effect of differing anticoagulant concentration.*

Use of other anticoagulants such as oxalate and citrate is associated with even greater fluid shifts.[6] Fluoride causes fluid shift and has an inhibitory effect on some enzymatic measurements of TC.[121] These anticoagulants are not recommended for use in lipid measurements. Heparin does not produce fluid shifts and appears to be an acceptable alternative for TC measurement;[122] however, heparin activates lipoprotein lipase both *in vitro* and *in vivo*.[123] TG concentration falls gradually in specimens containing heparin, including "serum" specimens from patients who are receiving therapeutic doses of heparin.[124] Most serum separator tubes do not appear to affect the results of lipid measurements.[121]

Capillary versus Venous Blood

The source of specimen appears to have some effect on TC concentration, which may be important in remote-site screening programs. Several studies have reported that TC is higher in capillary blood plasma than in venous serum, the average bias being approximately 3%.[125] Since heparin was utilized, this effect is not due to shifts in fluid due to use of anticoagulant. Other studies have reported no difference[126] or lower[127] concentrations of TC in capillary blood. It is unclear whether these conflicting results are due to variations in the site of collection, the puncture technique, the post-prandial state, or other variables. Further research is necessary to clarify this issue, since different measurement techniques may be a source of inaccurate classifications of patients' risk factors.[128]

Hemoconcentration

Hemoconcentration from posture or other causes increases the relative concentration of proteins and protein-bound substances, which would produce a predictable change in all lipid and apolipoprotein measurements. A patient who stands for 5 min will experience an apparent increase in lipid concentration of 9%, and a further increase to 16% after 15 min.[129] This can be minimized by having a patient remain seated for at least 15 min before venipuncture.[6, 129, 130]

Use of a tourniquet can also cause significant hemoconcentration. After 1 min there is no significant change in TC or protein concentration; however, after 2 min the apparent TC concentration increases up to 5%, and after 5 min apparent increases of 10–15% can occur.[6] If a tourniquet remains for 15 min during phlebotomy, lipid measurements increase by 20–40%.[131]

Posture changes have also become associated with changes in plasma, noradrenaline concentrations, and apparent sympathetic activity, which may also indirectly alter lipid concentration.[132]

Specimen Storage

Specimens for most lipid testing are stable at 0–4° C for up to 4 days, with HDL-C and LDL-C being less stable. Serum specimens show some spontaneous hydrolysis of phospholipids and TG which is inhibited in EDTA plasma.[6] For longer storage of lipids, specimen stability is better at −70° C than at −20° C. Apo B decreases slightly when frozen,[133] but other lipid components appear stable when frozen for up to 6 months.[121] Alterations in apo B and apo(a) measurements after storage are method dependent; if specimens are to be stored for long periods, the laboratory must verify that its method is not affected by specimen storage. Repeated freezing and thawing should be avoided, as it can affect results of lipid measurement after as few as two cycles.[134] This is a particular concern for specimens stored at −20° C, as most "frost-free" freezers regularly increase their temperatures to melt accumulated frost, which may result in specimen thawing.

RECOMMENDATIONS FOR MINIMIZING PREANALYTICAL VARIATIONS

The following recommendations are derived from NCEP Laboratory Standardization Panel (LSP) recommendations for minimizing the effect of preanalytical factors on lipid and lipoprotein testing.[135]

1. A subject's lipid and lipoprotein profile should only be measured when the individual is in a steady metabolic state.
2. Subjects should maintain their usual diet and weight for at least 2 weeks prior to the determination of their lipids or lipoproteins.
3. Multiple measurements should be performed within 2 months, at least 1 week apart, before making a medical decision about further action.
4. Subjects should not perform vigorous physical activity during the 24 h prior to testing.
5. Fasting or non-fasting specimens can be used for TC testing. However, a 12 h fasting specimen is required for TG and lipoproteins.
6. The subject should be seated for at least 5 min before specimen collection.
7. The tourniquet should not be kept on more than 1 min during venipuncture.
8. TC, TG, and HDL-C concentrations can be determined in either serum or plasma. When EDTA is used as the anticoagulant, plasma should be immediately cooled to 2–4° C to prevent changes in composition, and values should be multiplied by 1.03.
9. For TC testing, serum can be transported either at 4° C or frozen. Storage of specimens at −20° C is adequate for TC measurement. However, specimens must be stored frozen at −70° C or lower for TG and lipoprotein/apolipoprotein testing.
10. All blood specimens should be considered potentially infectious and handled accordingly.

No specific recommendations were made by the LSP concerning apolipoprotein testing. However, steps similar to those described should be taken to help minimize preanalytical sources of variation in apolipoprotein measurements.

CONCLUSION

Results of lipid measurements are affected by many biological, behavioral, and clinical factors, and by variability in sample collection and handling. Failure to control for these variables can lead to misclassification of patient risk. Therefore, standardization of preanalytical sources of variation to the fullest extent possible will improve the accuracy and utility of lipid testing in assessing the risk of CHD. ❖

REFERENCES

1. Harris EK. Statistical aspects of reference values in clinical pathology. Prog Clin Pathol 1981;8:45–66.
2. Ford RP. Essential data derived from biological variation for establishment and use of lipid analysis. Ann Clin Biochem 1989;26:281–5.
3. Mogadam M, Ahmed SW, Mensch AH, Godwin ID. Within-person fluctuations of serum cholesterol and lipoproteins. Arch Intern Med 1990;150:1645–8.
4. Smith SJ, Cooper GR, Myers GL, Sampson EJ. Biological variability in concentrations of serum lipids: sources of variation among results from published studies and composite predicted values. Clin Chem 1993;39:1012–22.
5. Cooper GR, Smith SJ, Myers GL, Sampson EJ, Magid E. Estimating and minimizing effects of biologic sources of variation by relative range when measuring the mean of serum lipids and lipoproteins. Clin Chem 1994;40:227–32.
6. Cooper GR, Myers GL, Smith SJ, Sampson EJ. Standardization of lipid, lipoprotein, and apolipoprotein measurements. Clin Chem 1988;34:B95–B105.
7. Cohn JS, McNamara JR, Schaefer EJ. Lipoprotein cholesterol concentrations in the plasma of human subjects as measured in the fed and fasted states. Clin Chem 1988;34:2456–9.
8. Gordon DJ, Trost DC, Hyde J, et al. Seasonal cholesterol cycles: the Lipid Research Clinics Coronary Primary Prevention Trial placebo group. Circulation 1987; 76:1224–31.
9. Kim HJ, Kalkhoff RK. Changes in lipoprotein composition during the menstrual cycle. Metabolism 1979;28:663–8.
10. Lussier-Cacan S, Xhignesse M, Desmarais JL, et al. Cyclic fluctuations in human serum lipid and apolipoprotein levels during the normal menstrual cycle: comparison with changes occurring during oral contraceptive therapy. Metabolism 1991;40:849–54.
11. Tangney C, Brownie C, Wu SM. Impact of menstrual periodicity on serum lipid levels and estimates of dietary intakes. J Am Coll Nutr 1991;10:107–13.
12. Reichel R, Widhalm K. Lipids and lipoproteins during pregnancy. In Widhalm K, Naito HK, eds. Recent aspects of diagnosis and treatment of lipoprotein disorders. New York: Alan R Liss, 1988:125–133.
13. Rifai N, Pham Q, McMurray RG. Serum lipid and apolipoprotein changes in normal pregnancy. Clin Chem 1989;35:1066.
14. Terstrat J, Hessel LW, Seepers J, Van Gent CM. The influence of meal frequency on diurnal lipid, glucose and cortisol levels in normal subjects. Eur J Clin Invest 1978;8:61–6.

15. Fraser CG. Biological variation in clinical chemistry. An update: collected data, 1988-1991. Arch Path Lab Med 1992;116:916–23.
16. Haarbo J, Christiansen C. Treatment-induced cyclic variations in serum lipids, lipoproteins, and apolipoproteins after 2 years of combined hormone replacement therapy: exaggerated cyclic variations in smokers. Obstet Gynecol 1992; 80:639–44.
17. Panteghini M, Pagani F. Pre-analytical, analytical and biological sources of variation of lipoprotein(a). Eur J Clin Chem Clin Biochem 1993;31:23–8.
18. Matuchansky C, Fabre J, Guillard O, Morechar-Beauchant M, Reinberg A. Effects of cyclic (nocturnal) total parenteral nutrition and continuous enteral nutrition on circadian rhythms of blood lipids, lipoproteins and apolipoproteins in humans. Am J Clin Nutr 1985;41:727–34.
19. Chandler WL, Loo SC. Lipoprotein (a) does not show circadian variations. Thromb Haemot 1990;63:151.
20. Summary of the second report of the National Cholesterol Education Program (NCEP) Expert Panel on Detection, Evaluation and Treatment of High Blood Cholesterol in Adults (Adult Treatment Panel II). JAMA 1993;269:3015–23.
21. Bookstein L, Gidding SS, Donovan M, Smith FA. Day-to-day variability of serum cholesterol, triglyceride, and high density lipoprotein cholesterol levels: impact on the assessment of risk according to the National Cholesterol Education Program guidelines. Arch Intern Med 1990;150:1653–7.
22. Strobl W, Widhalm K. The natural history of serum lipids and apolipoproteins during childhood. In: Widhalm K, Naito HK (eds). Detection and treatment of lipid and lipoprotein disorders of childhood. New York: Alan R. Liss, 1985:101–21.
23. Rifai N, Heiss G, Doetsch K. Lipoprotein(a) at birth, in blacks and whites. Atherosclerosis 1992;92:123–9.
24. Van Biervliet JP, Labeur C, Michiels G, Usher DC, Rosseneu M. Lipoprotein(a) profiles and evolution in newborns. Atherosclerosis 1991;86:173–81.
25. Siest, G, Henny J, Schiele F, Young DB (eds.). Interpretation of clinical laboratory tests. Foster City, CA: Biomedical Publications, 1985.
26. Berenson GS, Srinivasan SR, Cresanta JL, Foster TA, Webber LS. Dynamic changes of serum lipoproteins in children during adolescence and sexual maturation. Am J Epidemiol 1981;113:157–70.
27. Steinmetz J, Choukaife A, Visvikis S, Henny J, Siest G. Biological factors affecting concentration of serum LpAI lipoprotein particles in serum, and determination of reference ranges. Clin Chem 1990;36:677–80.
28. Jenner JL, Ordovas JM, Lamon-Fava S, et al. Effects of age, sex, and menopausal status on plasma lipoprotein(a) levels: the Framingham Offspring Study. Circulation 1993;87:1135–41.
29. Srinivasan SR, Freedman DS, Webber LS, Berenson GS. Black-white differences in cholesterol levels of serum high-density lipoprotein subclasses among children: the Bogalusa Heart Study. Circulation 1987;76:272–9.
30. Srinivasan SR, Dahlen GH, Jarpa RA, Webber LS, Berenson GS. Racial (black-white) differences in serum lipoprotein(a) distribution and its relation to parental myocardial infarction in children: the Bogalusa Heart Study. Circulation 1991; 84:160–7.
31. Tyroler HA, Heiss G, Schonfeld G, et al. Apolipoprotein A-I, A-II, and C-II in black and white residents of Evans county. Circulation 1980;62:249–54.
32. Webber LS, Harsha DW, Phillips GT, Srinivasan SR, Simpson JW, Berenson GS Cardiovascular risk factors in Hispanic, white, and black children: the Brooks County and Bogalusa Heart studies. Am J Epidemiol 1991;133:704–14.
33. Fulton-Kehoe DL, Eckel RH, Shetterly SM, Hamman RF. Determinants of total high density lipoprotein cholesterol and high density lipoprotein subfraction lev-

els among Hispanic and non-Hispanic white persons with normal glucose tolerance: the San Luis Valley Diabetes Study. J Clin Epidemiol 1992;45:1191–200.

34. Haffner SM, Gruber KK, Morales PA, et al. Lipoprotein(a) concentrations in Mexican Americans and non-Hispanic whites: the San Antonio Heart Study. Am J Epidemiol 1992;136:1060–8.
35. Cobbaert C, Kesteloot H. Serum lipoprotein(a) levels in racially different populations. Am J Epidemiol 1992;136:441–9.
36. Marcovina SM, Albers JJ, Jacobs DR Jr, et al. Lipoprotein(a) concentrations and apolipoprotein(a) phenotypes in Caucasians and African Americans: the CARDIA study. Arterioscler Thromb 1993;13:1037–45.
37. Lipid Research Clinics Population Studies Data Book. Volume 1. The Prevalence Study. Aggregate distributions of lipids, lipoproteins and selected variables in 11 North American Populations. NIH publication No. 80-1527. Bethesda, MD: Public Health Service, National Institutes of Health, July 1980.
38. Johnson CL, Rifkind BM, Sempos CT, et al. Declining serum total cholesterol levels among U.S. adults. JAMA 1993;269:3002–8.
39. National Center for Health Statistics, National Heart, Lung, and Blood Institute Collaborative Lipid Group. Trends in serum cholesterol levels among US adults aged 20 to 74 years: data from the National Health and Nutrition Examination Surveys, 1960 to 1980. JAMA 1987;257:937–42.
40. Cowan LD, Wilcosky T, Criqui MH, et al. Demographic behavioral, biochemical, and dietary correlates of plasma triglyceride. Arteriosclerosis 1985;5;466–80.
41. Slunga L, Asplund K, Johnson O, Dahlen GH. Lipoprotein (a) in a randomly selected 25–64 year old population: the Northern Sweden Monica Study. J Clin Epidemiol 1993;46:617–24.
42. Kloer HU. Diet and coronary heart disease. Postgrad Med J 1989;65:S13–21.
43. Grundy SM, Denke MA. Dietary influences on serum lipids and lipoproteins. J Lipid Res 1990;31:1149–72.
44. Harris WS, Connor WE, Illingworth DR, Rothrock DW, Foster DM. Effects of fish oil on VLDL triglyceride kinetics in humans. J Lipid Res 1990;31:1549–58.
45. Blankenhorn DH, Johnson RL, Mack WJ, El Zein HA, Vaillas LI. The influence of diet on the appearance of new lesions in human coronary arteries. JAMA 1990; 263:1646–52.
46. Ripsin CM, Keenan JM, Jacobs DR, et al. Oat products and lipid lowering: a meta-analysis. JAMA 1992;267:3317–25.
47. Kris-Atherton DM, Krommel D, Russell ME, et al. The effects of diet on plasma lipids, lipoproteins and coronary heart disease. J Am Diet Assoc 1988;88:1373–400.
48. Cooper GR, Myers GL, Smith SJ, Sehlant RC. Dietary oat fiber sources and blood lipids. JAMA 1992;268:986.
49. Sacks FM, Ornish D, Rosner B, et al. Plasma lipoprotein levels in vegetarians. JAMA 1985;524:1337–41.
50. McMurray MP, Cerqueira MT, Connor SL, Connor WE. Changes in lipid and lipoprotein levels and body weight in Tarahumara Indians after consumption of an affluent diet. N Engl J Med 1991;325:1704–8.
51. Wolf RN, Grundy SM. Influence of weight reduction on plasma lipoprotein in obese patients. Arteriosclerosis 1983;3:160–9.
52. Newman B, Selby JV, Quesenberry Jr., et al. Nongenetic influences of obesity on other cardiovascular disease risk factors: an analysis of identical twins. Am J Public Health 1990;80:675–8.
53. Kaplan NM. Obesity: location matters. Heart Dis Stroke 1992;1:148–50.
54. Despres JP, Moorjani S, Lupien PJ, et al. Regional distribution of body fat, plasma lipoproteins, and cardiovascular disease. Arteriosclerosis 1990:10:497–511.

55. McGill HC Jr. Potential mechanisms for the augmentation of atherosclerosis and atherosclerotic disease by cigarette smoking. Prev Med 1979;8:390–403.
56. Knoke JD, Hunninghake DB, Heiss G. Physiological markers of smoking and their relation to coronary heart disease. Arteriosclerosis 1987;7:477–82.
57. Yokode M, Kita T, Arai M, et al. Cholesteryl ester accumulation in macrophages incubated with low density lipoprotein pretreated with cigarette smoke extract. Proc Natl Acad Sci USA 1988;85:2344–8.
58. Craig WY, Palomaki GE, Haddow JE. Cigarette smoking and serum lipid and lipoprotein levels: an analysis of published data. Br Med J 1989;298:784–8.
59. Craig WY, Palomaki GE, Johnson M, Haddow JE. Cigarette smoking-associated changes in blood lipid and lipoprotein levels in the 8- to 19-year-old age group: a meta-analysis. Pediatrics 1990;85:155–8.
60. Willet W, Hennekens CH, Castelli W, et al. Effects of cigarette smoking on fasting: triglyceride, total cholesterol, and HDL cholesterol in women. Am Heart J 1983;105:417–21.
61. Steinberg D, Pearson TA, Kuller LH. Alcohol and atherosclerosis. Ann Intern Med 1991;114:967–76.
62. Haskell WL, Camargo C, Williams PT, et al. The effect of cessation and resumption of moderate alcohol intake on serum high density lipoprotein subfractions: a controlled study. N Engl J Med 1984;310:805–10.
63. Taskinen MR, Nikkila EA, Valimaki M, et al. Alcohol-induced changes in serum lipoproteins and in their metabolism. Am Heart J 1987;113:458–64.
64. Wilson DE, Schneibman PH, Brewster AC, Arky RA. The enhancement of alimentary lipemia by ethanol in man. J lab Clin Med 1970;75:264–74.
65. Taskinin MR, Nikkila EA, Valinmaki M, et al. Alcohol-induced changes in serum lipoproteins and in their metabolism. Am Heart J 1987:113;458–64.
66. Valimaki M, Laitinen K, Ylikahrit R, et al. The effect of moderate alcohol intake on serum apolipoprotein AI-containing lipoproteins and lipoprotein(a). Metabolism 1991;40:1168–72.
67. Solvoll K, Selmer R, Loken EB, Foss OP, Trygg K. Coffee, dietary habits, and serum cholesterol among men and women 35–49 years of age. Am J Epidemiol 1989;129:1277–88.
68. Thelle DS, Arnesen E, Forde OH. The Tromso Heart Study: Does coffee raise serum cholesterol? N Engl J Med 1983;308:1454–7.
69. Bak AAA, Grobbee DE. The effect on serum cholesterol levels of coffee brewed by filtering or boiling. N Engl J Med 1989;321:1423–7.
70. Zock PL, Katan MB, Merkus MP, Van Dusseldorp M, Harryvan JL. Effect of a lipid-rich fraction from boiled coffee on serum cholesterol. Lancet 1990;335: 1235–37.
71. Van Dusseldorp M, Katan MB, Demacker PNM. Effect of decaffeinated versus regular coffee on serum lipoproteins: a 12- week double-blind trial. Am J Epidemiol 1990;132:33–40.
72. Fried RE, Levine DM, Kwiterovich PO, et al. The effect of filtered-coffee consumption on plasma lipid levels. Results of a randomized clinical trial. JAMA 1992;267:811–5.
73. Berlin JA, Colitz GA. A meta-analysis of physical activity in the prevention of coronary heart disease. Am J Epidemiol 1990;132:612–28.
74. Ekelund LG, Haskell WL, Johnson JL, et al. Physical fitness as a prediction of cardiovascular mortality in asymptomatic North American men. N Engl J Med 1988;319:1379–84.
75. Wood PD, Stefanick ML, Williams PT, Haskell WL. The effects of plasma lipoproteins on a prudent weight reducing diet, with or without exercise, in overweight men and women. N Engl J Med 1991;325:461–6.

76. Lampman RM, Santinga JT, Hodge MF, Block WD, Flora JD, Bassett DR. Comparative effects of physical training and diet in normalizing serum lipids in men with type IV hyperlipoproteinemia. Circulation 1977;55:652–9.
77. Dufaux B, Assmann G, Hollmann W. Plasma lipoproteins and physical activity: A review. Int J Sports Med 1982;3:123–36.
78. Swank AM, Robertson RJ, Deitrich RW, Bates M. The effect of acute exercise on high density lipoprotein cholesterol and the subfractions in females. Atherosclerosis 1987;63:187–92.
79. Tucker LA, Friedman GM. Walking and serum cholesterol in adults. Am J Public Health 1990;80:1111–3.
80. Hardman AE, Hudson A, Jones PRM, Norgan NG. Brisk walking and plasma high-density lipoprotein cholesterol concentration in previously sedentary women. Br Med J 1989;299:1204–5.
81. McCann BS, Warnick GR, Knopp RH. Changes in plasma lipids and dietary intake accompanying shifts in perceived workload and stress. Psychosomatic Med 1990;52:97–108.
82. Dimsdale JE, Herd JA. Variability of plasma lipids in response to emotional arousal. Psychosom Med 1982;44:413–30.
83. Stoney CM, Matthews KA, McDonald RH, Johnson CA. Sex differences in lipid, lipoprotein, cardiovascular and neuroendocrine responses to acute stress. Psychophysiology 1988;25:645–56.
84. Genest JJ, McNamara JR, Ordovas JM, et al. Effect of elective hospitalization on plasma lipoprotein cholesterol and apolipoproteins AI, B and Lp(a). Am J Cardiol 1990;65:677–9.
85. Walton KW, Scott PJ, Dykes PW, Dawids JWL. The significance of alteration in serum lipids in thyroid dysfunction. II. Alteration of the metabolism and turnover of ^{131}I-low density lipoprotein in hypothyroidism and thyrotoxicosis. Clin Sci 1965;29:217–38.
86. Betteridge DJ. Lipids, diabetes and vascular disease: the time to act. Diabetic Med 1989;6:195–218.
87. Manolio TA, Savage PJ, Burke GL, et al. Association of fasting insulin with blood pressure and lipids in young adults. Arteriosclerosis 1990;10:430–6.
88. Hughes TA, Clements RS, Fairlough PK, Bell DSH, Segrest JP. Effects of insulin therapy on lipoproteins in non-insulin-dependent diabetes mellitus. Atherosclerosis 1987;67:105–14.
89. Ruotolo G, Zoppo A, Parlavecchia M, Giberti B, Micossi P. Apolipoprotein(a) levels in Type I and Type 2 diabetes mellitus. Acta Diabetologica 1991;28:158–61.
90. Haffner SM, Tuttle KR, Rainwater DL. Decrease of Lp(a) with improved metabolic control in subjects with insulin-dependent diabetes mellitus. Diabetes Care 1991;14:302–7.
91. Pollare T, Vessby B, Lithell H. Lipoprotein lipase activity in skeletal muscle is related to insulin sensitivity. Arterioscler Thromb 1991;11:1192–203.
92. Wheeler DC, Varghese Z, Moorhead JF. Hyperlipidemia in nephrotic syndrome. Am J Nephrol 1989;9(Suppl 1):78–84.
93. Rifai N, King ME, Sica D. Effects of long-term hemodialysis on lipid, lipoprotein and apolipoprotein levels of black patients with chronic renal failure. Ann Clin Biochem 1988;25:242–5.
94. Murphy BG, McNamee, Duly E, et al. Increased serum apolipoprotein(a) in patients with chronic renal failure treated with continuous ambulatory peritoneal dialysis. Atherosclerosis 1992;93:53–5.
95. Karadi I. Romics L, Palos G, et al. Lp(a) lipoprotein concentration in serum of patients with heavy proteinuria of different origin. Clin Chem 1989;35:2121–3.

96. Joven J, Villabona C, Vilella E, et al. Abnormalities of lipoprotein metabolism in patients with the nephrotic syndrome. N Engl J Med 1990;323:579–84.
97. Levy RI. Cholesterol, lipoproteins, apolipoproteins and heart disease: present status and future prospects. Clin Chem 1981;27:653–62.
98. Gore JM, Goldberg RJ, Matsumoto AS, et al. Validity of serum total cholesterol level obtained within 24 hours of acute myocardial infarction. Am J Cardiol 1984; 54:722–5.
99. Shephard MDS, Hester J, Walmsley RN, White GH. Variation in plasma apolipoprotein AI and B concentrations following myocardial infarction. Ann Clin Biochem 1990;27:9–14.
100. Rantapaa-Dahlqvist S, Wallberg-Jonsson S, Dahlen G. Lipoprotein(a), lipids and lipoproteins in patients with rheumatoid arthritis. Ann Rheum Dis 1991;50:366–8.
101. Mendez I, Hachinski V, Wolfe B. Serum lipids after stroke. Neurology 1987:37: 507–11.
102. Alvarez C, Ramos A. Lipids, lipoprotein and apolipoproteins in serum during infection. Clin Chem 1986;32:142–5.
103. Maeda S, Abe A, Makino K, Noma A, Kawade M. Transient changes of serum lipoprotein(a) as an acute phase reactant. Clin Chem 1989;89:145–50.
104. McKenney JM, Wright JT, Goodman RP, et al. The effect of low dose of hydrochlorothiazide on blood pressure, serum potassium and lipoproteins. Pharmacotherapy 1986;6:179–84.
105. Krone W, Nagele H. Effects of antihypertensives on plasma lipids and lipoprotein metabolism. Am Heart J 1988;116:1729–34.
106. Woodcock BG, Rietbrock N. Beta-blocker induced changes in the cholesterol: high density lipoprotein cholesterol ratio and risk of coronary heart disease. Klin Wochenschr 1984;62:843–9.
107. Goldsland IF, Crook D, Simpson R, Proudler T, Felton C. The effects of different formulations of oral contraceptive agents on lipid and carbohydrate metabolism. N Engl J Med 1990;323:1375–81.
108. Wahl P, Walder C, Knopp R, et al. Effects of estrogen/progestin potency on lipid/lipoprotein cholesterol. N Engl J Med 1983;308:862–7.
109. Bush TL, Fried LP, Barrett-Conner E. Cholesterol, lipoproteins and coronary heart disease in women. Clin Chem 1988;34:B60–B70.
110. Miller VT, Muesing RA, LaRosa JC, et al. Effects of conjugated estrogen with and without three different progestogens on lipoproteins, high density lipoprotein subfractions and apolipoprotein AI. Obstet Gynecol 1991;77:235–40.
111. Henriksson P, Angelin B, Berglund L. Hormonal regulation of serum Lp(a) levels. Opposite effects after estrogen treatment and orchidectomy in males with prostatic carcinoma. J Clin Invest 1992;89:1166–71.
112. Ilowite NT, Samuel P, Ginzler E, Jacobson MS. Dyslipoproteinemia in pediatric systemic lupus erythematosus. Arthritis and Rheumatism 1988;31:859–63.
113. Ballantyne CM, Podet EJ, Patsch WP, et al. Effects of cyclosporine therapy on plasma lipoprotein levels. JAMA 1989;262:53–6.
114. Farmer JA, Ballantyne CM, Franzier OH, et al. Lipoprotein(a) and apolipoprotein changes after cardiac transplantation. J AM Coll Cardiol 1991;18:926–30.
115. Markell MS, Friedman EA. Hyperlipidemia after organ transplantation. Am J Med 1989;87(suppl 5N):61–67.
116. Van Thiel DH, Iqbal M, Jaln I, Todo S, Starzl TE. Gastrointestinal and metabolic problems associated with immunosuppression with either CsA or FK 506 in liver transplantation. Transplant Proc 1990;22(Suppl 1):37–40.
117. Panteghini M, Pagani F. Serum concentrations of lipoprotein(a) during normal pregnancy and postpartum. Clin Chem 1991;37:2009–10.

118. Knopp RH, Walden CE, Wahl PW, et al. Effect of postpartum lactation on lipoprotein lipids and apoproteins. J Clin Endocrinol Metab 1985;60:542–7.
119. Rifai N, Merrill JR, Holly RG. Postprandial effect of a high fat meal on plasma lipid, lipoprotein cholesterol and apolipoprotein measurements. Ann Clin Biochem 1990;27:489–93.
120. Cloey T, Bachorik PS, Becker D, et al. Reevaluation of serum-plasma differences in total cholesterol concentration. JAMA 1990;263:2788–9.
121. Naito HK. Problems associated with lipid and lipoprotein analyses. In: Widhalm K, Naito HK, eds. Detection and treatment of lipid and lipoprotein disorders of childhood. New York: Alan R. Liss, 1985:19–60.
122. Lum G, Gambino R. A comparison of serum versus heparinized plasma for routine chemistry tests. Am J Clin Pathol 1974;61:108–13.
123. Ferlito S, Ricceri M, Ossino AM. Acute lipidemic effect of calcium heparin in normolipemic and hyperlipemic subjects. Int Angiol 1989;8:140–4.
124. Horton G, Cole TG, Gibson DW, Kessler G. Decreased stability of triglycerides and increased free glycerol in serum from heparin-treated patients. Clin Chem 1988;34:1847–9.
125. Greenland P, Bowley NL, Meiklejohn B, Doane KL, Sparks CE. Blood cholesterol concentration: fingerstick plasma vs. venous serum sampling. Clin Chem 1990; 36:628–30.
126. Ishikawa TT, Morrison J, Fallat R, Parsons D, Glueck CJ. Comparison of capillary and venous blood sampling for quantitation of plasma cholesterol. J Lab Clin Med 1974;84:281–6.
127. Kupke IR, Zeugner S, Gottschalk A, Kather B. Differences in lipid and lipoprotein concentrations of capillary and venous blood samples. Clin Chim Acta 1979;97: 279–83.
128. Naughton MJ, Luepker RU, Strickland D. The accuracy of portable cholesterol analyzers in public screening programs. JAMA 1990;283:1213–7.
129. Tan MH, Wilmshurst EG, Gleason RE, Soldner JS. Effect of posture on serum lipids. N Engl J Med 1973;289:416–19.
130. Miller M, Bachorik PS, Cloey TA. Normal variation of plasma lipoproteins: postural effects on plasma concentrations of lipids, lipoproteins, and apolipoproteins. Clin Chem 1992;38:569–74.
131. Page IH, Moinuddin M. The effect of venous occlusion on serum cholesterol and total protein concentration: a warning. Circulation 1962;25:651–2.
132. Howes LC, Krum H, Louis WJ. Plasma cholesterol levels are dependent on sympathetic activity. J Hypertension 1987;5(Supp):S361–3.
133. Brown SA, Epps DF, Dunn JK, et at. Effect of blood collection and processing on radioimmunoassay results for apolipoprotein B in plasma. Clin Chem 1990;36: 1662–6.
134. Sgoutas DS, Tuten T. Effect of freezing and thawing of serum on the immunoassay of lipoprotein(a). Clin Chem 1992;28:1873–7.
135. National Cholesterol Education Program. Recommendations for improving cholesterol measurement: A report from the Laboratory Standardization Panel of the National Cholesterol Education Program. NIH Publication No. 90–2964. Bethesda, MD: National Institutes of Health, 1990:28–9.

❖ CHAPTER 4 ❖

Measurement of Cholesterol Concentration

Joseph D. Artiss, Donald H. Feldbruegge, Martin H. Kroll, Matthew J. McQueen, Terry Pry, Bennie Zak, and Joachim Ziegenhorn

INTRODUCTION

❖ The National Cholesterol Education Program's (NCEP) Expert Panel on the Detection, Evaluation, and Treatment of High Blood Cholesterol in Adults has released its guidelines for the treatment of adults who have high blood serum cholesterol (TC) concentrations.[1] This report further establishes the direct relationship between increased concentrations of serum TC and coronary heart disease. For the first time, standardized cut-off concentrations have been recommended for blood serum TC: desirable, < 200 mg/dL (5.17 mmol/L); borderline, 200–239 mg/dL (5.17–6.19 mmol/L), and high, 240 mg/dL (6.20 mmol/L) and above.

Subsequent to this first report, the panel released its recommendations for the pediatric population (2–19 years of age).[2] Like the earlier report for the adult population, this report also recommends uniform cut-off values for serum TC: 170 and 200 mg/dL (4.39 and 5.17 mmol/L). In addition, both reports recommend that treatment decisions be based on low-density lipoprotein cholesterol (LDL-C) values. What is involved here, aside from the establishment of an accurate value for serum TC, is the determination of triglyceride and high-density lipoprotein cholesterol values for use in calculating an LDL-C value.

Concurrent with the above report, the Laboratory Standardization Panel of the NCEP released its report on the status of the measurement of serum TC in the United States.[3] For the first time, this report established guidelines for the accuracy and precision of serum TC determinations. As of 1992, laboratories were expected to be able to attain TC concentrations that are within 3% of the "true value" as defined by the definitive method.

The timing of the release of these reports was not coincidental. They were part of a well-planned and well-executed program to decrease serum TC concentrations in the general population of the United States. For the first time, the determination of cholesterol had become the subject of a great deal of interest, both professionally and, perhaps new to us all, publicly.

Although there is a reference procedure for the determination of serum TC, it is neither applicable nor transferable to most clinical laboratories. Therefore, it is the purpose of this chapter to provide a sound, nonproprietary enzymic procedure that, with the appropriate precautions, will yield quite acceptable TC results. The authors do not anticipate that laboratories will necessarily use this procedure as a matter of routine, but rather as a check when problems with their commercial system arise. However, the ease with which this proposed procedure can be carried out would not preclude its use internally in a routine laboratory in the same manner as the currently accepted reference procedure when used externally.

METHODOLOGIES

Definitive Method

A definitive method is one that has been subjected to a thorough investigation and evaluation for sources of inaccuracy, including nonspecificity.[4] The premise of such a method is that its level of bias and imprecision is of a magnitude that is compatible with the method's stated purpose. As might be inferred from its name, the mean value of the definitive method is accepted as being the "true value." At present, the definitive method for TC is an isotope dilution-mass spectrometric method developed by the National Institute of Standards and Technology (NIST) (formerly the National Bureau of Standards [NBS]).[5] Definitive methods tend to be too labor intensive and cost prohibitive for use in routine clinical laboratories. The method for TC is no exception to this rule, and therefore it is not recommended for use in most field laboratories.

Reference Method

A reference method is one whose accuracy, precision, and transferability, when compared to the definitive method, are sufficient, and whose incidence of susceptibility to known interferences is so thoroughly documented that the stated end purposes of the method can be achieved.[6] The currently accepted reference method for TC is a Centers for Disease Control and Prevention (CDC) modification[7, 8] of the Abell modification[9] of the earlier method of Sperry and Brand.[10] The substance of this method rests in the fact that the cholesterol esters are hydrolyzed prior to extraction and reaction with a Liebermann-Burchard reagent.[11, 12] It should be noted that the reference method has been demonstrated to yield results that are, on average, 1.6% higher than the definitive method.[13, 14]

The reference method is intended to be technically and financially less demanding than the definitive method. However, in the case of TC, it is still beyond the means of most clinical laboratories.

Enzymic Methods

Enzyme-based reagents for the determination of TC have largely supplanted the older (strong acid medium) chemical approaches to routine laboratory measurements. The change to enzymic methods is due mainly to the better specificity afforded by the enzymes, increasing use of automation, and the elimination of some of the hazardous materials used in the older chemical methods. However, as is often the case, the new enzymic methods frequently have problems with interferences that were not experienced previously. Several recent articles[15-17] have discussed the determination of cholesterol, but these articles have made only passing mention of interferences. There have also been reports of interference studies carried out on several chemistry analyzers[18-20] and a recent review of glucose determinations[21] that includes interference effects that may be applicable. Until recently,[22] there had not been a review of TC methodologies for over a decade.[23, 24]

Space does not permit a thorough review of enzymic methods and all potential interferences. However, a brief description of the common problems and concerns about enzymic TC procedures follows.

Enzyme Specificity

The specificity of the enzymes used in cholesterol determinations—or for that matter, any other analytical procedure—may be considered by some to be suspect. One argument that is almost certainly valid is that the enzymes cannot be characterized to the same degree as reagent chemicals. Furthermore, it is doubtful that we will ever be able to characterize enzymes to this degree in the foreseeable future. However, it may be that, unlike the wet chemical procedures, the enzyme preparations are much more forgiving and do not require the same degree of characterization. The reason that these systems are so robust is not at all clear. It may be simply because the enzymes behave as catalysts and therefore do not participate directly in the analytical process. Alternatively, perhaps it is that we have learned how to thoroughly characterize and control their behavior in any given system. The use of recombinant sources for enzymes may facilitate uniformity among preparations.

Cholesterol Esterase

In the past, the completeness of enzymic hydrolysis was considered to be a problem with totally enzymic procedures.[25] However, it has been relatively well established that this is no longer a major problem.[26-31] The significant difference between earlier reagents and newer ones is the incorporation of esterase(s) of microbial origin. Although the manufacturers of kits do not always wish to be totally specific in their description of the individual reagents, it would appear that the genus of this microbe is *Pseudomonas*[31, 32] and that the species is probably *fluorescens*. The Boehringer-Mannheim (Indianapolis, IN) claim of at least 99.5% hydrolysis[34] seems to have been verified.[25, 26] Bateson et al.,[32] in comparing enzymic to chemical hydrolysis using an esterase from the above-mentioned genus and species, reported the efficiency of the enzyme(s) to be in excess of 99%. Therefore, it seems reasonable to conclude that the new generation of esterases provides quite acceptable degrees of hydrolysis.

Cholesterol Oxidase

Like the esterases, it is not always easy to identify the microbial source of the oxidases. However, Deeg and Ziegenhorn[27] have reported the use of an enzyme from a *Streptomyces* species as being ideal for Boehringer's kinetic TC procedure. Likewise they utilize an enzyme from *Nocardia Erythropolis* for end-point determinations.[33] Bateson et al.[32] utilized an oxidase from a *Cellulomonas* genus; in this case it is not the authors but rather the enzyme producer who does not wish to exactly identify the specific species of microbe used to obtain the enzyme.

Almost certainly none of the above-mentioned sources of enzyme are totally specific to a 3,5-betahydroxysterol such as cholesterol, but it might be concluded that this lack of specificity is not particularly important. We can say this in light of the fact that all other reactive steroidal compounds that occur naturally in human blood are present in concentrations of at least several orders of magnitude less than cholesterol. Therefore, even if there were 100% cross-reactivity, the effects would not be appreciable.

Measurement

In order to complete the enzymic determination of cholesterol, one final step is necessary. That step involves the detection and measurement of the product of the cholesterol oxidase reaction as a means of quantifying the amount of cholesterol that was present. Three general approaches have been taken as a means of quantifying the cholesterol oxidase reaction. Each is discussed in the sections that follow.

Electrochemical Measurement of Oxygen Consumed

Methods have been described[35, 36] and made commercially available for measurement (by electrode) of oxygen consumption, in a manner similar to that used for some glucose oxidase methods. Noma and Nakayama[36] reported that bilirubin and ascorbic acid, at 10 mg/dL (171 µmol/L and 568 µmol/L respectively), were without effect on either the amount or the rate of oxygen consumption. The oxygen consumption methods are not easily automated and generally require a substantial amount of cholesterol oxidase. Thus, these methods have not become widely used.

Ultraviolet Spectrophotometric Measurement of Cholest-4-en-3-one Produced

Another approach for quantitating the cholesterol oxidase reaction has been to measure the cholestenone produced during the reaction. Flegg[37] measured the absorption of this compound at 240 nm, and also measured the absorption of the

2,4-dinitrophenylhydrazone derivative at 390 nm. The procedure is time consuming and difficult to perform and is probably of use only in research applications; even there, it seems to have little application when compared to the third approach to be discussed. Interference studies have not been reported, and it is quite unlikely that all other ultraviolet-absorbing species would be absent. Additionally, it almost certainly would not be able to achieve the analytical sensitivity of the peroxidase coupled reaction.

Measurement of Hydrogen Peroxide Produced by the Cholesterol Oxidase Reaction

Measurement of the hydrogen peroxide produced is the most common method of quantitating the cholesterol oxidase reaction. The measurement of hydrogen peroxide has been accomplished by use of several different approaches. Allain et al.[38] reported the first totally enzymic method of determining serum TC by measuring hydrogen peroxide. This method was based on the Trinder reaction previously used for the determination of glucose.[39] The hydrogen peroxide produced in the cholesterol oxidase reaction was used to effect the coupling of two chromogenic substrates by horseradish peroxidase (EC 1.11.1.7). In this original procedure, the chromogenic compounds used were phenol and 4-aminoantipyrine (4AAP). The product of the reaction is presumed to be a quinoneimine dye that can be measured photometrically at a wavelength of about 500 nm. Compounds other than 4AAP and phenol represent variations of the same analytical theme. The ease of use and ready applicability to automation has made numerous variations of this procedure very popular.

The list of substances that may interfere with the peroxidase coupled measurements have been reported to include: ascorbic acid,[40-44] bilirubin,[40-48] triglycerides,[41, 48-52] hemolysis,[42-44] and possibly unknown substances.[53] It is difficult to put absolute estimates on the degree of interference that may be exhibited. The magnitude of this difficulty is relatively obvious when one considers the numerous types of reagent formulations and instrument configurations that are available for use. Data on the effects of these substances on the following procedure are presented later in the chapter.

STANDARDIZATION

As reference ranges for TC have evolved from being based upon concentrations found in 95% of the population to "optimal" target values with specific cut-off points above which intervention is recommended, extraordinary attention has been focused on the accuracy with which the laboratory measurement of TC must be performed. It is no longer acceptable that 47% of laboratories in the College of American Pathologists (CAP) comprehensive Chemistry Survey submit TC results that are outside of 5% of the target value.[3] The initial goal put forward by the NCEP was that levels of bias and imprecision for routine TC measurement should not exceed ± 5%. By 1992, these levels were to be no greater than ± 3%, and the NCEP suggested that all cholesterol values should be traceable to the

Abell-Kendall reference method. To achieve this, a program of certified calibration materials has been proposed, in conjunction with a network of reference laboratories. Although initially it was thought that this level of inaccuracy and imprecision would be very difficult to achieve, it was demonstrated to be possible as early as 1990.[54, 55]

The need for this combination of low inaccuracy and low imprecision stems from the closeness of the cut-off points proposed by the NCEP. Furthermore, because the cut-off points are near the center of the population distribution, more patients would be erroneously diagnosed than if the cut-off points were near the 97.5 percentile used in the typical reference range.[56] A falsely low TC that leads to an unjustified assumption that an individual's TC is "normal" is just as unacceptable as an erroneously high result that leads to an attribution of increased risk and resultant aggressive dietary and/or drug intervention.

It is not enough that the laboratory know the accuracy of a method; it must also know the accuracy of the complete analytical system. Contributing to this issue are the patient specimen, analytical instrument, chemical measurement procedure, reporting mechanism, and interpretation by the clinician. The contribution of the reporting system can be illustrated by the use of SI units. In the United States cholesterol is currently reported in rounded-off increments of one milligram per deciliter and only in terms of free cholesterol.[22] In SI units, a similar increment can be maintained only if the TC level is reported to the second decimal place. Hence, the capability to do this should be a part of both manual and automated reporting systems.[57] This is a small distinction in itself, but it becomes more important when committing to an overall percentage of error of 5%, 3%, or even less, as has been suggested.[56]

Table 4–1 ❖ Limited List of Commercially Available Lipid Standards, Calibrators, and Controls and Their Traceability to CDC, NIST, or NBS SRM911

Manufacturer	*Name of Product*	*Comments*
Boehringer-Mannheim Diagnostics	Precical Calibrator	Traceable to NBS SRM 911 (information provided by vendor)
Bio Merieux	Lipid Unitrol	Not traceable
Diagnostic Chemicals Ltd.*	DC-IEP Serum Elevated	Claimed to be traceable to CDC
Dade	Lipid-Trol	Claimed to be traceable to NBS standards
Fisher	Serachem Elevated Lipids Control Serum	Not tracetable
Sigma	#1 Sigma Lipid Control Elevated	Not traceable
	#2 Sigma Lipid Control Normal	Not traceable
	#3 Cholesterol Standards	Traceable to CDC
	#4 Cholesterol Calibrators	Traceable to CDC
Quantimetrix Corp.	Serum Lipid Control	Not traceable

*Canadian Company, Price Edward Island

The need for appropriate lipid standards, calibrators, and controls has become paramount. To illustrate this, a limited sample of lipid controls and standards is shown in Table 4-1. (A more comprehensive listing is provided in Chapter 11.) In reviewing the table, the products listed appear to fall within three categories: those that are traceable to the CDC through the NBS SRM911; those that are similar in that they represent the results of determinations done on the materials by the CDC or the NIST, or by the manufacturers themselves; and those that are not traceable to a reference procedure or material. The latter class of material is most appropriate for measuring imprecision.

Cholesterol standardization at the level of the manufacturer and the individual laboratory is illustrated by the flowchart shown Figure 4-1. Standardization at the level of the manufacturer is essential, and the flowchart indicates three approaches that can be taken. The accuracy of the CDC reference method is traceable to the accepted NIST definitive method, an isotope-dilution-mass-spectrometry procedure with certified reference material (NBS SRM911). The SRM911 is a human-based lyophilized serum reference material that can be utilized to assess the accuracy of a particular assay system.

Figure 4-1 ❖ Cholesterol Standardization Flowchart

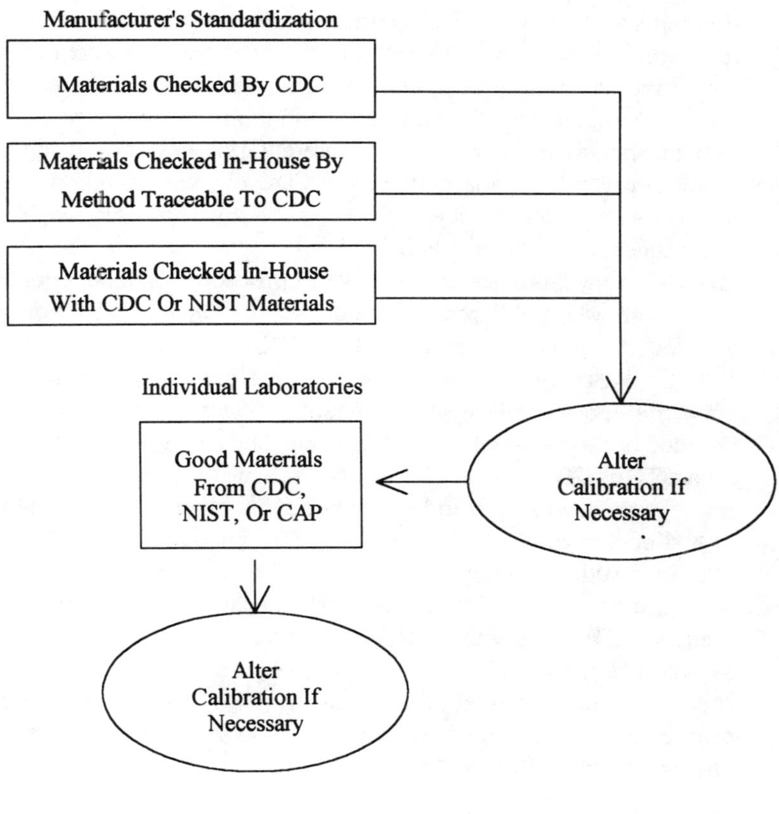

The average clinical laboratory should subscribe to at least one proficiency testing program. There are several commercial companies that sell materials traceable to the CDC reference method for standardization purposes. The CAP provides survey materials that are traceable to the CDC reference method as well as materials to test for linearity. Linearity checks are a sensitive technique to observe analytical problems before they have profound effects on accuracy.[58] It is important to note that the user may experience problems with this approach if the system being standardized is sensitive to matrix effects.

Even when a particular method, or a particular combination of method and instrument, is traceable to the CDC or NIST and has acceptable imprecision and inaccuracy, it is quite a different matter to say that the system in one's own laboratory meets the required limits. The NBS SRM911 and the CAP survey materials demonstrate marked matrix effects with some methods. As with all calibration and control materials, great care has to be exercised in extrapolating the values provided by the manufacturer to another situation unless all conditions are identical to those used by the manufacturer.

The control or calibrator material may show matrix effects when introduced into a different analytical system, even when the principles of the chemical procedure are the same in the manufacturer's system and in the system used in an individual laboratory. Lyophilization can cause a matrix effect with biases of 2–17%, depending on the instrument and reagent system.[59] The matrix effect is introduced into survey materials by the preparation and preservation of blood (serum). These matrix effects have had an impact on proficiency surveys.[60] Problems with matrix effects can interfere with the standardization process. Efforts are underway to try to better understand these effects in the biological reference materials that are used in the standardization of TC measurements.[61,62]

A complementary strategy for the laboratory is to send split aliquots from patient specimens to one of the reference laboratories using the Abell-Kendall method, which is standardized to the CDC reference method. This strategy is certainly of some value, but it is limited in that it does not really help enhance the day-today confidence in the results being issued. Also, the expense involved may limit the ability of many laboratories to follow this procedure on an adequate scale.

Even when different methods for cholesterol measurement have a bias within ± 3% when compared to the CDC reference method, these methods will still show a range of bias, some negative and some positive. This means that each method will still give significantly different percentages of the population deemed at risk based on the NCEP guidelines.[56] At the 50th percentile, where the mean TC is 200 mg/dL (5.17 mmol/L), a 3% positive bias would cause 6.7% more of the population to be diagnosed as having borderline-high cholesterol.[63] The effects of methodological bias for the measurement will be greater near the 50th and 70th percentiles than at the tails of the distribution.[63] These considerations should be taken into account when assigning allowable bias for methods for measuring TC, and they imply that the laboratory should keep the bias as small as possible in order to produce quality results. Further, it should be noted that the National Reference Method Laboratory (NRML) network laboratories have their own levels of bias and imprecision which could, in turn, be passed on to the routine clinical laboratories.[64,65]

RECOMMENDED METHOD

Materials

The following materials can be obtained from Genzyme Corporation, Cambridge, MA: cholesterol esterase (CHE), from *Pseudomonas sp.*, Sterol-ester acylhydrolase (EC 3.1.1.13); cholesterol oxidase (COX), from *Cellulomonus sp.*, cholesterol: oxygen oxidoreductase (EC 1.1.3.6); lipase, from *Chromobacterium viscosum*, Triacylglycerol acyl-hydrolase (EC 3.1.13); and peroxidase (HRP), from horseradish (donor: hydrogen-peroxide oxidoreductase) (EC 1.11.1.7).

Triton X-100, α-cyclodextrin (α-CD), magnesium acetate, and 4-aminoantipyrine may be obtained from Sigma Chemical Co., St. Louis, MO.

Potassium ferrocyanide (reagent grade) may be obtained from Spectrum Chemical Mfg. Co., Redondo Beach, CA.

The sodium salt of 2-hydroxy-3,5-dichlorobenzenesulfonic acid (may also be listed as 3,5-dichloro-2-hydroxybenzenesulfonic acid) (HDCBS) may be obtained from either Biosynth International, Skokie, IL or Research Organics, Inc., Cleveland, OH. (There are other sources of this material but we have not evaluated them.)

Cholesterol (NBS SRM911b) was obtained from NIST, the U.S. Department of Commerce, Gaithersburg, MD.

Specimen Collection and Storage

(Please refer to Chapter 3 for a detailed description of specimen collection.)

Blood is generally drawn from the patient's antecubital vein or other convenient arm vein.[66] Concentrations of TC in finger-stick samples of whole blood have been found to be comparable with venous plasma, providing the initial drop of blood which contains tissue fluid is discarded and excessive "milking" of the finger is minimized to eliminate hemolysis.[67, 68]

TC concentrations in serum or plasma have been shown to be stable when subjected to various storage conditions. Although it is standard practice to separate serum from the blood clot by centrifugation within 1 h, a delay in this procedure for 48 h produced no change in measured TC.[69] However, when blood was collected, clotted, centrifuged, and stored at 6–7° C in the same serum separator tube for seven days, a slight increase (1.8%) was observed.[70]

After using a secondary serum standard in the long-term Lipid Research Clinics Program, no evidence was uncovered to suggest that cholesterol in pooled serum deteriorates while stored at –20° C.[71] Cholesterol was also shown to be stable in either frozen or lyophilized forms of pooled serum during nearly 5 y of storage at –20° C.[72]

When plasma is to be used for TC analysis, anticoagulant-treated blood should be centrifuged as soon as possible to prevent hemolysis of red blood cells. No change in TC was noted after storage of heparinized plasma for 4 d at 25° C.[73]

The patient need not be fasting if the sample is to be used for TC only.[74] If other tests are to be performed, then a normal 12 h fast is required.

Reagents

The TC reagent is prepared in TRIS-HCl buffer (50 mmol/L, pH 7.6) to contain per liter 0.5 g of Triton X-100, 1.0 mmol of 4AAP, 2.0 mmol of HDCBS, 4.1 mmol of α-cyclodextrin, 50 μmol of potassium ferrocyanide, 400 U of CHE, 800 U of COX, 294 U (guaiacol) of HRP, and 200 kU of lipase.

Cholesterol standards are prepared with NBS SRM911b cholesterol which has been dried under vacuum over silica gel at 55° C for at least 12 h.[75] The appropriate amount of this material is dissolved in 2-methoxyethanol containing 200 g/L Triton X-100.

Procedure

Pipette 1.0 mL of reagent into the appropriate test tubes, add and mix 3 μL of sample, standard or control. Allow the reaction mixture to incubate for 12 min at 37° C. The absorbance is measured at 510 nm against a reagent blank.

Automated Procedure

In the event the reader wishes to automate the procedure, the parameters for the COBAS FARA are outlined in Table 4-2. It should be noted that the 3 μL sample volume stretches the limits of the FARA and that reproducibility may become a problem with some instruments. Therefore, these parameters are intended to serve only as a guideline for the reader.

Calibration

The following protocol is used by one of our laboratories for the periodic evaluation of their routine TC procedure.

Due to the matrix effects of lyophilized calibrators, it is necessary to standardize the COBAS FARA using a serum-based material. The calibrators are prepared by pooling non-lipemic patient and/or employee samples. The sample selection criteria is that the samples must have a TC of approximately 250 mg/dL. In addition, the samples are tested for HIV and hepatitis reactivity; only negative-reacting samples are included in the calibrator pool. After being collected, the pool is filtered through a 0.45 micron Gelman GN-6 Metricell filter. The pool is then mixed thoroughly and aliquoted into microtubes (about 150 μL per tube), after which the microtubes are frozen at –80° C for long-term storage. The pool is then allowed to age (stabilize) for 1 m, after which the TC content is determined by the Abell-Kendall method.

At random, 36 microtubes of the pool are selected and sent to a laboratory certified by the NRML network for determination of TC by the CDC reference method. (For a complete listing of NRML network laboratories, refer to Chapter 11.) The reference laboratory performs triplicate assays of the pool—each assay requires 4 microtubes, due to the large sample volume of the Abell-Kendall method—on each of 3 different days and informs the issuing laboratory of the

Table 4–2 ❖ Parameters for COBAS FARA*

General			Calculation	
Measurement Mode		Absorbance	Sample Check Limit	No
Reaction Mode		R1-I-S-A	Reaction Direction	Increasing
Calibration Mode		Slope Average	Check Direction	On
Reagent Blank		Reagent/Diluent	Conversion Factor	1.0000
Wavelength		510 nm	Slope	0.0000
Temperature		37.0° C	Test Range Low	0.0 mg/dL
Decimal Position		0	High	500.0 mg/dL
Units		mg/dL	Range Type	Normal Range
Analysis Conditions			Norm Range Low	Not Given
R1	Reagent 1	325 µL	High	Not Given
	Time Control	No	Number of Calc Steps	1
I	Incubation Time	30 s	Calc Step A	Endpoint
	M1 Reading	25 s	First Reading	M1 Reading
	M2 Reading	None	Last Reading	Reading 1
S	Sample Volume	3.0 µL	Calibration	
	Diluent (flush)	H_2O	Calibration Interval	Each Run
	Diluent Volume	40.0 µL	Number of Standards	1
	Cleaner Cycle	Off	Position of Standard	1
	Timer Control	None	Standard Value	259 mg/dL
A	Readings		Replicate Analysis	Single
	First Reading	720 s		
	Number/Reads	1		
	Interval	5 s		
	M3 Reading	None		
	M4 Reading	None		

*To convert values in mg/dL to mmol/L, multiply by 0.0259.

mean assayed value for these 9 determinations. This is now the assigned value of the calibrator: this value is confirmed by comparing 12 patient samples—3 samples from each of 4 separate runs—to the values of the Abell-Kendall reference method. If necessary, the assigned value of the pool will be readjusted to result in acceptable patient comparisons to the reference procedure.

Periodic patient-comparison studies have demonstrated that calibrators and controls prepared and stored in this manner are stable for up to 5 y.

Performance

The data collected with the manual procedure suggests that the within-run reproducibility of this reagent system should be less than 1.5% and that the between-run reproducibility should be less than 3%.[76] The reagent is linear to about 900 mg/dL (23 mmol/L) cholesterol. Interference studies conducted at the concentrations recommended by the National Committee for Clinical Laboratory Standards illustrate minimal interference from ascorbate,[76] hemolysis,[76] and lipemia.[77] Since this reagent was developed to clear turbidity, with lipemic samples those analyzers that make blank measurements within the first 1–2 min of the reaction will give erroneously low results. It is of note that NCCLS recommends evaluation of ascorbate at concentrations up to 30 mg/dL (1.7 mmol/L). Although this is well above concentrations that are commonly encountered, extremely increased concentrations of ascorbate will almost certainly interfere.[77]

Bilirubin interference is difficult to evaluate, for numerous reasons. Evidence would suggest that not all forms of bilirubin react and interfere equally. That is to say that the degree of interference is not the same for protein-bound bilirubin as compared to bilirubin glucuronide, and it is not likely that either are the same as commercial material made up in alkali that is used to "spike" normal serum for interference studies. However, it would appear that, at worst, 20 mg/dL (342 µmol/L) bilirubin will cause a negative interference of about 6% with this procedure.[76]

REFERENCE RANGES

Within the last decade, the reference ranges for serum TC filled the better part of two pages of most textbooks. There were different ranges for age (one per half-decade) and sex. In an attempt to alleviate the confusion and perhaps cynicism that this caused, the NCEP has recommended the use of two cut-off values for TC. Serum concentrations below 200 mg/dL (5.17 mmol/L) are considered "desirable," concentrations between 200 and 239 mg/dL (5.17–6.19 mmol/L) are considered "borderline," and concentrations of 240 mg/dL (6.20 mmol/L) and above are considered "high." Thus, the clinician's decision as to whether follow-up studies are necessary is based solely upon two numbers: 200 and 240 mg/dL (5.17 and 6.20 mmol/L). Decisions about treatment are based on the patient's medical history and concentration of LDL-C.

Earlier in the chapter we discussed transferability—or the lack of it—of the Abell-Kendall method to the routine laboratories in which the method might be attempted. The fact that more complex procedures such as Abell-Kendall are designated as "accurate" does not guarantee accuracy in any or all cases. The time and care required to perform the tests, as well as to make up a reagent whose matrix is so critical to the quality of the measurement, are beyond the scope of many laboratories. The procedure proposed here does not suffer from the same problems as the Abell-Kendall method, in that it is easily carried out in any laboratory and offers certain features that are not available in the field methods it closely resembles. A critical aspect of its reagent matrix is that severely hypertriglyceridemic specimens are more amenable to measurement, owing to the lipolysis that occurs simultaneously with the sequence of enzyme reactions leading to the measured color. In addition, the fatty acids split off by the lipase are trapped as the transparent guests of the host molecule, α-cyclodextrin. As a result, interference by turbidity of the sample is avoided, as has been suggested in the past for both the sample and the sample blanking. ❖

REFERENCES

1. Report of the National Cholesterol Education Program Expert Panel on the Detection, Evaluation, and Treatment of High Blood Cholesterol in Adults. Arch Intern Med 1988;148:36–68.
2. National Cholesterol Education Program (NCEP): Highlights of the report of the Expert Panel on Blood Cholesterol Levels in Children and Adolescents. Pediatrics 1992;89:495–501.
3. Current status of blood cholesterol measurement in clinical laboratories in the United States: A report from the Laboratory Standardization Panel of the National Cholesterol Education Program. Clin Chem 1988;34:193–201.
4. Dorsey DB. How does the National Reference System for the clinical laboratory standardize results? Pathologist 1984;May:307.
5. Cohen A, Hertz HS, Mandel J, et al. Total serum cholesterol by isotope dilution/mass spectrometry: A candidate definitive method. Clin Chem 1980;26: 854–60.
6. Tentative guidelines for the development of definitive methods in clinical chemistry for the National Reference System in Clinical Chemistry, NRSCC1-T. Villanova, PA: National Committee for Clinical Laboratory Standards, 1982.
7. Duncan IW, Mather A, Cooper GR. The procedure for the proposed cholesterol reference method. Atlanta, GA: Centers for Disease Control, 1982.
8. Cooper GR, Smith SJ, Duncan IW et al. Interlaboratory testing of the transferability of a candidate reference method for total cholesterol in serum. Clin Chem 1986;32:921–9.
9. Abell LL, Levy BB, Brodie BB, Kendall FE. Simplified methods for the estimation of total cholesterol in serum and demonstration of its specificity. J. Biol Chem 1951;195:357–66.
10. Sperry WM, Brand FC. The colorimetric determination of cholesterol. J Biol Chem 1943;150:315–24.
11. Liebermann C. Ueber des Oxychinoterpen. Ber Dtsch Chem Ges 1885;18:1803–9
12. Burchard H. Beitrage Zur Kenntnis des Cholesterins. Chem Zentralbl 1890;61:25–7.
13. Ellerbe P, Myers GL, Cooper GR, et al. A comparison of results for cholesterol in human serum obtained by the reference method and by the definitive method of the national reference system for cholesterol. Clin Chem 1990;36:370–5.
14. Bernert JT, Akins JR, Cooper GR, et al. Factors influencing the accuracy of the National Reference System total cholesterol reference method. Clin Chem 1991; 37:2053–61.
15. Naito HK. Reliability of lipid, lipoprotein, and apolipoprotein measurements. Clin Chem 1988;34(suppl):B84–94.
16. Current status of blood cholesterol measurement in clinical laboratories in the United States: a report from the Laboratory Standardization Panel of the National Cholesterol Education Program. Clin Chem 1988;34:193–201.
17. Cooper GR, Myers GL, Smith SJ, Sampson EJ. Standardization of lipid, lipoprotein, and apolipoprotein measurements. Clin Chem 1988; 34(suppl):B95–105.
18. Joseph JC, Konishi R, Peterson D. Interference studies on four chemistry analyzers [Abstract]. Clin Chem 1984;30:949–50.
19. Glick MR, Ryder KW, Jackson SA. Graphical comparisons of interferences in clinical chemistry instrumentation. Clin Chem 1986;32:470–5.
20. Glick MR, Ryder, KW. Analytical systems ranked by freedom from interferences. Clin Chem 1987;33:1453–8.
21. Burrin JM, Price CP. Measurement of blood glucose [Review]. An Clin Biochem 1985;22:327–42.

22. Zak B, Artiss JD. Some observations on cholesterol measurement in the clinical laboratory [Review]. Microchem J 1990;41:251–70.
23. Zak B. Cholesterol methodologies: a review. Clin Chem 1977;23:1201–14.
24. Witte DL, Brown LF, Feld RD. Enzymatic analysis of serum cholesterol and triglycerides: a brief review. Lab Med 1978;9:39–44.
25. Wiebe DA, Bernert JT. Influence of incomplete cholesterol ester hydrolysis on enzymic measurements of cholesterol. Clin Chem 1984;30:352–6.
26. Hassemer DJ, Koch DD, Wiebe DA. Four commercially-available cholesterol methods meet 1992 NCEP guidelines. Clin Chem 1988;34:1231.
27. Deeg R, Ziegenhorn J. Kinetic enzymatic method for automated determination of total cholesterol in serum. Clin Chem 1983;29:1798–802.
28. Lindback B, Ullsten E, Bjorkehem I. Is accuracy of clinical chemical routine analyses increasing? Scand J Clin Lab Invest 1988;48:375–8.
29. Katteerman R, Moller DJ, Moller G, et. al. Multicentre study of a new enzymatic method of cholesterol determination. J Clin Chem Clin Biochem 1984;22:245–51.
30. Bachorick PS, Virgil DG, Derby C, et al. Enzymatic analysis of total- and HDL-cholesterol: comparison with the standardized Liebermann-Burchard method used by the lipid research clinics program. Clin Chim Acta 1988;174:307–14.
31. McNamara JR, Schaefer EJ. Automated enzymatic standardized lipid analyses for plasma and lipoprotein fractions. Clin Chim Acta 1987;166:1–8.
32. Bateson J, Artiss JD, Zak B. Sensitive enzymic methods for HDL and HDL subclass cholesterol measurement. Clin Chem 1988;34:1230.
33. Siedel J, Rollinger W, Röschlau P, Ziegenhorn J. Total cholesterol, end-point and kinetic method. In: Bergmeyer HU, Bergmeyer J, Graβl, M, eds. Methods of enzymatic analysis, 3rd ed. Weinheim: VCH Verlagsgesellschaft mbH, 1985:139–48.
34. Siedel J, Hagele EO, Ziegenhorn J, Wahlefeld AW. Reagent for enzymatic determination of serum total cholesterol with improved lipolytic efficiency. Clin Chem 1983;29:1075–80.
35. Kumar A, Christian GD. Enzymatic assay of total cholesterol in serum or plasma by amperometric measurement of rate of oxygen depletion following saponification. Clin Chim Acta 1977;74:101–8.
36. Noma A, Nakayama K. Polarographic method for rapid microdetermination of cholesterol with cholesterol esterase and cholesterol oxidase. Clin Chem 1976; 22:336–40.
37. Flegg HM. An investigation of the determination of serum cholesterol by an enzymatic method. Ann Clin Biochem 1973;10:79–84.
38. Allain CC, Poon LS, Chan CSG, Richmond W, Fu PC. Enzymatic determination of total serum cholesterol. Clin Chem 1974;20:470–5.
39. Trinder P. Determination of glucose in blood using glucose oxidase with an alternative oxygen acceptor. Ann Clin Biochem 1969;6:24–7.
40. Pesce MA, Bodourian SH. Enzymic measurement of cholesterol in serum with the Centrifichem centrifugal analyzer. Clin Chem 1977;23:280–2.
41. Pesce MA, Bodourian SH. Interference with the enzymic measurement of cholesterol in serum by use of five reagent kits. Clin Chem 1977;23:757–60.
42. Deacon AC, Dawson PJG. Enzymic assay of total cholesterol involving chemical or enzymic hydrolysis: a comparison of methods. Clin Chem 1979;25:976–84.
43. Technicon SMAC Methodology Files, Method No. SGH-0065J81. Technicon Instruments Corp., Tarrytown, NY 10591.
44. Garber CC, Feldbruegge D. Evaluation of the performance of the automated enzymatic cholesterol method on the SMAC [Abstract]. Clin Chem 1978;24:1020.

45. Fingerhut B. Enzymic serum cholesterol measurement with a basic autoanalyzer and the DuPont ACA method. Clin Chem 1978;24:1624–7.
46. Witte DL, Brown LF, Feld RD. Effects of bilirubin on detection of hydrogen peroxide by use of peroxidase. Clin Chem 1978;24:1778–82.
47. McGowan MW, Artiss JD, Zak B. Spectrophotometric study on minimizing bilirubin interference in an enzyme reagent mediated cholesterol reaction. Microchem J 1982;27:564–73.
48. McGowan MW, Artiss JD, Zak B. Determination of free, esterified, and total serum cholesterol using a sensitive color reaction. Microchem J 1983;28:294–9.
49. Cobbledick R, Hinberg IH, Poon R, Cziraki S. Elimination of lipid interference with the COBAS-BIO procedure for cholesterol [Abstract]. Clin Chem 1985;31:1024.
50. Sharma A, Artiss JD, Strandbergh DR, Zak B. The turbid specimen as an analytical medium: hemoglobin determination as a model. Clin Chim Acta 1985;147:7–14.
51. Sharma A, Artiss JD, Zak B. A method for the sequential colorimetric determination of serum triglycerides and cholesterol. Clin Biochem 1987;20:167–72.
52. Poon R, Hinberg I, Schrader I, Wadhwani J. Measuring cholesterol without interference from lipemia. Clin Chem 1990;36:2149–50.
53. James DR, Price CP. Interference in colorimetric reactions for measuring hydrogen peroxide. Ann Clin Biochem 1984;21:398–404.
54. Tetrault GA, Miller WG, Chinchilli VM, et al. Regional interlaboratory standardization of determinations of cholesterol, high-density lipoprotein cholesterol, and triglycerides Clin Chem 1990;36:145–9.
55. McQueen MJ, Patten RL, Henderson AR, Krishnaw S, Wood DE, Webb J. Results of a Province-wide quality assurance program assessing the accuracy of cholesterol, triglycerides and high-density lipoprotein cholesterol measurements and calculated low-density lipoprotein cholesterol in Ontario, using fresh human serum. Arch Pathol Lab Med 1991;115:1217–22.
56. Kroll MH, Ruddel M, Elin RJ. Analytical bias for cholesterol and the percent of the population deemed at risk for coronary heart disease. Clin Chem 1988;34:2009–11.
57. The SI manual in health care, 2nd ed. Toronto: Metric commission of Canada, 1982:11–22.
58. Kroll MH, Emancipator K. A theoretical evaluation of linearity. Clin Chem 1993; 39:405–13.
59. Kroll MH, Chesler R, Elin RJ. Effect of lyophilization on results of five enzymatic methods for cholesterol. Clin Chem 1989;35:1523–6.
60. Naito HS, Kwak Y-S, Hartfield JL, Park JK, Travers EM, Myers GL et al. Matrix effects on proficiency testing materials. Arch Pathol Lab Med 1993;117:345–51.
61. Myers GL, Waymack PP. Matrix effects in biological reference materials used in standardization of cholesterol measurements. Fresenius J Anal Chem 1990;338: 538–42.
62. Ross JW, Myers GL, Gilmore BF, Cooper GR, Naito HR, Eckfeldt J. Matrix effects and the accuracy of cholesterol analysis. Arch Pathol Lab Med 1993;117:393–400.
63. Kroll MH, Ruddel M, Elin RJ. A model for the effect of bias for cholesterol on the population at risk. Clin Chem 1988;34:2256–9.
64. Bennett ST, Connelly DP, Eckfeldt JH. Assessment of split-sample proficiency testing for cholesterol by use of a computer simulation model. Clin Chem 1991; 37:497–503.
65. Bennett ST, Eckfeldt JH, Belcher JD, Connelly DP. Certification of cholesterol measurements by the national reference method laboratory network with routine clinical specimens: effects of network laboratory bias and imprecision. Clin Chem 1992;38:651–7.

66. Bachorik PS, Albers JJ, Ellefson RD, Kane JP, Wood, PD. Collection of blood samples for lipoprotein analysis. Clin Chem 1982;28(6):1375–8.
67. Alzofon J, Tilton KA, Haley NJ. Enzymatic determination of cholesterol in plasma obtained by fingerstick. Clin Chem 1985;31(1):168.
68. Kaplan SA, Yuceoglu AM, Strauss J. Chemical microanalysis: Analysis of capillary and venous blood. Pediatrics 1959;24:270.
69. Ono T, Kitaguchi K, Takehara M, Shiiba M, Hayami K. Serum constituents analyses: Effect of duration and temperature of storage of clotted blood. Clin Chem 1981;27(1):35–8.
70. Haider T, Per Foss O. The analytical variation and mean difference of serum lipid values in duplicate samples subjected to different times of storage. Scand J Clin Lab Invest 1983;43:439–43.
71. Hainline A, Karon JM, Winn CL, Gill JB. Accuracy and comparability of long-term measurements of cholesterol. Clin Chem 1986;32(4):611–15.
72. Kuchmak M, Taylor L, and Olansky AS. Suitability of frozen and lyophilized reference sera for cholesterol and triglyceride determinations. Clin Chim Acta 1982; 120:261–71.
73. Keller VH. Errors, resulting from storage, in the determination of eleven parameters in heparinized whole blood and plasma. Z Klin Chem Klin Biochem 1975; 6:217–24.
74. Manual of laboratory operations for the Lipid Research Clinics Program. In: Hainline A, Karon J, Lippel K, eds. Lipid and lipoprotein analysis, 2nd ed. Bethesda MD: National Heart, Lung and Blood Institute, National Institutes of Health, 1982:5.
75. Deacon AC, Dawson JG. Enzymic assay for total cholesterol involving chemical or enzymic hydrolysis: a comparison of methods. Clin Chem 1979;25:976–84.
76. Bateson JE, Artiss, JD, Zak B. The development of an enzymic reference equivalent method for total cholesterol. Clin Chem 1989;35:1071–2.
77. Bateson JE, Artiss JD, Zak B. Cholesterol measurement in grossly lipemic specimens. Clin Chem 1989;35:1072.
78. Peddicord CH, Barnes WA. Ascorbic acid interferogram for cholesterol in the Demand and TDx. Clin Chem 1988;34:773–4.

CHAPTER 5

Measurement of Triglyceride Concentration

Thomas G. Cole, Sigrid G. Klotzsch, Judith R. McNamara

INTRODUCTION AND CLINICAL SIGNIFICANCE

❖ Triglycerides (TG) are water-insoluble lipids, consisting of fatty acids linked to glycerol, which are transported in the blood as core constituents of lipoproteins. TG represent a concentrated source of metabolic energy, containing 9 kcal/g TG. A major source of TG is dietary fats, which are absorbed from the intestine, assembled into chylomicrons, and secreted into the lymph. Chylomicrons ultimately enter the blood compartment, where most of the TG are rapidly hydrolyzed by lipoprotein lipase to glycerol and fatty acids, which are absorbed by adipose tissue for storage or by other tissues requiring a source of energy. A peak of chylomicron-associated TG occurs within 3–6 h after ingestion of a fat-rich meal; however, the rate of absorption of fats is highly variable, depending upon the individual and the dietary composition of the fat.[1] The liver also produces the TG-rich very-low-density lipoproteins (VLDL), which transport TG synthesized in the liver from circulating fatty acids. The rate of VLDL synthesis is regulated by many factors, including substrate availability and hormonal status. The VLDL are precursors of the atherogenic low-density lipoproteins (LDL), some of which are formed as a consequence of VLDL TG hydrolysis by lipoprotein lipases and by the transfer of apolipoproteins to other lipoproteins.

TG are measured in the clinical laboratory for three major purposes: to establish triglyceridemic status; to assess risk of coronary heart disease (CHD), which is associated with abnormal concentrations of TG; and to determine LDL cholesterol (LDL-C) concentration. LDL-C concentration is often calculated in mg/dL through use of the Friedewald equation:[2]

LDL-C = Total Cholesterol – Total TG/5 – HDL Cholesterol (HDL-C).

For results in mmol/L, use Total TG/2.22.

TG concentrations are an important indicator of health. Fasting TG concentrations below 200 mg/dL (2.3 mmol/L) are considered normal. Concentrations between 200 and 400 mg/dL (2.3 and 4.5 mmol/L) are borderline high and concentrations above 400 mg/dL (4.5 mmol/L) are high; both situations require attention in the presence of other CHD risk factors. TG concentrations above 1000 mg/dL (11.3 mmol/L) can lead to abdominal pain and may be life-threatening due to chylomicron-induced pancreatitis.[3]

Although the relationship between TG concentrations and risk of CHD has not been firmly established, several studies have found increased TG concentrations to be positively correlated with increased risk of CHD.[4, 5] The relationship may be stronger in women than in men.[6] Whether TG concentrations represent an independent risk factor is not clear, since other independent risk factors, such as reduced concentrations of HDL-C, are often associated with increased TG.[6] Finally, if LDL-C is not determined directly, TG can be measured for use in the Friedewald equation by which LDL-C concentration is approximated. This approach makes certain assumptions about the relative proportions of cholesterol and TG mass in VLDL and is unreliable at TG concentrations above 400 mg/dL (4.5 mmol/L), when chylomicrons represent a sizable fraction of the TG-rich lipoproteins[7, 8] or in the presence of remnant lipoproteins associated with Type III hyperlipoproteinemia. For additional information on the limitations of the Friedewald calculation, refer to Chapter 7.

The recent introduction of a method for the direct measurement of LDL-C based on the immunoprecipitation of non-LDL lipoproteins may reduce the need for the Friedewald calculation using TG concentration, since the method is not affected by increased TG concentrations and does not depend on lipid ratios.

Clinical methods for the determination of TG concentration involve the enzymatic measurement of glycerol after hydrolysis of TG. Since glycerol is formed through normal metabolic processes, the measurement of TG concentrations will be overestimated if unesterified glycerol is not subtracted through the use of a glycerol blank. In normal individuals, endogenous glycerol represents the equivalent of 10 to 20 mg/dL (0.11 to 0.22 mmol/L) TG, which is a tolerable amount of error. In certain situations, however, endogenous, non-TG-associated glycerol concentrations may be much higher, confounding the interpretation of TG concentrations. Such situations include metabolic diseases such as diabetes mellitus, emotional stress, intravenous administration of drugs or nutrients containing glycerol, and contamination of blood collection devices by glycerol. In addition, prolonged storage of whole blood under non-refrigerated conditions can cause increased concentrations of free glycerol liberated from erythrocyte phospholipids. Although reagents and methods are currently available to eliminate the interference of glycerol, only about 5% of all American clinical laboratories blank TG measurements for endogenous glycerol.

REVIEW OF EXISTING METHODOLOGIES

Almost all clinical laboratories use enzymatic methods for the analysis of TG concentration. Although a multitude of variations have been developed, three basic steps are common to all the methods.[9, 10] These are illustrated in Figure 5–1. The first step uses lipases which are optimized for the hydrolysis of TG to glycerol and fatty acids. Earlier formulations were not always optimized to hydrolyze the TG commonly found in the circulation, i.e., those containing fatty acids of 16 or more carbon atoms. However, manufacturers are now aware of the importance of complete hydrolysis. Although phospholipids are not hydrolyzed by current lipase mixtures, mono- and diglycerides are hydrolyzed and contribute to the amount of TG quantified. Since mono- and diglycerides represent only about 3% of the total plasma neutral glyceride concentration, and since a portion of the mono- and diglycerides arise from *in vivo* and *in vitro* hydrolysis of TG, they are included as part of the TG measurement. All enzymatic methods then quantify the amount of glycerol present after hydrolysis.

The conversion step commonly uses an enzyme, such as glycerol kinase, to phosphorylate glycerol for further enzymatic reactions, or may generate an intermediate which can be used directly for the quantification of TG.

The final step leads to the formation of either a colored dye (commonly quinoneimine or formazin) or an ultraviolet light-absorbing chemical whose concentration can be measured spectrophotometrically (usually NAD^+ appearance or NADH disappearance) and which can then be related to the concentrations of TG in the specimen.

The calibration of TG analyses is confounded by the lack of validated definitive and reference methods. A "reference" chromotropic acid method as used by the Centers for Disease Control and Prevention (CDC) is considered the accuracy target for TG. Calibrators should be traceable to this method. Glycerol solutions, although suitable as standards, should not be used alone for calibration, because the hydrolysis step of the analytical process is not evaluated. Since TG are a heterogeneous mixture of glycerides with various fatty acid moieties, the choice of analytical standard becomes critical. The CDC uses a standard composed of triolein and tripalmitin (2:1, w/w), as this ratio approximates the distribution of fatty acids in lipoprotein-associated TG; however, values are reported as triolein equivalents. Due to the difference in molecular weights, a disparity of approximately 10% occurs when values are reported as tripalmitin equivalents. Truly unambiguous values would result if the Systeme International d'Unites units

Figure 5–1 ❖ Generalized Scheme of Enzymatic TG Analyses

Triglycerides —Hydrolysis→ Glycerol

Glycerol —Conversion→ Intermediate

Intermediate —Reporter Formation→ Colored Dye or Ultraviolet Chemical (mmol/L) were adopted, as recommended by the International Federation of Clinical Chemistry; however, such a change does not appear to be close at hand for the United States.

Since all enzymatic methods measure TG as the quantity of glycerol in a specimen, the endogenous glycerol will cause an overestimation of TG concentration if it is not taken into consideration. Two methods for the elimination of this interference have been devised. The first, the "external blanking" method, requires the use of two assay reagent solutions, one containing all components necessary for the measurement of TG and endogenous glycerol and one which is similar but lacks the lipase enzymes and therefore measures only the endogenous glycerol. In practice, a specimen is run twice, once with each reagent, and the difference in values represents the true TG concentration. The major advantage of this method is that it can be used with most analytical systems; one need only set up a separate test for each reagent solution. In addition, if the system is arranged so that each test provides a value before calculation of the true TG concentration, the concentration of the endogenous glycerol will be known, which may have medical significance in some situations. A disadvantage is that this method does require the duplicate analysis of a specimen using two complete sets of reagents, cuvettes, and other consumable supplies, thereby increasing the cost and the time of the analysis.

An alternative method, the "internal blanking" method, uses a single reagent which is split into two components: the first contains all enzymes necessary to consume the endogenous glycerol without the generation of a measurable light-absorbing compound or with subtraction of the resulting absorbance as blank; the second contains lipase and any other chemicals necessary for the measurement of TG. In practice, the specimen is reacted with the first component of the reagent for a period of time sufficient for the removal of all the endogenous glycerol. At this point a baseline reading is made, the second component of the reagent is added, and the complete reaction, including the hydrolysis of TG, is carried to completion. The value obtained from the second reaction represents the true or net TG concentration. Advantages of this method are that it requires only a single test on the analyzer to be devoted to the analysis, and that it is economical, since only a single, but divided, reagent and a single cuvette per analysis are required. Disadvantages of this method are that it requires a versatile instrument capable of the addition of the second component of the reagent at the proper time, and not all instruments have this capability; that the time required for analysis is prolonged due to the sequential reactions taking place in a single cuvette; and that the actual value of the endogenous glycerol is not available. The measurement of free glycerol concentration may be desirable in some situations, such as when a patient has a high level of endogenous lipoprotein lipase activity in response to heparin therapy.[11]

Much discussion has been devoted to the issue of why and when to glycerol-blank TG measurements.[12] Ideally, if economical and convenient methods were available, a glycerol blank should be included in every TG measurement. In reality, such is not the case; glycerol blanking is not utilized in most clinical laboratories. Complicating the decision is the fact that excessively high concentrations of endogenous glycerol is a problem in only a very small percentage of specimens;[13] however, in these situations an inappropriate medically significant decision may be made. Therefore, if all specimens are not glycerol blanked, decisions must be made as to which specimens are to be blanked. Unfortunately, the lack of communication between the laboratory and the ordering physician often precludes proper decision-making, and any system that depended upon it probably would be in error a large part of the time. To reduce the potential for errors, the following recommendations are made:

1. Clinical laboratories should use only systems and reagents with glycerol blanking capabilities and should have the reagent available at all times to be used as necessary.
2. All hospital in-patient specimens should be glycerol-blanked, due to the higher incidence of endogenous glycerol in this population.
3. Specimens from hospital out-patients or from other sources generally need not be glycerol-blanked, unless specimens are from patients who may be expected to have increased concentrations of glycerol, such as patients from diabetes clinics or other specialty clinics.
4. All specimens must be kept refrigerated until analyzed. Improper storage or transport of specimens may cause an increase in free glycerol. At room temperature, lipases may continue to hydrolyze TG. In addition, unrefrigerated whole blood may have high concentrations of free glycerol from erythrocyte phospholipid hydrolysis.
5. Any specimen with TG concentration > 200 mg/dL (2.3 mmol/L) should be glycerol-blanked using a "reflex" ordering system.
6. Any suspicious specimen, such as one that has high TG concentrations but is not turbid, should also be glycerol-blanked.

Grossly lipemic specimens are often associated with other potential sources of error. One error is caused by the "clearing effect" action of the lipase as it hydrolyzes TG. Light scattering due to large-sized, TG-rich lipoproteins causes an artificially high baseline absorbance measurement in systems which utilize a serum blank rather than a reagent blank. As the analysis progresses, the turbidity is reduced as the lipase hydrolyzes TG and reduces the size of lipoproteins. The overall effect is a slight underestimation of total TG concentration in an up-reaction, or an overestimation in a down-reaction. Reagent-blanked systems are not affected by this phenomenon. Several methods have been devised to minimize the error. The inclusion of proper detergents in the reagent mixture will eliminate the problem in some, but not all, cases.[14] Fortunately, since the magnitude of the error is small and occurs only in specimens with increased TG concentrations, the error is usually acceptable.

In addition, grossly lipemic specimens, particularly those with a high chylomicron content, may become heterogenous as chylomicrons float to the top of the sample cup in an analyzer. Therefore, grossly lipemic specimens should be mixed thoroughly prior to analysis and the analysis must be carried out with minimal delay. The large amounts of fatty acids liberated by the hydrolysis of TG, particularly in lipemic specimens, can interfere in the analysis due to the formation of turbidity and the inactivation of some lipases by product inhibition. This situation is avoided by the inclusion in the reaction buffer of a chemical reservoir for the liberated fatty acids. Generally, bovine serum albumin or α-cyclodextrin is used for this purpose.[13]

As for all analytes, TG concentration should be measured as soon as possible after blood is drawn. In certain specimens containing high values of lipase activities, such as those obtained from heparinized patients, TG will become hydrolyzed upon standing.[10, 11] Over time, the glycerol blank concentration in these specimens will increase while the true TG concentration will decrease; however, total TG concentration will remain stable. This process is temperature-dependent and persists at 4° C, but is minimized if the specimen is frozen at −20° C.

REFERENCE AND DEFINITIVE METHODS

At this time, no true definitive or reference method has been established for the measurement of TG. The National Institute of Standards and Technology (NIST) is in the final stages of publication of two definitive methods for the measurement of TG. Both methods are based on isotope dilution/mass spectrometry. One method measures only TG and the other measures tri-, di-, and monoglycerides as well as free glycerol. The chemical method the CDC uses as its "in-house reference method" has been accepted by The National Committee for Clinical Laboratory Standards as the interim reference method for the U.S. National Reference System. The CDC uses this method to maintain the accuracy base for the CDC-NHLBI (National Heart, Lung and Blood Institute) Lipid Standardization Program, which requires precision consistent with a coefficient of variation of 5% and accuracy within 5% of the target values set on frozen serum specimens by the CDC. The method involves silicic acid-chloroform extraction, alkaline hydrolysis, and color formation with chromotropic acid.[15–18] Due to several technical factors, including the difficulty of transferring the method between laboratories, the CDC is investigating modifications of the method, as well as alternative candidate methods.

RECOMMENDED CLINICAL METHODS

Given the lack of true definitive and reference methods for the analysis of TG, it is difficult to recommend any particular method over another. However, the following are examples of two systems that meet either the criteria of the Lipid Standardization Program or of the College of American Pathologists (CAP) Proficiency Program. It must be emphasized that many other systems also meet these criteria, and that these are offered only as examples of systems that incorporate the two common methods of glycerol blanking.

Example of an External Glycerol-Blanked System

The Core Laboratory of the Lipid Research Center of Washington University uses the Technicon RA-1000 instrument and Miles/Technicon reagents for the analysis of

TG. The assay is based on an endpoint measurement of the generation of a stable quinoneimine dye. The RA-1000 is a random-access instrument which allows multiple analyses to be conducted on the same specimen. In practice, two independent tests are defined and each specimen is run by each test: total TG using reagent T01-1868-01 and glycerol blanking using reagent T01-2013-01, which lacks lipase. Calibration is carried out on the total TG test using Technicon Setpoint, as described by the manufacturer. Technicon provides a system-specific calibrator value. At lot number change, this value is verified in-house by overlap with the previous calibrator. The blank TG is calibrated using a 50 mg/dL (5.43 mmol/L) glycerol standard which is a 1:5 dilution of the Sigma Diagnostics 250 mg/dL (27.15 mmol/L) glycerol standard (G-1394).

Aliquots of specimen, 200 µL minimum, are poured into specimen cups and placed in the instrument. Initial absorbance readings at 500 nm are taken on 350 µL of reagent pipetted into the cuvette tray, followed by the addition of 3.5 µL of specimen. After 5 min, the final absorbance reading is made and the calculated concentration of TG or glycerol is printed for each test.

Although the calculation of true TG concentrations can be made internally by the instrument, we prefer to make the calculation manually, thereby allowing the value of the glycerol blank to be recorded. Internal and external quality control for the system is maintained through use of the lyophilized pools purchased from Miles. Relative to the interim reference method used by the CDC, the described system has an average absolute positive bias of approximately 3% and a coefficient of variation of 1.8%.

Example of an Internal Glycerol-Blanked System

The Clinical Laboratories of Barnes Hospital of the Washington University Medical Center uses the Hitachi 747 instrument and Boehringer Mannheim reagents for the analysis of TG. The assay is based on the endpoint measurement of the generation of a stable phenazone dye. The Hitachi 474 is a random-access analyzer that allows multiple reagent additions to be added to the reaction cuvette during analysis. The instrument is calibrated daily following the manufacturer's instructions using Hitachi calibrators, Precical and Precimat. Precical is a human serum based calibrator with a calibration value set by the manufacturer. Precimat is a 21 mg/dL (2.28 mmol/L) glycerol solution which is diluted 1:21 with 0.9% NaCl prior to use (equivalent to 9.5 mg/dL [0.11 mmol/L] TG). It is assigned a value of 0 to compensate for the nonspecific background reaction in the reagent. The assay is started by mixing 150 µL of Reagent 1 and 3 µL of specimen in a cuvette. During this time the endogenous glycerol is consumed. After 5 min, 150 µL of Reagent 2, which contains lipase, is added to the same cuvette. After 5 min, a final absorbance reading is taken at 505 nm. Results represent true TG concentrations and are calculated based upon the difference in absorbance readings. Internal quality control for the system is maintained through the use of a commercially available control material (DADE Monitrol Level I and II). External quality control is maintained through the CAP Proficiency Program.

Table 5–1 ❖ Recommended Cutpoints for Evaluation of Fasting Triglyceridemic Status		
Classification	*mg/dL*	*mmol/L*
Normal	TG < 200	TG < 2.3
Borderline High	TG 200–400	TG 2.3–4.5
Hypertriglyceridemic	TG > 400	TG > 4.5
High Risk for Pancreatitis	TG > 1000	TG > 11.3

REFERENCE RANGES

The National Institutes of Health Consensus Development Conference on Treatment of Hypertriglyceridemia has set limits for evaluation of triglyceridemic status.[19] These limits, shown in Table 5–1, have been accepted by the National Cholesterol Education Program Expert Panel on Detection, Evaluation, and Treatment of High Blood Cholesterol in Adults,[20] with slight modification.[21] TG values of North American males and females are presented in the appendix to this book.

INTERFERING SUBSTANCES

The sources of potential interfering substances in the analysis of TG have been recently reviewed in depth[9] and such reviews should be referred to for more detailed information. The limits of acceptable analytical performance in the presence of ascorbic acid, bilirubin, hemolysis, and gross lipemia need to be evaluated for each system.[22] In many situations, the inclusion of a serum blank may eliminate the effects of the interfering substances.

Glycerol

As discussed above, endogenous glycerol causes an overestimation of all TG measurements. Methods are available to eliminate this interference and should be used whenever the inclusion of the endogenous glycerol concentration may lead to errors in medical judgment. However, when blanking is instituted, it is important to avoid conditions that would cause *in vitro* hydrolysis prior to analysis, since this would result in a falsely underestimated TG concentration.

Ascorbic Acid

As an antioxidant, ascorbic acid can interfere with oxidation/reduction reactions. Specimens from patients ingesting megadoses of Vitamin C may have improper analysis of TG and other analytes. Concentrations of ascorbic acid up to 20 mg/dL (1135 μmol/L) may be acceptable for most systems.

Bilirubin

Both chemical and spectral interference can be a serious problem with high concentrations of bilirubin in colorimetric methods. Concentrations of bilirubin up to 20 mg/dL (340 µmol/L) may be acceptable for most systems.

Hemolysis

Slight to moderate hemolysis may cause no interference, but gross hemolysis is unacceptable. The effects of hemoglobin interference will vary depending on the oxidative state of the hemoglobin, the type of reaction being performed, and the type of spectrophotometric system in use. In addition, the lysis of erythrocytes results in the dilution of lipid constituents. ❖

REFERENCES

1. Cohn JS, McNamara JR, Cohn SD, Ordovas JM, Schaefer EJ. Postprandial plasma lipoprotein changes in human subjects of different ages. J Lipid Res 1988;29: 469–79.
2. Friedewald WT, Levy RI, Frederickson DS. Estimation of the concentration of low-density lipoprotein cholesterol in plasma, without use of the preparative ultracentrifuge. Clin Chem 1972;18:499–502.
3. Frederickson DS, Goldstein JL, Brown MS. The familial hyperlipoproteinemias. In Stanbury JB, Wyngaarden JB, eds. Metabolic basis of inherited diseases, 4th ed. New York: McGraw-Hill, 1978:604.
4. Brunzell, JD, Albers JJ, Chait A, Grundy SM, Groszek E, McDonald GB. Plasma lipoproteins in familial combined hyperlipidemia and monogenic familial hypertriglyceridemia. J Lipid Res 1983;24:147–55.
5. Genest J, Sniderman A, Cianfeone K, et al. Hyperapobetalipoproteinemia. Plasma lipoprotein responses to oral fat load. Arteriosclerosis 1986;6:297–304.
6. Austin MA. Plasma triglyceride and coronary heart disease. Arterioscler Thromb 1991;11:2–14.
7. Warnick GR, Knopp RH, Fitzpatrick V, Branson L. Estimating low-density lipoprotein cholesterol by the Friedewald equation is adequate for classifying patients on the basis of nationally recommended cutpoints. Clin Chem 1990;36:15–19.
8. McNamara JR, Cohn JS, Wilson PWF, Schaefer EJ. Calculated values for low-density lipoprotein cholesterol in the assessment of lipid abnormalities and coronary disease risk. Clin Chem 1990;36:36–42.
9. Klotzsch SG, McNamara JR. Triglyceride measurements: a review of methods and interferences. Clin Chem 1990;36:1605–13.
10. Naito HK, David JA. Laboratory considerations: Determination of cholesterol, triglycerides, phospholipids and other lipids in blood and tissues. In: Story JB, ed. Lipid research methodology. New York: Alan R. Liss, 1984:31.
11. Hortin GL, Cole TG, Gibson DW, Kessler G. Decreased stability of triglycerides and increased free glycerol in serum for heparin-treated patients. Clin Chem 1988; 34:1847-9.
12. Cole TG, Glycerol blanking in triglyceride assays: Is it necessary? Clin Chem 1990;36:1267–8.
13. Jessen RH, Dass CJ, Eckfeldt JH. Do enzymatic analyses of serum triglycerides really need blanks of free glycerol? Clin Chem 1990;36:1372–5.

14. Artiss JD, Strandbergh DR, Zak B. Elimination of glycerol interference in a colorimetric enzymic triglyceride assay. Clin Chim Acta 1989;182:109-16.
15. Carlson LA, Wadstrom LB. Determination of glycerides in blood serum. Clin Chim Acta 1959;4:197–205.
16. Carlson LA. Determination of serum triglycerides. J Athero Res 1963;3:334–6.
17. Van Handel E. Zilversmit DB, Micromethod for the direct determination of triglycerides. J Lab Clin Med 1957;50:152–7.
18. Lofland Jr HB. A semiautomated procedure for the determination of triglycerides in serum. Anal Biochem 1964;9:393–400.
19. Consensus Development Conference. Treatment of hypertriglyceridemia. JAMA 1984;251:1196–1200.
20. National Cholesterol Education Program. Report of the National Cholesterol Education Program Expert Panel on Detection, Evaluation, and Treatment of High Blood Cholesterol in Adults. Arch Intern Med 1988;148:36–69.
21. National Cholesterol Education Program. Summary of the second report of the National Cholesterol Education Program Expert Panel on Detection, Evaluation, and Treatment of High Blood Cholesterol in Adults. (Adult Treatment Panel II). JAMA 1993;269:3015–23.
22. Glick JR, Ryder KW. Interferographs. User's guide to interferences in clinical chemistry instruments. Indianapolis: Science Enterprises, Inc, 1987.

CHAPTER 6

Measurement of High-Density Lipoprotein Cholesterol Concentration

Donald A. Wiebe and G. Russell Warnick

INTRODUCTION

❖ In recent years, high-density lipoprotein cholesterol (HDL-C) has become an accepted component of the lipoprotein profile [total cholesterol (TC), triglycerides (TG), HDL-C, and low-density lipoprotein cholesterol (LDL-C)] used to assess an individual's risk of developing coronary heart disease (CHD). Numerous epidemiologic studies have demonstrated the inverse relationship between HDL-C and CHD: individuals with low concentrations of HDL-C have increased incidence of CHD, while those with high concentrations of HDL-C (as in familial hyperalphalipoproteinemia) seldom present with symptoms of CHD. The fact that pre-menopausal females tend to have higher HDL-C and less incidence of CHD compared to males supports the protective role of HDL-C. Furthermore, in post-menopausal women HDL-C concentrations decrease while their incidence of CHD increases. Thus, there has been substantial interest in HDL-C measurements as a means to assess potential risk for CHD, and most clinical laboratories routinely perform HDL-C analysis.

The National Cholesterol Education Program (NCEP) has had a substantial impact on the need to perform accurate and precise laboratory measurements of TC. NCEP activities have resulted in increased awareness by physicians and the public of the association between CHD risk and increased concentrations of TC. In response to the NCEP recommendation that every individual should know his or her TC value, TC screening programs have become common in such non-traditional sites as shopping malls.

In 1988, the NCEP Adult Treatment Panel (ATP) developed uniform treatment guidelines for individuals with increased TC. HDL-C had a limited role in the initial ATP guidelines, with concentrations < 35 mg/dL (0.91 mmol/L) considered to be a risk factor for CHD.[1] Updated guidelines developed by ATP II and released in 1993 recognized HDL-C as an independent risk factor and placed greater importance on HDL-C by recommending its measurement together with TC during the initial screen.[2] An HDL-C concentration of < 35 mg/dL (0.91 mmol/L) is still considered indicative of high risk for CHD, while a concentration ≥ 60 mg/dL (1.56 mmol/L) is considered protective. Thus, for reliable classifications of patients, clinical laboratories must have reliable methods for HDL-C measurement.

Physicians generally treat patients whose lipid patterns are associated with increased risk of developing CHD. Occasionally, patients have normal concentrations of cholesterol (< 200 mg/dL [5.18 mmol/L]) with low HDL-C (< 35 mg/dL [0.91 mmol/L]). Among the therapeutic approaches to increasing HDL-C, exercise, stopping smoking, and weight loss are perhaps the most effective. More often, patients present with increased TC or LDL-C concentrations, in which case intervention programs are primarily oriented toward lowering cholesterol through diet therapy and, if necessary, drug therapy. Some of the common cholesterol-lowering drugs, such as niacin and the fibrates, tend to increase HDL-C. (For more information on diet and drug therapy for dislipidemia, refer to Chapter 2.)

DEFINITION

HDL, the smallest in size of the lipoproteins, comprises a complex family of lipoprotein particles that exists in a constant state of dynamic flux as the particles interact with other HDL particles and with low-density lipoprotein (LDL) and very-low-density lipoprotein (VLDL) particles. HDL has the highest proportion of protein relative to lipid compared to other lipoproteins, comprising > 50% protein. The major HDL proteins are designated apolipoproteins (apo) AI and AII; apo CI, CII, CIII, E, AIV, and D exist in smaller amounts. Phospholipids are the principal lipid component of HDL, with lesser amounts of cholesterol esters, unesterified cholesterol, and TG (see Table 6–1). Since cholesterol esters are hydrolyzed to unesterified or free cholesterol in most analytical procedures, the esterified portion is usually quantified as unesterified cholesterol.

Classically, HDL-C refers to the fraction of TC (both free cholesterol and cholesterol esters) associated with HDL as defined by ultracentrifugation (hydrated density), although in common practice the fractions separated by chemical precipitation or electrophoretic (mobility) properties are also referred to as

Table 6–1 ❖ Lipid Composition of HDL Particles

Lipid	*% of Total Lipid*
Phospholipids	50
Cholesterol esters (as free cholesterol)	30
Cholesterol	10
Triglyceride	10

HDL-C. Therefore, HDL-C is defined by the method used to isolate the lipoprotein and includes a family of similar particles that vary in size, chemistry, and composition.

REVIEW OF EXISTING METHODS

Ultracentrifugation

Lipoproteins can be readily separated on the basis of their differing hydrated densities using ultracentrifugation techniques. The proportion of lipid, especially TG, associated with the proteins in a particular lipoprotein adds to the buoyancy of the total complex, allowing separation of the major classes by either equilibrium or rate methods. The nomenclature commonly used in the literature classifies lipoproteins by their relative densities. Therefore, by definition, VLDL are those particles whose density is < 1.006 Kg/L, the background density of serum, while LDL are those whose density is 1.006–1.063 Kg/L and HDL are those whose density is 1.063–1.210 Kg/L. These classes are approximately comparable to electrophoretic fractions designated pre-beta, beta, and alpha lipoproteins, respectively.

Fractionation of the lipoproteins can be accomplished using ultracentrifugation following adjustment of the specimen density with salts such as NaBr or KBr.[3] The following procedure describes the 1.063 Kg/L separation of VLDL and LDL (top fraction) from HDL (bottom fraction), as illustrated in Figure 6–1.

The reliability of lipoprotein quantitation following separation by ultracentrifugation techniques depends upon both the performance of the analytical method, such as cholesterol analysis, and the skills of the technologist in performing accurate recovery and transfer of the lipoprotein fractions from the ultracentrifuge tube. Further adjustment of the 1.063 Kg/L bottom fraction to a density of 1.210 Kg/L followed by ultracentrifugation can be used to isolate the HDL from other serum proteins. However, for quantitative analysis, HDL is usually considered simply to be the fraction of density > 1.063 Kg/L.

Ultracentrifugation has long been considered the ultimate comparison method for the isolation and quantitation of lipoproteins. Unfortunately, ultracen-

Figure 6–1 ❖ Ultracentrifugation at 1.063 Kg/L.

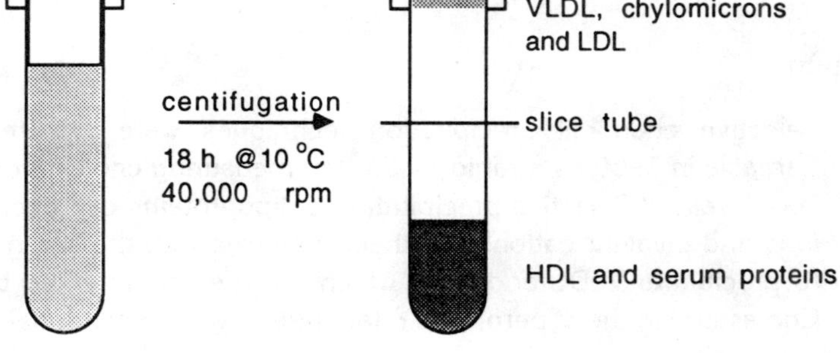

trifugation cannot meet the stringent requirements for a reference method. Achieving complete and reproducible recovery is difficult, even for experienced technologists, and fractions may become cross-contaminated. In addition, fractions isolated by ultracentrifugation are heterogeneous in that they contain other functional particles. For example, the HDL fraction with density between 1.063–1.210 Kg/L may contain considerable amounts of lipoprotein(a) [Lp(a)].

An interesting use of the ultracentrifuge for quantitation of lipoprotein fractions is the single-spin vertical autoprofiler (VAP) approach.[4] An aliquot (1.3 mL) of plasma adjusted to a density of 1.21 Kg/L with solid KBr is overlaid with 3.5 mL of normal saline and centrifuged in a vertical rotor (DuPont-Sorvall rotor TV865) for 45 min at 64,000 rpm. The bottom of the centrifuged sample tube is then punctured and the cholesterol concentration of the effluent monitored continuously. A computer is used to analyze the profile and generate values for cholesterol associated with HDL, LDL, VLDL, and Lp(a). The VAP approach is capable of accurate and reproducible lipoprotein quantitation.

Electrophoresis

Historically, lipoproteins were isolated using electrophoretic techniques, visualized with lipophilic dyes, and named on the basis of mobility by comparison to mobilities of common serum proteins. Thus, lipoprotein classes were commonly referred to as alpha, pre-beta, and beta lipoproteins (see Figure 6–2). Lipoproteins can be separated using a variety of electrophoretic media, such as paper, agarose gel, cellulose acetate, and polyacrylamide gel, with numerous buffers.[5] Lipid stains, such as Oil Red O, Fat Red 7B, or Sudan Black, have been used to detect lipoproteins after electrophoresis. Alpha lipoproteins, which have mobility comparable to alpha-proteins, are approximately comparable to the HDL class.

Electrophoresis has proved primarily useful for qualitative analysis of lipoproteins and has not been considered appropriate for quantitation. Since the lipophilic dyes are not specific for a class of lipid, such as cholesterol, TG, or phospholipids, they cannot be used to accurately quantitate lipoproteins. One interesting approach involves staining electrophoretic plates with specific enzyme reagents, such as cholesterol esterase and oxidase for cholesterol coupled with a peroxidase indicator.[6] However, the system must be reproducible to achieve the precision or accuracy necessary for HDL-C quantitation in clinical applications. Recent studies suggest that newer automated electrophoretic systems can be made suitable for accurate HDL-C quantitation.[7]

Precipitation

Selective chemical precipitation techniques were reported by Burstein and Samaille in 1960 as a rapid method of measuring cholesterol associated with the lipoproteins.[8] Selective precipitation of lipoproteins can occur by mixing polyanions and divalent cations or other chemicals with the serum/plasma specimens to precipitate VLDL and LDL, which are then sedimented by centrifugation.[9-11] Cholesterol in the supernate is quantitated by a conventional enzymic or other as-

Figure 6–2 ❖ Comparison of Electrophoresis Media

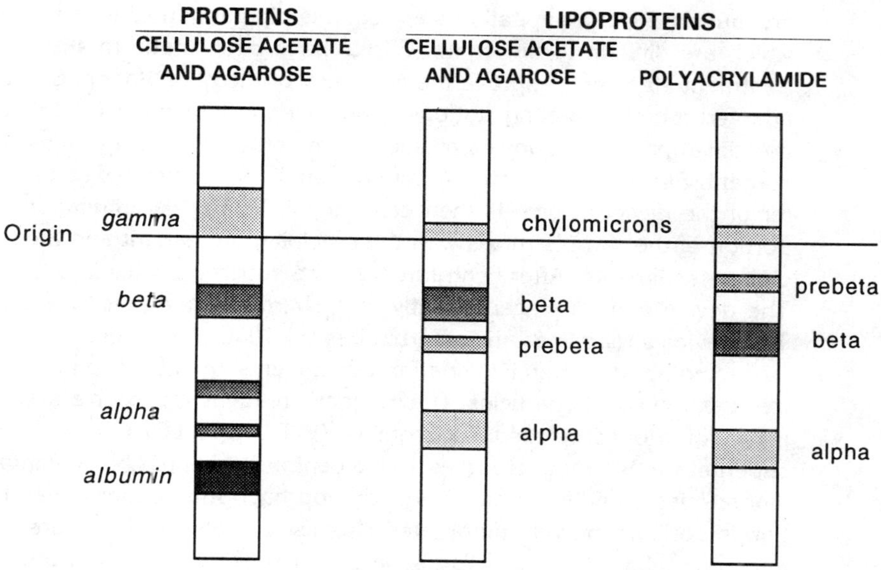

say. Following are several of the common reagents available for selective isolation of lipoproteins:[12-16]

- ❖ heparin–manganese chloride[12]
- ❖ heparin–calcium chloride[13]
- ❖ dextran sulfate–magnesium chloride[14]
- ❖ sodium phosphotungstate-magnesium chloride[15]
- ❖ polyethylene glycol[16]

Each of the chemical precipitation methods has several modifications or variations that reportedly improve the selectivity or performance of the system. For example, heparin-manganese chloride methods are reported with 1.0 and 2.0 mol/L $MnCl_2$, where the lower concentration is the preferred reagent with serum and the higher is preferred with EDTA plasma. The higher concentration with plasma is required to compensate for the chelation of some of the divalent cations by EDTA and avoid incomplete precipitation of apo-B-containing lipoproteins. Dextran sulfate-$MgCl_2$ precipitation has been reported with both 500,000 and 50,000 molecular weight materials; the HDL-C values are significantly lower with the first reagent. The sodium phosphotungstate procedures have been used with or without $MgCl_2$, adjusted for pH and at several concentrations. Similarly, a variety of polyethylene glycol concentrations and pH conditions have been reported for HDL-C analysis.

The cholesterol analysis can be performed on automated chemistry analyzers, but the HDL precipitation step requires pipetting, mixing, and centrifugation, which are difficult to automate. Efforts have been made to streamline the pretreatment steps and improve the efficiency of the quantitation. Some users have adapted robotic pipetting stations to achieve these tasks. A new alternative commercial approach employs a plastic device containing pre-measured precipitation reagent. Simply, this involves pouring a sufficient volume of serum into the center of the device, which is then centrifuged. During centrifugation, a measured portion of the sample mixes with the precipitation reagent and the resultant precipitate sediments. After centrifugation, the supernate rises into the central well. The device can be placed directly in a chemistry analyzer for direct sampling. This device streamlines and standardizes the HDL separation.

Another commercial variation circumvents centrifugation by using magnetized dextran sulfate particles. The magnetic reagent approach allows for the separation of the complexed lipoproteins (VLDL and LDL) from HDL-containing supernate with a magnet rather than a centrifuge—a possible advantage for small laboratories. In addition, the magnetic approach may afford streamlined separation in conjunction with automated chemistry analyses. There are also compact analyzers with integrated HDL separations schemes built into the system, thus eliminating a pre-treatment step. These types of systems will be attractive to clinics that lack centrifuges to perform the more traditional HDL-C assays. (For further discussion of these analyzers, refer to Chapter 8.)

Quantitation of HDL-C in routine laboratories is done almost exclusively by one of the precipitation techniques. There are several reasons for this:

- ❖ There is high demand for routine quantitation of HDL-C as a CHD risk factor.
- ❖ Expensive equipment, such as an ultracentrifuge, is not required.
- ❖ Cholesterol analysis is performed directly on supernate.
- ❖ The technique can be partially automated for high-volume laboratories.

Therefore, HDL-C was more easily quantitated by the relatively simple and inexpensive selective precipitation techniques compared to either ultracentrifugation or electrophoresis methods.

Selection of the most suitable HDL reagent for a given laboratory is not a simple decision. Inordinate effort would be required to perform an in-depth evaluation of each method to assess all the possible performance characteristics and optimize the procedures to ensure that the cholesterol in the supernate accurately represents the patient's HDL-C. Most laboratories rely on commercial reagents, with the majority using either dextran sulfate or sodium phosphotungstate procedures (see Figure 6-3). Described in some detail below are two methods which are representative of typical assays with these two reagents.

Figure 6–3 ❖ *Laboratory Use of HDL-C Reagents, by Year*

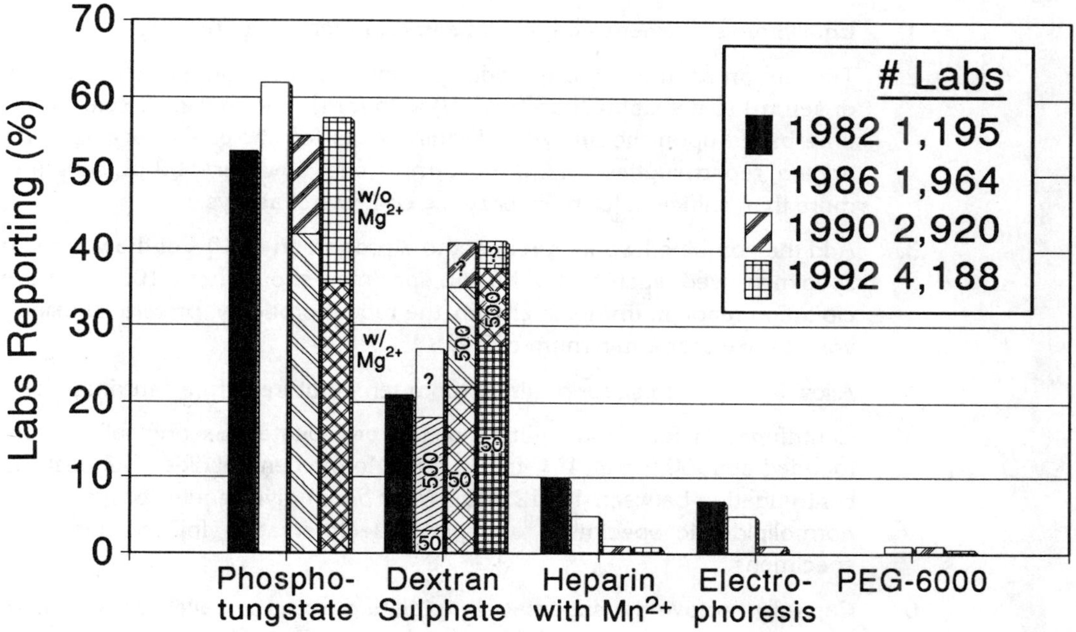

From CAP Proficiency Survey, Comprehensive Chemistry C-A

Commonly Used HDL-C Methods

Dextran Sulfate[14]

This method was reported as a selected method for clinical chemistry.[17]

Reagents: Dextran sulfate (50,000 MW) is manufactured by Sochibo (Meudon-LA-Foret, France) and distributed in the United States by Genzyme (Boston, MA). Reagent-grade magnesium chloride hexahydrate crystalline material is available from several sources.

Solutions: Stock dextran sulfate solution is prepared with 2.0 g dextran sulfate in 100 mL deionized water. Stock 1.0 mol/L magnesium chloride is prepared with the dry hexahydrated salt using 20.3 g $MgCl_2$ in 100 mL deionized water. Combined working solution is prepared by mixing equal volumes of the two stock solutions to obtain a 10 g/L solution of dextran sulfate and 0.5 mol/L $MgCl_2$. Alternatively, the combined reagent can be prepared directly. With the addition of 0.05% sodium azide (NaN_3), this reagent is stable for at least 1 y at room temperature.

Precipitation sequence:

1. Equilibrate specimens and reagents to room temperature.
2. Transfer an accurate and reproducible volume of specimen (either plasma or serum) to a suitable tube (e.g., 10 × 75 mm). The user may select a volume based upon the analytical requirements for cholesterol analysis. The original report suggests a 2.0 mL sample size; however, 500 µL should be more than sufficient for most enzymic cholesterol assays.
3. Add the combined working reagent to a proportion of 10% of the volume of specimen used, such that a 1.0 mL specimen should have 100 µL of precipitation reagent. Immediately mix the tube thoroughly, preferably using a vortex mixer, for a minimum of 3 s.
4. Allow the tubes to sit for 5 min at room temperature before centrifugation.
5. Centrifugation for sedimentation of the precipitate was originally recommended at 1500 × g at 4° C for 30 min. More recent studies[18] suggest that centrifugation between 10–12,000 × g for 5 min gives equivalent results on normolipidemic specimens and better sedimentation for hyperlipidemic specimens.
6. Carefully remove the supernate by pipetting from the pellet for cholesterol analysis. The use of conical-shaped rather than round-bottom tubes may facilitate the removal of the supernate without disturbing the pellet.

   Note: Incomplete separation of apo B-containing lipoproteins may be observed by the presence of cloudy or turbid supernates in hypertriglyceridemic specimens. Cloudy supernates may be cleared by ultrafiltration with a 0.22 µm filter or by repeat precipitation procedure with the original specimen diluted with saline at a ratio of 1:1. Alternatively, the dilution can be made subsequently on turbid supernates. Simply mix equal volumes of the turbid supernate with an equal volume of saline-containing precipitation reagent at the same concentration originally added to the specimen.
7. Perform the cholesterol analysis on the supernate.

   Note: The cholesterol assay should be optimized for high performance in the low cholesterol concentration range. The cholesterol result is corrected for dilution by multiplying by 1.1. In the case of turbid supernates diluted with saline, multiply by the additional dilution factor times 2.

Sodium Phosphotungstate[15, 19]

Reagents: Reagent grade phosphotungstate free acid crystals and magnesium acetate (tetrahydrate) are available from several chemical suppliers.

Solutions: Stock 4% sodium phosphotungstate is prepared with 10.0 g of the free acid dissolved in 125 mL deionized water mixed with 40 mL 1.0 mol/L NaOH.

The resulting solution is adjusted to pH 7.6 and brought to a total volume of 250 mL. The pH of the stock solution should be checked for stability and adjusted as required with dilute NaOH. Stock 2.0 mol/L magnesium acetate [$(CH_3CO_2)_2Mg \cdot 4H_2O$] is prepared with 42.89 g of magnesium acetate dissolved in 100 mL deionized water in a volumetric flask. The magnesium acetate solution should be checked by the laboratory's routine Mg assay to ensure proper concentration (see discussion below). A combined working reagent is prepared by mixing 4 parts of the 4% sodium phosphotungstate with 1 part 2.0 mol/L magnesium acetate.

Analysis to confirm the concentration of the stock 2.0 mol/L magnesium acetate solution: Transfer 1.0 mL of the stock solution to a 250 mL volumetric flask and bring to volume with deionized water for a 0.008 mmol/L solution that contains 19.45 mg/dL Mg. Analysis of the diluted magnesium acetate solution by the laboratory's routine Mg procedure should range from 19.25–19.65 mg/dL. The stock magnesium acetate can be easily diluted with deionized water or adjusted with solid magnesium acetate to compensate for overly concentrated or dilute reagent, respectively.

Note: With this reagent, it is important that preparation be consistent from batch to batch.

1. Prepare fresh combined working reagent for a batch of specimens.
2. Allow the specimens to equilibrate to room temperature. Label the conical centrifuge tubes with appropriate identification marks for the specimens and controls.
3. Using a precision dispenser/dilutor, transfer 500 µL of serum and 63 µL of combined working reagent to a conical-shaped centrifuge tube.
4. Vortex the resultant mixture for 15 s.
5. Centrifuge the specimens at 1500 × g (2900 rpm with the Beckman Accu-Spin FR) at 4° C for 45 min.

   Note: To achieve reproducibility with this reagent, it is important that the timing and temperature of the measuring, mixing, and centrifugation steps be standardized and consistent.
6. Carefully remove the supernate and transfer it to a proper container for cholesterol analysis without disturbing the pellet.
7. Analyze the supernatant for cholesterol content and correct the result for dilution by multiplying by the factor 1.126.

CDC Reference Method

The following procedure is used at the Centers for Disease Control and Prevention (CDC) to assign HDL-C target values for human-based serum pools.[20] Although there is no validated reference method for HDL-C now, the CDC HDL-C procedure is currently considered the best accuracy target because most of the major epidemiologic/population studies have used CDC pools as the reference target for analyses. Since the accepted NCEP cutpoints for HDL-C are derived from such population studies, appropriate patient classification is dependent on obtaining routine results in agreement with the CDC method. Therefore, the CDC HDL-C method is recommended as the reference method for calibrating and checking the accuracy of routine methods.

The three key components to the CDC HDL-C method are outlined below and illustrated in Figure 6-4.

Figure 6–4 ❖ *CDC HDL-C Reference Method*

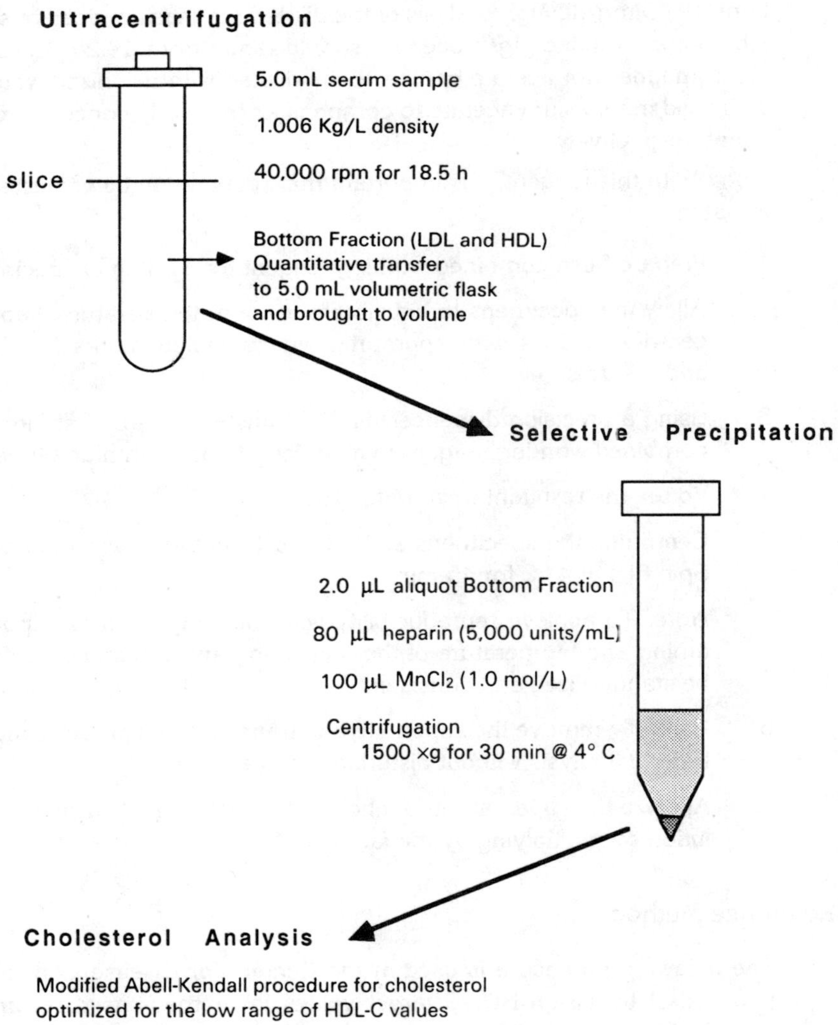

1. Ultracentrifugation at a density of 1.006 Kg/L to isolate HDL and LDL from chylomicrons and VLDL.[21] Selective precipitation procedures for HDL may be falsely increased when increased TG-rich lipoproteins (chylomicrons and VLDL) prevent their sedimentation. Therefore, ultracentrifugation eliminates these lipoproteins as potential interferences.

2. Selective precipitation of LDL with heparin/$MnCl_2$. CDC uses the same precipitation reagent as that used for the Lipid Research Clinics (LRC) studies.

   Reagents: Pharmaceutical-grade heparin, Lipo-Hepin (40,000 units/mL) from Riker Laboratories (Minneapolis, MN) and reagent-grade $MnCl_2 \bullet 4H_2O$, available from several chemical suppliers, are used for the HDL procedure.

   Solutions: Working heparin solution (5,000 units/mL). Mix 1.0 mL heparin with 7.0 mL saline (0.15 mol/L NaCl). This mixture will be stable for at least 1 m with storage at 4° C. A 1.0 mol/L $MnCl_2$ solution is prepared by dissolving 197.91 g of the tetrahydrate in water and bringing to volume in a 1.0 L volumetric flask.

   Precipitation: Add 80 µL heparin solution and 100 µL $MnCl_2$ to 2.0 mL of 1.006 Kg/L bottom fraction in an ice bucket, mixing thoroughly after each addition. Allow to stand in the ice bucket for 30 min, then centrifuge at 1500 × g for 30 min at 4° C. Recover supernates by pipetting for subsequent cholesterol analysis.

3. Analysis of cholesterol in the HDL supernate by the CDC reference method, a modified Abell-Kendall assay. The Abell-Kendall reference method for TC is the accepted accuracy target for any cholesterol tests in specific lipoproteins, including HDL-C. The HDL supernate aliquot required for cholesterol analysis is twice the amount required for TC to increase the sensitivity of the procedure in the lower HDL-C range.

Note: Because this three-step method is cumbersome, a direct precipitation method using dextran sulfate and magnesium, similar to the one described in this chapter, is under investigation as a "designated comparison method" (see Chapter 11).

EVALUATION OF HDL PRECIPITATION PROCEDURES

Regardless of which method a laboratory selects to measure HDL-C on patient specimens, the procedure should be capable of achieving acceptable analytical performance. Analytical performance goals for precision, accuracy, and total error in HDL-C measurement have been developed by the NCEP (see Table 6–2). Most of the common precipitation techniques can provide acceptable results when performed with an accurate and reproducible cholesterol assay. Precision can be assessed by replicate analysis of commercial controls or frozen serum pools. Currently, the only reliable approach to ensuring the accuracy of a routine HDL-C assay is a direct comparison of patient specimens with a laboratory using

Table 6-2 ❖ NCEP Analytical Goals for HDL-C	
Precision	*Accuracy**
INTERIM GOALS	
≥42 mg/dL (1.09 mmol/L): CV ≤6% <42 mg/dL: SD ≤2.5 mg/dL (0.065 mmol/L)	Bias ± 10%
1998 GOALS	
≥42 mg/dL (1.09 mmol/L): CV ≤4% <42 mg/dL: SD ≤1.7 mg/dL (0.044 mmol/L)	Bias ± 5%
*Bias compared to CDC three-step reference method.	

the CDC's HDL-C reference procedure or an equivalent method with well-established accuracy. Reference materials for HDL-C that behave as patient material for use in transferring accuracy from laboratory to laboratory are not available through the usual commercial sources.

Standardization of TC measurements has been improved by the activities of the Cholesterol Reference Method Laboratory Network, which was organized by CDC to make reference methods readily available for direct comparison on fresh patient specimens. The Network is in the process of establishing a similar program for standardizing HDL-C measurement. Network laboratories will be available to assist manufacturers and clinical laboratories in verifying the accuracy of their HDL-C methods. A detailed protocol for conducting a comparison study is available through the Network program.

Manufacturers and laboratories may wish to perform additional validation studies. The following set of universal evaluation and performance considerations is presented to enable a laboratory to validate its own HDL procedure.

1. Optimize cholesterol method:
   a. Linearity from 0–120 mg/dL (0–3.11 mmol/L).
   b. Reproducibility with 1 SD = 2 mg/dL (0.05 mmol/L) or less.

2. Establish reproducible specimen and reagent transfers using precise pipettes. Methods may be automated using a programmable dispensing apparatus. For example, the Beckman Accu-Prep and Micromedic systems are capable of high precision. High-volume laboratories can automate this process to the extent of a "walk-away" system.

3. Establish specific conditions for specimen processing:
   a. length of time for mixing (vortexing)
   b. time and temperature for incubation
   c. time, temperature, and force of centrifugation

4. Compare HDL results from fresh patient specimens with a reliable laboratory using an accurate method:

a. The Cholesterol Reference Method Laboratory Network sponsored by CDC is available for such comparisons. (Phone 404-488-4126 for additional information.)

b. Patient specimens should be distributed throughout the assayed range of 20–100 mg/dL (0.52–2.59 mmol/L).

c. Specimens stored or shipped under similar conditions should be analyzed at about the same time to minimize confounding effects of specimen deterioration.

The following experiments can be performed to assess the specificity and separation characteristics of any given HDL-C procedure:

1. Lipoprotein electrophoresis of the supernates and resolubilized precipitates provide an indication of the specificity of the separation. The HDL precipitates should be washed with a saline/precipitation reagent solution and dissolved with 0.6 mol/L NaCl prior to electrophoresis. The supernates should have an HDL or alpha band only and no beta or pre-beta band, and the precipitate should have no HDL band. The sensitivity of this technique depends on the characteristics of the stain used with the particular electrophoresis method.
2. Analysis of both the supernates and resolubilized precipitates for apo AI and B can assess the possible inadequate separation of lipoproteins. For example, apo B in the supernate suggests non-HDL contamination, while high concentrations of apo AI in the precipitate suggests precipitation of HDL.
3. Select several patient specimens with varying amounts of TG to compare the performance of the HDL-C assay at low and increased concentrations of TG. Some HDL-C methods experience incomplete sedimentation and supernatant turbidity when TG concentrations exceed 400–500 mg/dL (4.52–5.65 mmol/L). Specimens with chylomicrons can give erroneous values, often yielding cloudy supernates and overestimation of HDL. Such specimens should be re-precipitated after dilution, or the supernates should be filtered to obtain reliable results.

Reference Ranges

Data from an LRC prevalence study[22] provides HDL-C values for plasma using the heparin/$MnCl_2$ precipitation procedure and Technicon AutoAnalyzer cholesterol method. Keep in mind that the LRC HDL-C method required the use of 0.046 mol/L $MnCl_2$ and heparin with EDTA plasma specimens. If the study were repeated today, the method would have used a higher (0.092 mol/L) concentration of $MnCl_2$ and would have slightly lower HDL-C values. Refer to the appendix for the LRC HDL-C values in various North American populations.

NECP guidelines for interpreting HDL-C data have simplified this task for clinicians. Simply, serum HDL-C values < 35 mg/dL (0.91 mmol/L) are associated with increased risk of CHD for both males and females, while HDL-C values ≥ 60

mg/dL (1.56 mmol/L) are a positive indicator of patient health. As HDL-C values increase above 35 mg/dL (0.91 mmol/L), ratios of TC to HDL-C tend to decrease, which is also associated with CHD risk. In general, ratios, whether HDL or otherwise, have interesting statistical significance in determining a patient's risk of developing a given disease. However, ratios are unitless values, and NCEP guidelines suggest that the physician make decisions based on the individual's TC or LDL-C and HDL-C values rather than the ratios.

SUMMARY

HDL-C analysis has been routine for almost 20 years, yet there is still a need to improve analytical performance to ensure that HDL-C results meet clinical requirements. The analytical goals for accuracy and precision developed by the NCEP provide a performance target for clinical laboratories. The CDC cholesterol reference method network program, which has been effective in standardizing TC, will soon provide the same service for HDL-C, which will facilitate improvements in accuracy. The standardization process would be made more efficient with the development and widespread availability of HDL-C reference materials that behave as patient specimens (without matrix effects) for use in comparing other HDL-C assays to the consensus reference procedure. Priority should be given to encouraging manufacturers to take advantage of the opportunity to validate and standardize their methods, as this will contribute to improving the performance of many laboratories.

The importance of accurate HDL-C measurements was emphasized by Robert Levy, M.D., past director of the National Heart, Lung and Blood Institute, when he proposed the following performance requirement in an article in *Clinical Chemistry*:[23]

> One has to be able to measure accurately in the range below 50 mg/dL (1.30 mmol/L) a difference of less than 5 mg/dL (0.13 mmol/L) if the HDL determination is to be used to define an individual's risk.

In order to achieve this goal, manufacturers and individual laboratories must take steps to improve the accuracy and reproducibility of their methods. We have a long road to travel before these objectives are met, and our journey will be successful only with a coordinated effort. ❖

REFERENCES

1. Current status of blood cholesterol measurements in clinical chemistry laboratories in the United States: A report from the Laboratory Standardization Panel of the National Cholesterol Education Program. Clin Chem 1988;34:193–201.
2. The Expert Panel. Summary of the second report of the National Cholesterol Education Program (NCEP) expert panel on detection, evaluation, and treatment of high blood cholesterol in adults (Adult Treatment Panel II). JAMA 1993;269: 3015–23.

3. Lindgren FT, Jensen LC, Hatch FT. The isolation and quantitative analysis of serum lipoproteins. In: Nelson GF, ed. Blood lipids and lipoproteins: quantitation, composition, and metabolism. Wiley-Interscience, 1972:181–274.
4. Cone JT, Segrest JP, Chung BH, Ragland JB, Sabesin SM, Glasscock A. Computerized rapid high resolution quantitative analysis of plasma lipoproteins based upon a single vertical spin centrifugation. J Lipid Res 1982;23:923–35.
5. Lewis LA, Opplt JJ, eds. CRC handbook of electrophoresis, Vol 1. Boca Raton, FL: CRC Press, 1980.
6. Conlon DR, Blankstein LA, Pasakarins PA, Steinberg CM, D'Amelio JE. Quantitative determination of high-density lipoprotein cholesterol by agarose gel electrophoresis. Clin Chem 1979:25:1965–9.
7. Baldwin R, Warnick R, Leary T. Rapid, precise automated HDL cholesterol analysis with the Helena Rep system. [Abstract] Clin Chem 1992;38:1059.
8. Burstein M and Samaille J. Sur un dosage rapide du cholesterol lie aux a- et aux b-lipoproteins du serum. Clin Chem Acta 1960;5:609.
9. Levin SJ. High-density lipoprotein cholesterol: review of methods. American Society of Clinical Pathologists. Core Chemistry No. PTS 89-2 (PTS-36) 5/2, 1989.
10. Warnick GR, Cheung MC, Albers JJ. Comparison of current methods for high-density lipoprotein cholesterol quantitation. Clin Chem 1979;25:596–604.
11. Denmacker PNM, Hijmans AGM, Vos-Janssen HE, Van't Laar A, Jansen AP. Measurement of high-density lipoprotein cholesterol in serum: comparison of six isolation methods combined with enzymic cholesterol analysis. Clin Chem 1980;26:1780–6.
12. Warnick GR, Albers JJ. A comprehensive evaluation of the heparin-manganese precipitation procedure for estimating high density lipoprotein. J Lipid Res 1978;19:65–76.
13. Srinivasan SR, Radhakrishnamurthy B, Berenson GS. Studies on the interaction of heparin with serum lipoproteins in the presence of Ca^{++}, Mg^{++}, and Mn^{++}. Arch Biochem Biophys 1975;170:334–40.
14. Warnick GR, Benderson J, Albers JJ. Dextran sulfate-Mg^{2+} precipitation for quantitation of high-density lipoprotein cholesterol. Clin Chem 1982;28:1379–88.
15. Lopes-Virella MF, Stone P, Ellis S, Colwell JA. Cholesterol determination in high-density lipoproteins separated by three different methods. Clin Chem 1977;23: 882–4.
16. Brigg CJ, Anderson D, Johnson P, Deegan T. Evaluation of the polyethylene glycol precipitation method for the estimation of high-density lipoprotein cholesterol. Ann Clin Biochem 1981;18:177–81.
17. Warnick GR, Benderson J, Albers JJ. Dextran sulfate-Mg^{2+} precipitation procedure for quantitation of high density lipoprotein cholesterol. In: Cooper GR, ed. Selected methods of clinical chemistry, Vol. 10. Washington, DC: American Association of Clinical Chemistry, 1983:91–99.
18. Nguyen T, Warnick GR. Improved methods for the quantitation of total HDL and subclasses. Clin Chem 1989;35:1086.
19. Davies C, Fahie-Wilson MN. Use of magnesium acetate in high-density lipoprotein cholesterol assay. Clin Chem 1979;25:1510.
20. Gary Myers, Centers for Disease Control, Atlanta GA., personal communication, Jan 1990.
21. Hainline A, Karon J, Lippel K., eds. Manual of laboratory operations. In: Lipids Research Clinics Program, Lipid and lipoprotein analysis (2nd ed). Bethesda, MD: U.S. Dept. Health and Human Services, 1982.
22. The Lipid Research Clinics population studies data book. Vol.1: The prevalence study. NIH Publication No. 80-1527. Bethesda, MD: National Institutes of Health, 1980.
23. Levy RI. Cholesterol, lipoproteins, apoproteins, and heart disease: present status and future prospects. Clin Chem 1981;27:653–62.

CHAPTER 7

Measurement of Low-Density Lipoprotein Cholesterol Concentration

John D. Belcher, Judith R. McNamara, Gregory F. Grinstead, Nader Rifai, G. Russell Warnick, and Paul Bachorik

INTRODUCTION

❖ The positive relationship between total serum cholesterol (TC) concentrations and coronary heart disease (CHD) is the most consistent finding to emerge from the observational and experimental epidemiological studies of the last 15 years. Low-density lipoprotein cholesterol (LDL-C) accounts for approximately two-thirds of the serum TC and constitutes the primary atherogenic fraction of serum cholesterol. Other potentially atherogenic particles present usually constitute only a small proportion of serum TC, including very-low-density lipoproteins (VLDL), chylomicron remnants, intermediate-density lipoproteins (IDL), and lipoprotein(a) [Lp(a)].[1-4]

LDL consists of a hydrophobic lipid core composed of cholesterol ester and triglyceride (TG). The lipid core of the LDL particle is surrounded by an amphipathic coat composed of phospholipid, unesterified cholesterol, and apolipoproteins. Each LDL particle contains on its surface one molecule of apolipoprotein B-100 (apo B-100) and variable amounts of apo CIII (0.34–1.90 molecules) and apo E (0.08–0.74 molecules).[5] On a mean weight basis, LDL (1.006–1.063 kg/L) is composed of 38% cholesterol ester, 22% phospholipid, 21% protein, 11% TG, and 8% unesterified cholesterol.[6]

Our understanding of the role that LDL-C plays in the atherosclerotic process has increased significantly since the discovery of LDL receptors by Brown and Goldstein in the 1970s.[7] Knowledge gained from experiments and observations in molecular, cellular, animal, and human biology has provided a large body of evidence implicating LDL-C as an important risk factor for the development of

CHD. Undoubtedly there will be refinements to the evidence. Current debate centers on the predictive power of LDL particle number, which can be assessed by measuring LDL apo B-100, as opposed to LDL-C, which represents the amount of cholesterol transported by LDL and may not reflect the number of LDL particles to the same extent in all patients.[8] In addition, the cost/benefits of measuring total and LDL apo B-100 and Lp(a) for CHD risk assessment must be carefully considered. Clearly, international standardization of apo B-100 and Lp(a) measurements must be achieved before these measurements can replace or supplement LDL-C[9] in CHD risk assessment.

Other promising areas of research include the role of oxidized LDL in promoting atherosclerosis[10] and the relationship of low serum concentrations of LDL-C to cancer, hemorrhagic stroke, mental functioning, and infections.[11]

Despite the importance of LDL-C measurement to CHD prediction, we must recognize that CHD is a polygenetic, multi-factorial disease and that LDL-C is but one of the factors in the equation for determining CHD risk. Low concentrations of high-density lipoprotein cholesterol (HDL-C) and high concentrations of Lp(a), as well as cigarette smoking, hypertension, diabetes, obesity, clotting factors, male gender, and family history, give the clinician an overlapping but more complete mosaic of a patient's risk for CHD. In addition, coronary angiography, ultrasound, and exercise-treadmill tests with electrocardiographic measurements can provide confirmatory evidence in patients with suspected CHD.

Given the powerful association between increased LDL-C and CHD, a national consensus has emerged for the treatment of high serum LDL-C. The National Cholesterol Education Program Adult Treatment Panel (NCEP-ATP) recommended in 1988 that LDL-C be used as the primary criterion for treatment decisions by physicians for patients with hypercholesterolemia.[12] ATP treatment recommendations were based on each patient's LDL-C value and risk factor profile. These recommendations were revised in 1993;[13] the revised guidelines still use LDL-C concentration and risk factor profile as the primary treatment criteria, but additional emphasis has been placed on measurement of HDL-C. (See Chapter 2 and the revised NCEP guidelines for adults[13] and classification for children and adolescents[14] in Table 7–1.) These recommendations make accurate measurement of TC, TG, LDL-C, and HDL-C a national public health imperative.

Reference ranges for plasma LDL-C for North American men and women, compiled from the Lipid Research Clinics' (LRC) prevalence study,[15] are provided in the appendix. However, it should be noted that average LDL-C concentrations in the United States have declined slightly since the LRC prevalence study was conducted.

REVIEW AND DISCUSSION OF EXISTING METHODOLOGIES

Accurate measurement of LDL-C depends on *separation* of LDL particles in serum from other lipoproteins, namely VLDL and HDL, followed by the *measurement* of LDL-C. LDL can be separated from VLDL and HDL by methods based on physical characteristics such as density, size, charge, or apolipoprotein composi-

Table 7-1 ❖ NCEP Guidelines for Adults and Classification for Children and Adolescents

*Revised NCEP Guidelines for Adults: Initiation of Therapy**		
	Initiation Concentration	*Goal of Therapy*
Without 2 or more CHD risk factors	LDL-C ≥ 160 mg/dL (4.14 mmol/L)	LDL-C < 160 mg/dL (4.14 mmol/L)
With 2 or more CHD risk factors	LDL-C ≥ 130 mg/dL (3.36 mmol/L)	LDL-C < 130 mg/dL (3.36 mmol/L)
*Revised NCEP Guidelines for Adults: Initiation of Drug Therapy after Diet Therapy**		
Without 2 or more CHD risk factors	LDL-C ≥ 190 mg/dL (4.92 mmol/L)	LDL-C < 160 mg/dL (4.14 mmol/L)
With 2 or more CHD risk factors	LDL-C ≥ 160 mg/dL (4.14 mmol/L)	LDL-C < 130 mg/dL (3.36 mmol/L)
With CHD	LDL-C ≥ 130 mg/dL (3.36 mmol/L)	LDL-C < 100 mg/dL (2.59 mmol/L)
NCEP Classification for Children and Adolescents		
Desirable: LDL-C < 110 mg/dL (2.85 mmol/L)		
Borderline: LDL-C 110–129 mg/dL (2.85-3.34 mmol/L)		
High: LDL-C > 130 mg/dL (3.36 mmol/L)		

*See Chapter 2 for a list of risk factors.

tion. Traditionally, LDL have been defined as all lipoproteins with densities of 1.019–1.063 kg/L. However, in common practice this definition has been broadened to include IDL, or VLDL remnants, some of which are precursors of LDL and are found between the densities of 1.006 and 1.019 kg/L. In addition, since Lp(a) overlaps the LDL and HDL density ranges and is precipitated along with VLDL, IDL, and LDL during HDL precipitation assays, all or a portion of this lipoprotein is also present in the LDL fraction. Thus, in clinical profiles that utilize precipitation methods to separate HDL, LDL encompasses all of the lipoproteins with densities of 1.006–1.063 kg/L plus Lp(a), or:

$$\text{LDL} = \text{LDL} + \text{IDL} + \text{Lp(a)}$$

This definition has been used as the basis for cutpoints defined by the NCEP-ATP, and it is this definition to which this chapter is confined. Future assessment of CHD risk will likely take into account the lipoprotein heterogeneity present within the LDL fraction once the relative significance of various LDL subfractions and other lipoproteins found in the LDL density range have been clearly established and routine standardized methods have been developed for their measurement. Such methods are likely to be immunological in nature, so as to separate LDL subfractions based on their apolipoprotein composition and epitope expression.

Several ultracentrifugation methods have been developed to separate serum lipoproteins. Analytical ultracentrifugation was developed in the 1950s[16–18] and continues to be used today in some research laboratories. In this technique, lipoproteins are separated in the analytical ultracentrifuge and quantitated by optical refraction. This method of quantitation measures lipoprotein mass but does not give any information about lipid or protein composition. Also, analytical ultracentrifugation requires expensive equipment and has a low sample throughput. For these reasons, analytical ultracentrifugation has rarely been used in the clinical laboratory.

Sequential ultracentrifugation was developed by Havel, Eder and Bragdon[19] in 1955 to overcome some of the limitations of analytical ultracentrifugation. In this technique, lipoproteins are separated by repeated ultracentrifugation after progressively increasing the sample density.[19, 20] Lipoproteins can be isolated within any desired density interval and in sufficient quantities to allow for multiple chemical analyses. Sequential ultracentrifugation continues to be used today for preparative isolation of lipoproteins, although its clinical utility is limited by its requirement for lengthy ultracentrifugation and its low sample throughput compared to other methods currently available.

Discontinuous density gradient equilibrium ultracentrifugation evolved from sequential ultracentrifugation.[21-27] In this method, serum samples are layered in an ultracentrifuge tube with salt solutions of different densities. During ultracentrifugation, lipoproteins in the sample migrate to their isopycnic (equilibrium) density and the separated lipoproteins are then fractionated for analysis. The major advantage of discontinuous density gradient equilibrium ultracentrifugation is that only one centrifugation step is required to separate all of the lipoprotein classes. Also, sample volumes of 0.2 mL or less can be used with sensitive microenzymatic cholesterol and TG assays.[27] The disadvantages of the technique are the labor required to prepare the density gradients and fractionate the ultracentrifuge tubes, and the limited sample throughput, which is largely dictated by the number of sample slots per rotor and the centrifugation time.

Several types of rotors have been used for density gradient ultracentrifugation: swinging-bucket,[21–27] vertical,[28–31] and fixed-angle.[27] The swinging-bucket rotor has been used most widely because it generally provides better resolution of lipoprotein subfractions than the vertical and fixed-angle rotors. However, the vertical and fixed-angle rotor methods require much shorter centrifugation times (45 min to 3 h) than swinging-bucket rotor methods (22 to 66 h), with only a small loss in resolution. Moreover, fixed-angle and vertical rotors are available with 2–3 times more sample wells than the swinging-bucket rotors. The shorter centrifugation times of the fixed-angle and vertical centrifugation methods are due to the shorter sample separation zones in these rotors compared to swinging-bucket rotors. Also, both the vertical and fixed-angle density gradient methods rely on rate zonal rather than isopycnic separations. Lipoproteins separated by rate zonal methods do not reach their equilibrium density positions; therefore, assignment of lipoprotein cholesterol values is based on the rates of migration of the different lipoprotein fractions, and the data analysis is usually based on computer-assisted deconvolution of the cholesterol profile rather than the equilibrium densities of the lipoproteins.[27, 29]

In routine clinical application, the ultracentrifugation methods for measuring LDL-C concentrations to assess lipoprotein abnormalities and CHD risk are too expensive and time consuming. Currently the majority of laboratories estimate LDL-C by the Friedewald formula.[32] The NCEP will soon publish laboratory guidelines that stress the need for reliable methods for direct measurement of LDL-C that can replace the Friedewald formula. Direct LDL methods employed by clinical laboratories should measure all of the atherogenic lipoproteins contained within the traditional definition of LDL [i.e., LDL + IDL + Lp(a)]. An immunologic approach for measuring LDL directly and the Friedewald method for estimating LDL-C (along with certain precautions) are discussed in later sections of this chapter.

Other methods for separating LDL-C include size-exclusion and other types of chromatography, electrophoresis, and precipitation. The size-exclusion chromatography methods include agarose column chromatography,[33] high-performance gel filtration column chromatography,[34, 35] and fast-flow gel filtration chromatography.[36] The time required for analysis (24 h) has been the major difficulty with agarose column chromatography. Also, difficulty in obtaining homogeneous lipoprotein fractions has been reported.[35] The development of high-performance liquid chromatography and fast-flow gel filtration methods[34-36] has reduced the analysis time but has increased the cost and complexity of the procedure. Affinity chromatography using anti-LDL antibodies, heparin, or dextran sulfate linked to Sepharose gels has also been used to isolate LDL.

Electrophoresis methods, which can separate lipoproteins according to their charge and size, have been commonly used in clinical laboratories. Traditionally, electrophoresis was used for qualitative assessment of the various types of hyperlipoproteinemias. More recent applications, also available as prepackaged commercial kits, employed agarose and polyacrylamide gel electrophoresis at pH 8.6, followed by visualization using lipophilic stains such as Oil Red O or Sudan Black, or by enzymatic reagents for quantitative analysis. Lipoprotein concentrations were estimated by densitometry based on the color intensity of the separated bands. Because of imprecision and other factors, electrophoresis has been considered a semi-quantitative technique. Other potential problems are differential staining of lipoprotein lipids by lipophilic dyes, which can result in misdiagnoses. Serum from patients with biliary cirrhosis can appear normal on electrophoresis, despite the presence of unesterified cholesterol concentrations above 1000 mg/dL (25.9 mmol/L).[37] In addition, certain lipoproteins have atypical mobility. For example, IDL and chylomicron remnants can smear between LDL and VLDL; Lp(a) generally migrates on agarose with slow pre-beta with, or near, VLDL; and beta-VLDL, when present, migrates with LDL. Therefore, lipoprotein electrophoresis is not commonly used for routine clinical quantitation of LDL cholesterol. A recent report on an automated electrophoresis system suggests the potential for reliable quantitation.[38]

Several methods for selective chemical precipitation of LDL have been reported.[39-45] These precipitation methods quantitate LDL-C as the difference between TC and the soluble VLDL + HDL-C. When serum TG values are low, precipitation methods are precise and produce reasonably accurate results relative to ultracentrifugation methods. However, when samples with high TG concentrations are analyzed, most investigators have found that the precipitation methods are plagued with systematic errors.[39, 40, 44-48] Thus, the current LDL precipitation methods generally exhibit the same shortcomings as the Friedewald formula—i.e., interference from increased TG—and therefore do not appear to have substantial advantages for routine quantification of LDL-C. Immunochemical methods, in which LDL particles may be separated not on the basis of buoyant density but by their characteristic apolipoprotein composition, offer promise for the future. However, since most of the epidemiological data is based on the traditional definition of LDL [i.e., LDL + IDL + Lp(a)], new methods will need to define which LDL subfractions are being measured. It is conceivable that "LDL" as defined above could predict CHD risk better than any of the LDL subfractions alone. Before LDL subfractions will become widely accepted for clinical use, additional epidemiological studies are needed to determine the relative benefits of measuring LDL subfractions compared to "LDL" as it is currently defined by the NCEP.

REFERENCE METHOD FOR LDL CHOLESTEROL DETERMINATION

There is no consensus-approved and validated reference method for LDL-C as exists for TC. The accuracy target for national population studies has been the Centers for Disease Control and Prevention (CDC) reference method, which utilizes a single ultracentrifugation of plasma at its own non-protein density of 1.006 kg/L. The VLDL and chylomicrons float in the supernate after centrifuging at 39,000 rpm (135,000 ×g) for 18 h at 10° C in a Beckman 40.3 or 50.3 fixed-angle rotor (Beckman Instruments, Palo Alto, CA). The LDL, including IDL and Lp(a), and the HDL remain below in the infranate. The supernatant and infranatant fractions are generally separated by a tube-slicing technique and are quantitatively removed and reconstituted to a known volume. The cholesterol concentrations of serum, the 1.006 kg/L infranate, and an HDL supernate, obtained by heparin-manganese precipitation of the 1.006 kg/L infranate (see Chapter 6), are measured by the Abell-Kendall reference method (see Chapter 4). VLDL-C is then calculated as TC minus the 1.006 kg/L infranatant (LDL+HDL) cholesterol.

This method of quantitating VLDL-C is preferable to direct measurement of VLDL-C in the 1.006 kg/L supernate because the already low concentration of cholesterol in VLDL is lowered further due to losses that occur during tube slicing. LDL-C is calculated as the 1.006 kg/L infranatant cholesterol (LDL+HDL) minus HDL-C (determined by precipitation). Note that the CDC HDL-C reference method (see Chapter 6) uses the 1.006 kg/L infranate (VLDL-free serum) instead of whole serum for the separation of HDL using a precipitation method. Use of the 1.006 kg/L infranate for HDL determination eliminates interference caused by increased TG in some samples.

Some technical notes are worthy of mention. A similar method with HDL precipitation from whole plasma was adopted for use by the LRC[49] and other lipoprotein research laboratories. Variations of the method, with HDL separation by other precipitation reagents and enzymic lipid analysis, have become the method of choice for most diagnostic lipoprotein analyses. This LDL reference procedure was called "beta quantification"[49] since LDL lipoproteins are called "beta-migrating lipoproteins" in electrophoretic terminology.

Accurate volume measurement is essential for accurate LDL-C values. Typically, 5.0 mL of serum or plasma is pipetted into a 40.3 or 50.3 Ultraclear™ or polyallomer ultracentrifuge tube (Beckman Instruments) using a volumetric pipette. The serum is then carefully overlaid with 1 mL of 1.006 kg/L saline solution (0.15 mol/L NaCl). The tube is sealed with an aluminum cap and filled to capacity with the saline solution through the fill hole in the aluminum cap, and the fill hole is then sealed shut with a metal set screw. It is very important to squeeze the tube to be sure there are no leaks in the cap. A leak in the centrifuge tube during ultracentrifugation can lead to spurious results. Other brands of centrifuge tubes with alternative sealing methods are available, such as heat seals (Beckman Instruments), screw seals (Seton, Palo Alto, CA), or plug seals (Nalge, Rochester, NY).

The newer Tabletop TL100 Ultracentrifuge (Beckman Instruments) offers an alternative centrifugation method with smaller specimen volumes and shorter separation times. In a typical method, 2 mL of serum or plasma is added to a thick-walled 4 mL polycarbonate ultracentrifuge tube for the 100.3 fixed-angle rotor (Beckman Instruments). The specimen is overlaid with 1.6 mL of 1.006 kg/L saline. Capping of the tube is not required, which is an advantage. In addition, centrifugation time is reduced to 2.5 h at 100,000 rpm (541,000 ×g). LDL-C values determined by this method appear to be the same as those obtained with the 40.3 rotor.[50] Another micro variation using a 200 µL specimen in a 7 × 20 mm thick-walled tube for the TLA100 rotor (Beckman Instruments) gave results equivalent to the conventional macro method.[51]

After ultracentrifugation, both the 40.3/50.3 rotor method and the 100.3/TLA100 rotor methods require a special tube-slicing apparatus (Beckman Instruments) to separate the VLDL-containing 1.006 kg/L supernate from the LDL- and HDL-containing 1.006 kg/L infranate. The tube should be placed in the slicing apparatus so that the knife slices through the middle of the clear zone between the white supernate and yellow infranate. After slicing, the VLDL-containing supernate can be discarded unless it is needed for further biochemical analyses, such as agarose electrophoresis to diagnose Type III hyperlipoproteinemia.[52] Alternatively, the supernate can be analyzed to assess overall recovery of cholesterol.

It is very important that the 1.006 kg/L infranate containing LDL and HDL be *quantitatively* transferred to a volumetric flask, preferably of the same or similar volume as the starting serum volume. Quantitative transfer of the infranate requires careful and patient washing of the centrifuge tube with saline solution. The protein plug in the bottom of the tube can trap lipoproteins and must be dissolved. Also, lipoproteins can stick to the sides of the centrifuge tube and must be removed with careful washing. All of the washing must be done with a small amount of saline so as not to dilute the sample too much.

Once the LDL- and HDL-containing infranate has been reconstituted inside the volumetric flask, it should be gently but thoroughly mixed to homogeneity. Aliquots can be used for HDL-C determination by precipitation (see Chapter 6) and for cholesterol analysis (see Chapter 4).

In the proposed reference method, all cholesterol values are determined by the Abell-Kendall reference method. For routine clinical samples that require LDL determination by ultracentrifugation (see discussion below), a standardized enzymatic method can be used for the measurement of cholesterol to reduce the analysis costs.

POTENTIAL IMMUNOLOGIC METHODS FOR DIRECT LDL DETERMINATION

Metabolically speaking, immunologic methods directed against apolipoproteins may be the most specific methods available for separating lipoproteins since the apolipoprotein composition largely determines lipoprotein catabolism and atherogenicity. Apolipoprotein composition differs both between and within each lipoprotein class.

Practically speaking, immunologic methods directed against apolipoproteins inherently have a great deal of flexibility for designing lipoprotein separation procedures. Antibodies against apolipoproteins could be used to remove lipoproteins selectively by incorporating the antibodies into liquid phases that will precipitate immune complexes, or by attaching the antibodies to solid phases such as microtiter plates, latex beads, magnetic beads, agarose gels, or glass fibers. Also, immunologic methods are non-destructive to lipoproteins, since antibodies can bind lipoproteins without altering the lipoprotein's cholesterol content. However, the antibodies must be of sufficient affinity and concentration to quantitatively bind all of the desired lipoprotein in the sample.

The similarities and differences in the apolipoprotein composition of lipoproteins can be exploited using immunologic techniques to isolate the lipoprotein class or subclass desired. For example, in fasting serum, apo AI is found only on

Table 7-2 ❖ Concentration and Lipoprotein Distribution of Plasma Apolipoproteins in Normal Fasting Humans[1]

	Plasma Concentration		*Distribution Between Lipoproteins (mol%)*[2]			
Apolipoprotein	mg/dL	mmol/L	VLDL	IDL	LDL	HDL
AI	130	46.40	–	–	–	100
AII	40	23.00	–	–	–	100
B-100	80	1.44	2	8	90	–
CI	6	9.05	2	1	–	97
CII	3	3.37	30	10	–	60
CIII	12	13.60	20	10	10	60
D	10	4.55	–	–	–	100
E	5	1.46	20	20	10	50

[1]Adapted from Gotto AM Jr, Pownall HJ, Havel RJ. Introduction to the plasma lipoproteins. In: Segrest JP, Albers JJ, eds. Methods in enzymology, vol. 128. New York: Academic Press, 1986:9.
[2]For each apolipoprotein

HDL particles and apo B-100 is found only on VLDL, IDL, LDL, and Lp(a) particles (see Table 7–2).

There is at least one apo AI molecule on every HDL particle[53] and only one apo B-100 on every VLDL, IDL, LDL, and Lp(a) particle.[54] Therefore, antibodies to apo AI or apo B-100 could be used to isolate or remove either of these two lipoprotein fractions selectively.

The situation becomes more complex when one attempts to separate LDL, IDL, and Lp(a) from VLDL by immunologic methods, because all four of these lipoproteins contain apo B-100. This separation is desirable for any direct LDL method attempting to emulate the beta-quantification method which measures indirectly the combined LDL, IDL, and Lp(a) fractions. Theoretically, this separation may be possible using monoclonal antibodies to differentially expressed epitopes on apo B-100. However, this approach has not been successful in laboratories working with monoclonal antibodies to apo B-100.[55] All of the antibodies to VLDL apo B-100 developed to date cross-react with LDL apo B-100. Success with this approach would depend on developing monoclonal antibodies with the correct specificity, and on having those epitopes expressed universally in the population.

Perhaps an easier and more universal immunologic approach would be to utilize polyclonal antibodies to specific apolipoproteins to selectively remove lipoproteins. Using each apolipoprotein's molar concentration and percentage of distribution among lipoproteins in Table 7–2, we have calculated the average molar ratio of each apolipoprotein relative to apo B-100 on VLDL, IDL, and LDL (see Table 7–3). Since there is only one apo B-100 on each VLDL, IDL, and LDL particle, these ratios are equivalent to the average number of apolipoprotein molecules on each lipoprotein particle. In order to mimic the beta-quantification procedure, VLDL must be separated from IDL and LDL. In order to do this immunologically,

Table 7–3 ❖ Average Number of Apolipoprotein Molecules per Particle*

Apolipoprotein	VLDL	IDL	LDL
B-100	1.0	1.0	1.0
CI	6.3	0.8	0.0
CII	35.1	2.9	0.0
CIII	94.4	11.8	1.0
E	10.1	2.5	0.1

*The average number of apolipoprotein molecules on a VLDL, IDL, or LDL particle is equal to the ratio of the apolipprotein to apo B-100 since there is one and only one apo B-100 per particle. The ratios are calculated from Table 7–2 as follows:

$$\frac{\text{mmoles apo CI, CII, CIII or E (on VLDL, IDL, or LDL)}}{\text{mmoles apo B-100 (on VLDL, IDL, or LDL)}} = \frac{\text{serum concentration of apo CI, CII, CIII, or E} \times \text{\% distribution (on VLDL, IDL, or LDL)}}{\text{serum concentration of apo B-100} \times \text{\% distribution (on VLDL, IDL, or LDL)}}$$

where the serum concentrations are in mmol/L.

one must use epitopes or apolipoproteins expressed on VLDL that are not expressed on IDL and LDL, or vice versa.

Unfortunately, none of the apolipoproteins is found exclusively on VLDL (Table 7–3). Apo CI comes closest to fulfilling this criterion, since there are 6.3 molecules of apo CI on each VLDL particle, 0.8 molecules on each IDL, and none on LDL. Apo CII is the next closest, with 35 molecules per VLDL particle, 2.9 molecules per IDL, and none on LDL. Apo E is the third best candidate, with an average of 10 molecules per VLDL particle, 2.5 per IDL, and 0.1 per LDL. Thus, an immunoadsorbent containing antibodies directed against apo CI would be expected to remove all of the VLDL particles, approximately 80% of the IDL particles, and none of the LDL. In contrast, antibodies directed against apo E would remove all of the VLDL and IDL and approximately 10% of the LDL. Hence, because of the loss of IDL containing apo CI, immunoadsorption or immunoprecipitation with antibodies directed against apo CI should yield an IDL and LDL fraction that is somewhat lower in concentration than that measured in the beta-quantification procedure.

On average, IDL contains approximately 13% of TC and 18% of the combined IDL plus LDL cholesterol.[56] If antibodies directed against apo CI remove 80% of the IDL, then IDL+LDL cholesterol values would be, on average, approximately 14% (80% × 18%) lower in an immunologic method than in the beta-quantification method. In reality however, incomplete precipitation of HDL could partially offset the removal of IDL from the LDL fraction. Thus, compensating errors could make a procedure appear more accurate than it really is.

Theoretically, it may be possible to improve the yield of IDL with an immunologic method by gently dissociating apo CI or apo CII from IDL by changing the temperature, pH, or ionic strength, or by adding mild detergents or lipid emulsions. Apo CI and CII should be easier to displace from IDL than VLDL because of the apparent higher binding affinity apo CI and CII have for VLDL. This is probably due to VLDL's larger size and higher TG content. Apo CI and CII probably dissociate from IDL *in vivo* when the particle reaches a critical size or TG content.

Monospecific polyclonal sheep antibodies to human apo CI have been successfully used by Anderson, Gry, and Kostner[57] to precipitate VLDL and IDL for routine measurement of serum LDL-apo B. This method requires only 25 μL of human serum and has a linear correlation with LDL-apo B values measured after ultracentrifugation over a wide range of serum cholesterol (150–373 mg/dL [3.89–9.66 mmol/L]) and TG (50–200 mg/dL [0.57–2.26 mmol/L]) concentrations. Regression analyses show the values are distributed evenly around the line of identity with a constant coefficient of variation (CV) of 11.4 percent. The CV for the comparison method, ultracentrifugation, contributes 5–10% to this latter value. No reduction of Lp(a) is observed after immunoprecipitation with the anti-apo-CI antibody.

There is some apo CI on HDL particles. The average molar ratio of apo CI to apo AI in HDL is approximately 0.19. Addition of anti-apo-CI antibodies to human sera co-precipitates some, but not all, HDL with the VLDL and IDL.[57] This is not a problem if one is attempting to remove both VLDL and HDL in order to measure LDL. Over 20 years ago it was recognized that antibodies to apo AI, but not to apo AII, could precipitate the HDL fraction.[58, 59] It is now recognized that there are two types of particles in HDL: particles containing both apo AI and AII, usually in a 2:1 molar ratio, and particles containing AI but no AII.[53] Thus, antibodies to apo AI could be used to selectively remove HDL from human sera.

A direct LDL method could be envisioned as a three-step immunoprecipitation or immunoadsorption method. First, monospecific polyclonal or monoclonal antibodies to apo AI and apo CI are added to human serum either directly or attached to latex or magnetic beads. Second, the resulting immune complexes containing VLDL and HDL are removed either by filtration, centrifugation, or magnetic attraction. Third, the LDL-C is measured by a standardized enzymatic assay. In this scenario, VLDL and HDL are removed together. If both HDL-C and LDL-C concentrations are desired, then HDL can be removed first with antibodies against apo AI. HDL-C is calculated as the difference between TC and VLDL-C plus LDL-C remaining in the supernatant. Next, LDL-C is measured directly after the addition of antibodies against apo CI to remove VLDL. Additional steps may be necessary to selectively dissociate apo CI from IDL.

Immunoprecipitation or immunoadsorption as a direct method for LDL-C determination has many advantages:

1. Simplicity: Assays could be performed in clinical labs, physicians offices, or worksites.
2. Throughput: Many samples could be processed rapidly.
3. Economy: No expensive equipment would be required beyond what is needed for HDL and TC measurement, and labor costs would be minimal.
4. Potential for HDL-C: Immunoprecipitation or immunoadsorption of apo AI particles would be a legitimate candidate for an HDL-C method.
5. Specificity: The method would be highly specific for apolipoprotein composition.
6. Low sample volume: The sample volume requirement for enzymatic cholesterol methods would be approximately 0.1–0.2 mL.
7. Automation: Such methods could be readily automated.
8. Measurement of LDL-C in non-fasting samples: Post-prandial lipoproteins, chylomicrons, and chylomicron remnants contain apo AI and apo CI and would therefore be precipitated or adsorbed by antibodies to these apolipoproteins. However, it should be noted that LDL-C may be slightly underestimated in postprandial blood samples due to transient decreases in LDL that occur after fat ingestion.

There are also several disadvantages to using antibody-based methods to separate LDL from other lipoproteins:

1. The inability of anti-apolipoprotein antibodies to distinguish completely between VLDL and IDL, as discussed above.

2. Lot-to-lot antibody specificities would need to be carefully monitored and controlled to maintain the method's accuracy and precision.
3. The large amounts of antibodies, particularly apo AI, required are expensive to produce and would increase the cost per test.
4. Patients with hyperlipoproteinemias and other abnormal lipoprotein profiles would need to be studied to determine assay characteristics when lipoprotein concentrations are increased or when lipoprotein composition is abnormal.

Currently, there is a direct LDL-C measurement kit available (Genzyme Diagnostics, Cambridge, MA, and Sigma Chemical Company, St. Louis, MO). This kit uses anti-apolipoprotein antibodies attached to latex beads. The manufacturer has not published the apolipoprotein binding specificity of their antibodies. When the Genzyme Immunoseparation Reagent (GIR) was compared with beta quantification (BQ) on 74 fasting subjects with LDL-C of 27–241 mg/dL (0.70–6.24 mmol/L), TG of 26–936 mg/dL (0.29–10.85 mmol/L), and HDL-C of 12–95 mg/dL (0.31–2.46 mmol/L), the relationship by linear regression was: GIR = 1.05BQ + 4.8; r = 0.96.[60] Additional studies with unreported lipid concentrations, yielded correlation coefficients of r = 0.98 in both fasting and non-fasting subjects.[60]

A COMMON METHOD FOR ROUTINE ANALYSIS

As discussed earlier, the NCEP-ATP has concluded that the Friedewald formula is too variable to meet medical usefulness criteria due to the combined variation in the three underlying measurements, and has recommended development of appropriate methods for direct measurement of LDL-C. The NCEP will soon release guidelines for LDL-C measurement which are expected to express performance targets in terms of total error of less than 11.8% (combining bias and imprecision), consistent with a bias of 4% or less and an imprecision of 4% or less. The usual variability in the routine clinical laboratories in the measurement of TC, TG, and HDL-C will lead to variability in calculated LDL-C that exceeds the performance goal. On the other hand, the specificity and performance characteristics of the currently available direct LDL-C method has not been well established in long-term use. We therefore recommend a cautious approach to widespread replacement of the common Friedewald estimation with a direct LDL-C method until the latter has been proven specific and reliable.

The Friedewald equation has been considered reasonably accurate provided that three conditions are met:

1. The patient's TG concentration is less than 400 mg/dL (4.66 mmol/L).[61, 62]
2. Chylomicrons are not present.
3. The sample does not contain beta-VLDL ("floating" beta-lipoprotein, characteristic of Type III hyperlipoproteinemia).

Specimens that do not meet these three conditions can be referred to lipid specialty laboratories that use ultracentrifugation methods for accurate LDL-C determination, such as the LRC beta-quantification method described earlier.

Several points related to the use of the Friedewald formula are worth mentioning. First, TG concentrations of 400 mg/dL (4.66 mmol/L) are above the 95th percentile for both men and women. Second, chylomicrons can usually be detected as a layer of floating fat on top of plasma or serum that has been refrigerated overnight. And third, Type III hyperlipoproteinemia is relatively uncommon, occurring in the general population in approximately 1–2 individuals out of 1000. Thus, the Friedewald equation has been considered suitable for the majority of LDL-C measurements, provided underlying measurements are accurate and precise.

LDL-C is estimated from the Friedewald formula as:

$$\text{LDL-C} = \text{TC} - \text{HDL-C} - (\text{TG}/5)$$

where TG/5 is an estimate of VLDL-C and all concentrations are expressed in mg/dL. When values are expressed in mmol/L, VLDL-C is estimated as TG/2.22.

The main advantage of this procedure for estimating LDL-C is its simplicity; only measurements of TC, HDL-C, and total TG are required. Warnick et al.[61] evaluated the calculation in a standardized lipid laboratory and found that when TG were $<$ 200 mg/dL (2.33 mmol/L), 90% of estimated LDL-C values were within $\pm$ 10% of values measured by the beta-quantification method described above. At TG concentrations of 200–400 mg/dL (2.33–4.66 mmol/L) and 400–600 mg/dL (4.66–6.99 mmol/L), only 72% and 39%, respectively, of the estimates were within $\pm$ 10% of measured values. LDL-C values derived from an alternate formula[63] that estimated VLDL as TG divided by 6 when values are expressed in mg/dL (0.356 when expressed in mmol/L) were less accurate.[62] Nevertheless, at TG concentrations $\leq$ 500 mg/dL (5.83 mmol/L), 88% of LDL-C classifications based on NCEP cutpoints of 130 and 160 mg/dL (3.36 and 4.14 mmol/L) LDL-C were concordant with those by measured LDL-C.[61] McNamara et al.,[62] also in a standardized lipid laboratory, reported that at TG concentrations $\leq$ 400 mg/dL (4.66 mmol/L), 86% of LDL-C classifications based on NCEP cutpoints were concordant with those by measured LDL-C. In both cases it was concluded that the Friedewald formula gave acceptable LDL-C estimation for more individuals with TG $\leq$ 400 mg/dL (4.66 mmol/L) than with any of the other algorithms tested. The NCEP cutpoints (see Chapter 2 and the NCEP guidelines provided at the end of this chapter) were, in fact, derived from the second National Health and Nutrition Examination Survey,[64] which used the Friedewald formula to estimate LDL-C.

A note of caution is justified when using the Friedewald formula on hospitalized patients. Warnick et al.[61] subtracted the free glycerol from their TG measurements while McNamara et al.[62] and Friedewald et al.[32] did not. The three authors reached similar conclusions because the contribution of free glycerol to serum is small in most healthy individuals (see Chapter 5). However, serum free glycerol can be increased substantially in patients who are receiving glycerol-containing hyperalimentation fluids or heparin, and in patients who are critically ill.[65] Therefore, for such patients it is most prudent to measure LDL-C by ultracentrifugation, even if a glycerol-blanked TG assay is available.

Another important limitation of the Friedewald equation for estimating LDL-C is the requirement for fasting serum. Chylomicronemia or fasting VLDL TG above 400 mg/dL (4.66 mmol/L) can falsely overestimate VLDL-C and therefore underestimate LDL-C. Similarly, subjects with Type III hyperlipoproteinemia have abnormal VLDL, enriched with cholesterol relative to TG, which can result in an underestimation of VLDL-C and an overestimation of LDL-C. This can be especially troublesome since a Type III subject can be misdiagnosed as having Type II hyperlipoproteinemia, and the treatments for these disorders differ.

Detailed procedures for measuring TC, TG, and HDL-C are presented in Chapters 4, 5, and 6, respectively. In order to obtain adequately precise and accurate estimation of LDL-C, these underlying measurements must be made with a high degree of accuracy and precision. Because of the difficulty inherent in measuring all three analytes accurately and precisely, the NCEP-ATP has recommended the development of direct LDL-C methods. However, to date, there has been no formal effort to standardize LDL-C measurements, in spite of fact that the NCEP guidelines make LDL-C the major lipoprotein decision analyte. Informal proficiency surveys of beta-quantification methods in specialty lipid laboratories suggest wide variability between laboratories with biases greater than the NCEP target of 4%. An LDL-C standardization program is urgently needed to enable physicians and laboratories to implement NCEP recommendations.

In summary, LDL-C is an important analyte for patient classification and for monitoring response to treatment of hyperlipidemia, and standardization of LDL-C quantitation is essential. The reference method and the common separation method in specialty laboratories involves separation by ultracentrifugation. Estimation by the Friedewald equation, common in routine clinical laboratories, will likely be superseded by direct-LDL methods. ❖

REFERENCES

1. Austin MA. Plasma triglyceride as a risk factor for coronary heart disease. The epidemiological evidence and beyond. Amer J Epi 1989;129:249–59.
2. Simons LA, Dwyer T, Simons J, et al. Chylomicrons and chylomicron remnants in coronary artery disease: a case-control study. Atherosclerosis 1987;65:181–9.
3. Krauss RM, Williams PT, Brensike J, et al. Intermediate-density lipoproteins and progression of coronary artery disease in hypercholesterolaemic men. Lancet 1987;July 11:62–6.
4. Sandkamp M, Funke H, Schulte H, Kohler E, Assmann G. Lipoprotein(a) is an independent risk factor for myocardial infarction at a young age. Clin Chem 1990;36:20–3.
5. Lee DM, Alaupovic P. Apolipoproteins B, C-III and E in two major subpopulations of low-density lipoproteins. Biochim Biophys Acta 1986;879:126–33.
6. Chapman J. Comparative analysis of mammalian plasma lipoproteins. Methods in Enzymology 1986;128:70–143.
7. Goldstein JL, Brown MS. Binding and degradation of low density lipoproteins by cultured human fibroblasts. J Biol Chem 1974;249:5153–62.

8. Sniderman AD, Silberberg J. Is it time to measure apolipoprotein B? Arteriosclerosis 1990;10:665–7.
9. Vega GL, Grundy SM. Does measurement of apolipoprotein B have a place in cholesterol management? Arteriosclerosis 1990;10:668–71.
10. Steinberg D, Parthasarathy S, Carew T, et al. Beyond cholesterol: modifications of low-density lipoprotein that increase its atherogenicity. N Engl J Med 1989; 320:915–24.
11. Conference on Low Blood Cholesterol Levels: Disease Associations. Sponsored by the National Heart, Lung and Blood Institute, October 9–10, 1990.
12. The Expert Panel. Report of the National Cholesterol Education Program Expert Panel on Detection, Evaluation, and Treatment of High Blood Cholesterol in Adults. Arch Intern Med 1988;148:36–69.
13. The Expert Panel. Summary of the second report of the National Cholesterol Education Panel (NCEP) Expert Panel on Detection, Evaluation, and Treatment of High Blood Cholesterol in Adults (Adult Treatment Panel II). JAMA 1993;269: 3015–23.
14. NCEP Expert Panel on Blood Cholesterol Levels in Children and Adolescents. National Cholesterol Education Program (NCEP): Highlights of the report of the Expert Panel on Blood Cholesterol Levels in Children and Adolescents. Pediatrics 1992;89:495–501.
15. The Lipid Research Clinics Population Studies Data Book, Volume I, The Prevalence Study. US Department of Health and Human Services, Public Health Service, National Institutes of Health. 1–81.
16. Lindgren FT, Elliott HA, Gofman JW. The ultracentrifugal characterization and isolation of human blood lipids and lipoproteins, with applications to the study of atherosclerosis. J Phys & Colloid Chem 1951;55:80.
17. Lewis LA, Green AA, Page IH. Ultracentrifuge lipoprotein pattern of serum of normal, hypertensive and hypothyroid animals. Am J Physiol 1952;171:391.
18. de Lalla OF, Gofman JW. Ultracentrifugal analysis of serum lipoproteins. Methods of Biochemical Analysis 1954;1:459–78.
19. Havel RJ, Eder HA, Bragdon JH. The distribution and chemical composition of ultracentrifugally separated lipoproteins in human serum. J Clin Invest 1955;34: 1345–53.
20. Schumaker VN, Puppione DL. Sequential flotation ultracentrifugation. Methods in Enzymology 1986;128:155–70.
21. Lindgren FT, Nichols AV, Upham FT, et al. Subfractionation of the Sf 20–10^5 lipoproteins in a swinging bucket rotor. J Phys Chem 1962;66:2007–11.
22. Lossow WJ, Lindgren FT, Murchio JC, et al. Particle size and protein content of six fractions of the Sf > 20 plasma lipoproteins isolated by density gradient centrifugation. J Lipid Res 1969;10:68–76.
23. Hinton RH, Kowalski AJ, Mallinson A. Choice of conditions for the gradient flotation of serum lipoproteins in swing-out rotors. Clin Chim Acta 1973;44:267–70.
24. Redgrave TG, Roberts DCK, West CE. Separation of plasma lipoproteins by density-gradient ultracentrifugation. Anal Biochem 1975;65:42–49.
25. Foreman JR, Karlin JB, Edelstein C, et al. Fractionation of human serum lipoproteins by single-spin gradient ultracentrifugation: quantification of apolipoproteins B and A-I and lipid components. J Lipid Res 1977;18:759–67.
26. Nilsson J, Mannickarottu V, Edelstein C, et al. An improved detection system applied to the study of serum lipoproteins after single-step density gradient ultracentrifugation. Anal Biochem 1981;110:342–48.
27. Belcher JD, Egan JO, Bridgmen G, Baker R, Flack J. A micro-enzymatic method to measure cholesterol and triglyceride in lipoprotein subfractions separated by

density gradient ultracentrifugation from 200 microliters of plasma or serum. J Lipid Res 1991;32:359–70.

28. Chung BH, Wilkinson T, Geer JC, et al. Preparative and quantitative isolation of plasma lipoproteins: rapid, single discontinuous density gradient ultracentrifugation in a vertical rotor. J Lipid Res 1980;21:284–91.
29. Cone JT, Segrest JP, Chung BH, et al. Computerized rapid high resolution quantitative analysis of plasma lipoproteins based upon single vertical spin centrifugation. J Lipid Res 1982;23:923–35.
30. Campos E, McConathy WJ. Distribution of lipids and apolipoproteins in human plasma by vertical spin ultracentrifugation. Arch Biochem Biophys 1986;249:455–63.
31. Chung BH, Segrest JP, Ray MJ, et al. Single vertical spin density gradient ultracentrifugation. Methods in Enzymology 1986;128:181–209.
32. Friedewald WT, Levy RI, Fredrickson DS. Estimation of the concentration of low density lipoprotein cholesterol in plasma without use of the ultracentrifuge. Clin Chem 1972;18:449–502.
33. Rudel LL, Lee JA, Morris MD, et al. Characterization of plasma lipoproteins separated and purified by agarose-column chromatography. Biochem J 1974;139:89–95.
34. Williams MC, Stenoien CG, Kushwaha RS. Rapid method for measuring plasma low density lipoprotein turnover using high performance gel exclusion chromatography. J Chromatogr 1986;375:233–43.
35. Carroll RM, Rudel LL. Lipoprotein separation and low density lipoprotein molecular weight determination using high performance gel-filtration chromatography. J Lipid Res 1983;24:200–7.
36. März W, Siekmeier R, Scharnagl H, Seiffert UB, Gross W. Fast lipoprotein chromatography: New method of analysis for plasma lipoproteins. Clin Chem 1993;39:2276–81.
37. Levy RI. Cholesterol, lipoproteins, apoproteins, and heart disease; present status and future prospects. Clin Chem 1981;27:653–62.
38. Warnick GR, Leary ET, Goetsch J, Hicks D. Electrophoretic quantitation of LDL-cholesterol using the Helena REP. Clin Chem 1993;39(6):1122.
39. Weiland H, Seidel D. A simple specific method for precipitation of low density lipoproteins. J Lipid Res 1983;24:904–9.
40. Kersher L, Schiefer S, Draeger B, Maier J, Ziegenhorn J. Precipitation methods for the determination of LDL cholesterol. Clin Biochem 1985;18:118–225.
41. Maier J, Draeger B, Wehmeyer G, Gloger M, Ziegenhorn J. Method for the quantitation of serum low density lipoprotein cholesterol (LDL-CH) [Abstract]. Clin Chem 1983;29:1173.
42. Assmann G, Jabs H-U, Kohnert U, Nolte W, Schriewer H. LDL-cholesterol determination in blood serum following precipitation of LDL with polyvinyl sulfate. Clin Chim Acta 1984;140:77–83.
43. Moss MA, Wong CSY, Tan MH, et al. Determination of low density lipoprotein cholesterol (LDL-C) in serum by Biomerieux cholesterol/-phospholipids polyanion precipitation method and comparison with preparative ultracentrifugation. [Abstract]. Clin Chem 1986;32:1096–7.
44. Mainard F, Madec Y. Are "precipitated LDL" really low density lipoproteins? Clin Chim Acta 1987;162:141–6.
45. Shaikh M, Miller NE. Evaluation of a commercial reagent for precipitating human serum low-density lipoprotein. Clin Chim Acta 1985;152:213–17.
46. Demacker PN, Hijmans AG, Brenninkmeijer BJ, et al. Five methods for determining low-density lipoprotein cholesterol compared. Clin Chem 1984;30:1797–1800.

47. Assmann G, Jabs HU, Nolte W, Schriewer H. Precipitation of LDL with sulphopolyanions: a comparison of two methods for LDL cholesterol determination. J Clin Chem Clin Biochem 1984;22:781–5.
48. Mulder K, van Leeuwen C, Schouten JA, et al. An evaluation of three commercial methods for the determination of LDL-cholesterol. Clin Chim Acta 1984;143:29–35.
49. Lipid Research Clinics Program. Manual of laboratory operations, Vol. 1. Lipid and lipoprotein analysis. DHEW Publication No. (NIH) 75-628. Washington, DC: U.S. Government Printing Office, 1974.
50. Frantz I Jr. (Minneapolis, MN), personal communication.
51. Wu LL, Warnick GR, Wu JT, Williams RR, et al. A rapid micro-scale procedure for determination of the total lipid profile. Clin Chem 1989;35(7):1486.
52. Knopp RH, Walden CE, Heiss G, et al. Prevalence and clinical correlates of beta-migrating very-low-density lipoprotein: Lipid Research Clinics Program prevalence study. Am J Med 1986;81:493–502.
53. Cheung MC, Albers JJ. Characterization of lipoprotein particles isolated by immunoaffinity chromatography: particles containing A-I and A-II and particles containing A-I but no A-II. J Biol Chem 1984;259:12201–9.
54. Patterson BW, Miller NH, Fisher WR. Structural studies of apolipoprotein B: physical properties of the protein in guanidine hydrochloride. Biochimica et Biophysica Acta 1987;920:266-76.
55. Yves Marcel (Ottawa, Canada) and Elaine Krul (St. Louis, MO), personal communications.
56. Havel RJ, Eder HA, Bragdon JH. The distribution and chemical composition of ultracentrifugally separated lipoproteins in human serum. J Clin Invest 1955;34: 1345–53.
57. Andersen GE, Gry H, Kostner GM. A new micromethod for routine measurement of serum LDL-apolipoprotein B. J Lip Res 1988;29:377–80.
58. Albers JJ, Aladjem F. Precipitation of I-labeled lipoproteins with specific polypeptide antisera. Evidence for two populations with differing polypeptide compositions in human high density lipoproteins. Biochemistry 1971;10:3436–42.
59. Kostner G, Depisch A, Petek W, Holasek A. Untersuchung der beiden in der agarose gel-elektrophorese auftretenden α-lipioproteinbanden. Hoppe-Seylers Z. Physiol Chem 1971;352:1440–4.
60. Leary ET, Tjersland G, Warnick GR. Evaluation of the Genzyme Immunoseparation Reagent for direct quantitation of LDL cholesterol. [Abstract] Clin Chem 1993;39:1124.
61. Warnick GR, Knopp RH, Fitzpatrick V, et al. Estimating low-density lipoprotein cholesterol by the Friedewald equation is adequate for classifying patients on the basis of nationally recommended cutpoints. Clin Chem 1990;36:15–19.
62. McNamara JR, Cohn JS, Wilson PWF, et al. Calculated values for low-density lipoprotein cholesterol in the assessment of lipid abnormalities and coronary disease risk. Clin Chem 1990;36:36–42.
63. DeLong DM, DeLong ER, Wood PD, et al. A comparison of methods for the estimation of plasma low- and very low-density lipoprotein cholesterol: the Lipid Research Clinics prevalence study. JAMA 1986;256:2372–7.
64. National Center for Health Statistics, Division of Health Examination Statistics. Unpublished data from the second National Health and Nutrition Examination Survey, 1976–1980.
65. Jessen RH, Dass CJ, Eckfeldt JH. Do enzymatic analyses of serum triglycerides really need blanking for free glycerol? Clin Chem 1990;36:1372–5.

❖ CHAPTER 8 ❖

Lipid and Lipoprotein Analysis with Desk-Top Analyzers

Paul S. Bachorik

INTRODUCTION

❖ Cholesterol is transported in the circulation by plasma lipoproteins, primarily by low-density lipoproteins (LDL), which account for about two-thirds, and high-density lipoproteins (HDL), which carry about 20–25% of the circulating cholesterol in humans. An increased concentration of LDL cholesterol (LDL-C) is a well-established risk factor for the development of premature coronary heart disease (CHD),[1] and over the past few years it has been established that lowering plasma LDL-C concentrations reduces the risk of CHD.[2-7] Total cholesterol (TC) is an effective surrogate measure of LDL-C in adults, and therefore TC measurement has assumed increasing importance in identifying patients with high LDL-C concentrations. TC measurement is performed in various kinds of screening operations and in follow-up analyses used to evaluate the progress of patients being treated for hypercholesterolemia. A low concentration of HDL cholesterol (HDL-C) is independently associated with increased risk of CHD,[8-10] although it has not yet been demonstrated that raising HDL-C reduces risk.

The National Cholesterol Education Program Adult Treatment Panel (NCEP-ATP) recently updated[11] its original[12] recommendations concerning the diagnosis and treatment of hypercholesterolemia in adults. (Refer to Chapter 2 for additional information.) While the original recommendations focused primarily on TC and LDL-C, the updated report places increased emphasis on HDL-C by recognizing a high HDL-C concentration as a protective factor and recommending the measurement of HDL-C whenever TC is measured.

Technical developments over the past decade have increased the speed and accuracy of clinical measurements of lipids and lipoproteins and, to some extent, have moved the performance of these measurements from the conventional laboratory to the physician's office and to other nontraditional settings. The first of these developments was the introduction into the laboratory of completely enzyme-based methods for cholesterol and triglyceride (TG) measurement.[13–16] Such methods have virtually replaced the older, more involved and technically demanding chemical methods. The second development was the adaptation of this methodology to small, portable analyzers, commonly called desk-top analyzers. These analyzers were originally designed to be used in the physician's office, but they have also been widely used for TC screening in non-medical settings.[17-20]

In this chapter we discuss the use of desk-top analyzers for lipid and lipoprotein measurements in non-laboratory settings.

PRINCIPLES OF OPERATION

Total Cholesterol

Virtually all of the desk-top analytical systems currently available are based on the same kinds of enzymatic methods used in laboratory settings. For measurement of TC, they employ a single, combined reagent mixture which contains all of the necessary enzymes, buffers, and other cofactors required for the test. An aliquot of serum or plasma, usually 10–30 μL, is mixed with the combined reagent and allowed to incubate under controlled conditions for a short period, generally about 3–8 min, during which a colored reaction product forms. The intensity of the color is related to the cholesterol concentration. The results are read from a digital display on the instrument or from a printout on a paper tape. The chemical reactions involved are as follows:

$$\text{1. Cholesteryl esters} + H_2O \xrightarrow{\text{cholesteryl esterase}} \text{cholesterol} + \text{fatty acid}$$

$$\text{2. Cholesterol} + O_2 \xrightarrow{\text{cholesteryl oxidase}} \text{cholestanone} + H_2O_2$$

$$\text{3. } H_2O_2 + \text{dye} \xrightarrow{\text{peroxidase}} \text{colored dye product}$$

Cholesteryl esters are hydrolyzed and then cholesterol is oxidized to produce cholestanone and hydrogen peroxide. The amount of hydrogen peroxide produced is proportional to the amount of cholesterol in the sample. The concentration of H_2O_2 is then measured through a peroxidase catalyzed reaction in which a colored dye product is formed.

Triglycerides

TG are similarly measured through a series of enzyme reactions.

$$4.\ \text{Triglyceride} + H_2O \xrightarrow{\text{bacterial lipase}} \text{fatty acid} + \text{glycerol}$$

$$5.\ \text{Glycerol} + \text{ATP} \xrightarrow{\text{glycerokinase}} \text{glycerophosphate} + \text{ADP}$$

The glycerophosphate formed is most commonly measured using glycerophosphate oxidase.

$$6.\ \text{Glycerophosphate} + O_2 \xrightarrow{\text{glycerophosphate oxidase}} \text{dihydroxyacetone} + H_2O_2$$

The H_2O_2 produced is generally measured as in reaction 3 above.

In general, all of the reactions occur once the sample has been placed into the instrument. The instrument controls the time and conditions of the incubation, reads the color intensity, converts the readings to units of concentration, and displays the results. The entire analysis can be accomplished in about 10 min.

HDL Cholesterol

Within the past few years, desk-top analyzer tests have also become available for HDL-C measurement. These tests rely on the removal of the apolipoprotein B (apo B)-containing lipoproteins: very-low-density lipoproteins (VLDL), LDL, and Lipoprotein(a) [Lp(a)]. This is generally done by precipitation with a polyanion in the presence of a divalent cation, under conditions in which HDL remains soluble. Precipitants such as phosphotungstate or dextran sulfate-Mg^{++} have been used. The cholesterol content of the HDL-containing fraction is then measured as described above.

HDL tests have been difficult to develop, in part because the cholesterol measurement itself must be several times more sensitive than is required for TC measurement. With some desk-top analyzers, the apo-B-containing lipoproteins must be removed before the test is run, while in others both the precipitation and cholesterol measurement steps are performed automatically.

LDL Cholesterol

For analyzers that measure TC, TG, and HDL-C, LDL-C can be calculated using the empirical relationship of Friedewald et al.:[21]

$$\text{LDL-C} = \text{TC} - \text{HDL-C} - \text{TG}/5$$

where the factor TG/5 is an estimate of VLDL-C concentrations when given in mg/dL. The factor TG/2.22 is used when concentrations are expressed in mmol/L.

INSTRUMENTS

The variety of desk-top instruments available differ in their simplicity and ease of use. These instruments can be separated into two broad categories: those based on dry reagent chemistry and those for which reagents are furnished in a liquid form. Here we discuss the principles of operation of several of the more commonly used analyzers.

One of the simplest of the instruments to use is the Reflotron (Boehringer-Mannheim Diagnostics, Indianapolis, IN). This is one of the so-called whole-blood analyzers, capable of making measurements using either serum, plasma, or whole blood. The sample (30 µL) is applied to the sample application zone of a reagent strip. The sample moves downward into an absorbant pad through a glass fiber mesh which removes the red cells. The cell-free sample then moves into the reagent-containing zone, an area of the strip that has been impregnated with the enzymes and other reagents needed for the test. The reagents dissolve in the sample, allowing the reaction to proceed.

Once the sample is applied, the strip is inserted into the instrument. The reaction proceeds for about 3 min, during which time the color develops. At the end of this period, the intensity of the color is measured by reflectance photometry. A light beam is shined on the sample and the amount of light reflected is inversely proportional to the TC concentration in the sample. The reflectance is converted to units of concentration and the results are displayed on a small screen on the front of the instrument. Results can be expressed either in mg/dL or in mmol/L. The strip is then removed and the next test can be performed. The instrument is capable of analyzing one sample at a time.

The identity and conditions for the test and the calibration curve are coded magnetically on the back of the strip, and this information is read by the instrument when the strip is inserted. This configuration makes the test very simple to use, since the operator need only verify the operation of the optics of the instrument and then perform the tests. Because of its simplicity, the Reflotron has been widely used for large-scale TC screening operations in various field settings.[17–20, 22]

Reflotron test strips are available for TC, TG, and HDL-C. For HDL-C, 30 µL EDTA plasma (not whole blood) are applied to the strip. The strip then moves through a layer impregnated with dextran sulfate (M_r 50,000) and magnesium acetate, which precipitates the apo-B-containing lipoproteins. The HDL-containing fraction then moves to the reagent zone, where HDL-C is measured.

Since the introduction of the Reflotron, this technology has been developed further to allow several tests to be performed simultaneously on a single sample. This is accomplished by placing the reagent zones to the side of the cell-removal filter and providing a channel through which the cell-free blood filtrate moves from the filter to the reagent zones. The channel is split to allow portions of the sample to move to different test zones. One such analyzer is the Cholestech LDX (Cholestech Corporation, Hayward, CA) which can simultaneously measure TC and HDL-C. The prototype for a similar system (Biotrack, Ciba Corning Diagnostics, Mountain View, CA) capable of measuring a variety of analytes has recently been described.[23]

The Kodak DT60 (Eastman Kodak, Rochester, NY) is another example of a dry chemistry analyzer. With this instrument, however, the configuration of the test is somewhat different. The DT60 is something of a cross between a sequential analyzer and a batch analyzer. Reagents for the test are contained on a test slide in which the reagents are impregnated into a pad immediately below the sample application zone. The slide is also printed with a bar code that identifies the test and test conditions.

The slide is first inserted into the instrument, after which the sample (10 μL) is injected onto the slide through a port in the top of the instrument using a small, battery-powered, motorized pipet. After the sample has been applied, the slide moves automatically to an incubation chamber, where color development takes place. The reflectance of the sample is then measured and converted to units of concentration, and the results are printed on a paper tape. The entire analysis takes about 8 min. The incubation chamber, however, can accommodate a number of slides, so that samples can be continuously fed into the instrument. Once in operation, the results are printed about every 30 s. Test slides are available for TC, TG, and HDL-C. The HDL-C assay requires the preliminary manual removal of apo-B-containing lipoproteins. The method uses dextran sulfate (M_r 50,000) and $MgSO_4$ as the precipitating reagent; the reagent is provided with the test kit.

The DT60 differs from the Reflotron in several other respects. First, it is not a whole-blood analyzer; blood cells must be removed before the sample is analyzed. Second, the instrument is calibrated by the user at intervals of about 30 d using lyophilized reference sera provided by the manufacturer. Three levels of reference sera are used, and the calibration curves are calculated and stored automatically by the instrument.

The Abbott VISION (Abbott Laboratories, North Chicago, IL) is an example of a wet-chemistry, batch-type whole blood analyzer. The VISION analyzer uses a somewhat different principle of analysis. First, the reagent mixture is contained in a small plastic cassette. A sample of whole blood is injected into the cassette with the aid of a pipet. The cassette is then placed into the instrument, where it is centrifuged to sediment red cells. The position of the cassette is then automatically changed such that an accurately measured volume of the plasma spills into the chamber containing the reagent and is mixed. After an appropriate period, the absorbance of the mixture is measured and the absorbance readings are converted to units of concentration and displayed on a printed tape. The instrument can accommodate up to 10 cassettes at one time, and the entire analysis takes about 10 min. Test cassettes are available for measuring TC, TG, and HDL-C.

Because the instrument incorporates a centrifuge, it is physically larger than the Reflotron and the DT60 and is therefore less mobile. The analyzer differs from the Reflotron in that centrifugation rather than adsorption is used to remove the cells. Furthermore, the absorbance rather than the reflectance of the sample is measured.

Type of Sample Used

When using these instruments, it is necessary to consider the kind of sample that is used. The analyzers can be used with either venous or capillary samples, although in mass-screening settings, capillary samples are usually used. In general, the results obtained with the two kinds of samples are similar but not identical. In some studies, the values in capillary samples appear to be lower than those in venous samples; in other studies, the opposite was observed.[24-26] On average, the differences between venous and capillary samples are on the order of a few percentage points. However, the values in capillary samples from individual patients can vary considerably from those in venous samples collected at the same time.[17] The reasons for these differences have not been investigated in any detail, but possibilities may include the temperature of the patient's finger, differential effects of posture on capillary and venous samples, and the ease with which blood flow occurs from the finger puncture, among other factors.

Serum-plasma differences are another consideration. Certain anticoagulants exert an osmotic effect, resulting in the movement of water from the cells to the plasma. This causes dilution of the plasma lipoproteins, the degree of which depends on the osmotic activity of the anticoagulant used.[27] This difference was estimated to be about 3%,[28] and, according to the NCEP-ATP guidelines,[12] multiplication of plasma values by the factor 1.03 is recommended when evaluating patients. It must be kept in mind, however, that the term "plasma" used in this context refers specifically to EDTA plasma. Some agents, such as citrate, exert much larger osmotic effects, and cholesterol should not be measured in samples that contain such anticoagulants. Heparin, on the other hand, exerts little or no osmotic effect in concentrations used for blood collection, and TC values in heparinized plasma are indistinguishable from those in serum.[29] This is important because capillary blood samples are generally collected into heparinized capillary tubes, and measurements in capillary blood need not be converted to serum values. Nonetheless, it is well to bear in mind possible differences between venous and capillary samples, since the NCEP-ATP guidelines are based on serum.[11, 12] Finally, the factor 1.03 was determined at a time when blood-drawing tubes containing EDTA were supplied with sufficient EDTA to produce an anticoagulant concentration of 1 mg/mL. Tubes supplied currently, however, contain 1.5 mg/mL EDTA, and the hema-dilution has been estimated at about 4.5%.[29]

RELIABILITY OF LIPID AND LIPOPROTEIN-CHOLESTEROL MEASUREMENTS

The NCEP Lipid Standardization Panel (LSP) established the following criteria for acceptable TC measurement in the United States.[30] These criteria are given in Table 8–1. These guidelines are consistent with a total analytical error of ≤ 10%.

Over the past several years, the NCEP Lipoprotein Measurement Working Group has developed guidelines for reliable TG, HDL-C, and LDL-C measurements. These are indicated in Table 8–2. As of this writing, these guidelines have not yet been published by NCEP, but they have been accepted in principle by the

Table 8–1 ❖ NCEP Guidelines for Accuracy and Precision of Cholesterol Measurements	
Bias*	< 3%
Coefficient of variation (CV)	< 3%

*With respect to CDC reference method.

NCEP Coordinating Committee. The recommendations are cast in terms of total error; the larger the method bias, the more precise the methods must be to satisfy the recommendations for total error.

The NCEP guidelines do not distinguish between analyses performed in the laboratory and those performed in non-laboratory settings, nor do they establish separate guidelines for desk-top analyzers, since the impact of such measurements on the patient and the medical community is generally the same, regardless of how the measurements are made. A number of studies have evaluated the reliability of TC measurements with desk-top analyzers. In general, desk-top analyzers give fairly accurate measurements on average, but with individual samples measurements tend to be somewhat more variable than those obtained with laboratory-based methods. The use of desk-top instruments with finger-stick blood samples apparently contributes to this variability. In one recent study, the distribution of the difference between measurements of finger-stick samples with the Reflotron analyzer and with standardized laboratory measurements of venous samples taken at the same time was about twice as great as that for venous samples analyzed with both methods.[17] The reasons for the difference are not clear,

Table 8–2 ❖ Forthcoming NCEP Recommendations for Reliable Triglyceride, HDL-C and LDL-C Measurements

Analyte	*Total error*	*Consistent with*[1]
Triglyceride	≤ 15%	bias ≤ 5% CV2 ≤ 5%
HDL-C[3]	≤ 23%	bias ≤ 10% CV ≤ 6% at HDL-C ≥ 42 mg/dl (1.09 mmol/L) SD ≤ 2.5 mg/dL (1.065 mmol/L) at HDL-C < 42 mg/dL (1.09 mmol/L)
LDL-C	≤ 12%	bias ≤ 4% CV ≤ 4%

[1]The primary recommendations are cast in terms of total error. The bias and CV shown for each analyte is one example that will satisfy the primary criterion for that analyte.
[2]CV = coefficient of variation.
[3]Interim recommendations. More stringent provisional criteria (total error ≤ 13%) are recommended to be adopted by 1998, providing that suitable technology is available by that time.

but may derive, at least in part, from physiological differences in venous and capillary samples.

It has also been found that desk-top analyzers provide more reliable results when operated by trained laboratory technologists than by operators who lack formal laboratory training.[31] Desk-top analyzers operated under field conditions, in which the instruments must be transported frequently and may be operated in climates of varying temperature and humidity, generally provide more variable results than those operated at a fixed location.

In individual samples, however, even when venous samples are used, there can be substantial differences between measurements made with desk-top and laboratory-based systems. A comparison is given in Table 8–3. The data were collected in the author's laboratory in a split-sample comparison conducted over a 9-month period. The analyses were conducted on the Kodak DT60 using venous EDTA plasma. The comparison measurements were standardized according to the criteria established by the Centers for Disease Control and Prevention (CDC)–National Heart, Lung and Blood Institute (NHLBI) for lipid and HDL standardization and were made on a Hitachi 704 analyzer (Boehringer-Mannheim Diagnostics, Indianapolis, IN). There were no significant differences in group mean between the DT60 and laboratory measurements for TC, TG, or HDL-C. However, 15% percent of the TC values, 20% of the HDL-C values, and 26% of the TG values obtained from the DT60 differed by more than 10% from those derived from laboratory measurements. For TC, TG, and HDL-C values, 5–6% of the DT60's measurements differed from the laboratory measurements by more than 20%. For TC and HDL-C, there was no relationship between the magnitude of the differences and concentrations; for TG, the differences tended to decrease at higher concentrations ($p = 0.003$). Such differences can be of concern when dealing with individual patients, particularly those whose lipid and lipoprotein concentra-

Table 8–3 ❖ Comparison of Total Cholesterol (TC), Triglycerides (TG), and HDL-cholesterol (HDL-C) Values Measured with a Desk-top Analyzer and Standardized Laboratory Procedures[1]

	CONCENTRATION MEAN (SD) (mg/dL)[2]			DIFF.		PERCENTAGE OF SAMPLES WITH DIFFERENCE EXCEEDING[3]	
Analyte	n	*DT60*	*Laboratory*	%	p	*10%*	*20%*
TC	115	210.5 (53.7)	216.3 (53.3)	−2.7	NS[4]	15%	5%
TG	117	142.7 (79.7)	142.3 (76.2)	+0.3	NS	26%	6%
HDL-C	114	47.3 (13.1)	46.0 (13.5)	+0.9	NS	20%	5%

[1]Desk-top analyzer measurements were made with the Kodak DT60. Laboratory measurements were standardized through the CDC-NHLBI Lipid Standardization Program.
[2]To convert to mmol/L, divide TC or HDL-C by 38.7 and TG by 88.5.
[3] Refers to the absolute values of the differences between methods.
[4]NS = not significant ($p > 0.05$, Mann-Whitney Rank Sum Test).

tions are not already known; in such cases, the error might not be suspected and the patient might be misclassified.

The higher variability of desk-top analyzers can also be observed in several other ways. Table 8–4 indicates the variation in TC, TG, HDL-C, and LDL-C values as reflected by measurements in quality-control pools over a 1-y period using laboratory methods, and over a 7-m period using the DT60 analyzer. The control pools used with the DT60 were those supplied by the manufacturer. In all cases except one, the coefficients of variation (CVs) obtained with the desk-top analyzer were several times higher than those obtained from laboratory-based measurements. The one exception was a CV of 1.5% for the TG measurements obtained in one of the manufacturer's pools.

The data in Table 8–5 illustrate the results of TC, TG, and HDL-C analyses performed in survey pools distributed by the College of American Pathologists for two commonly used desk-top analyzers. For comparison, data for the same survey pools for several common laboratory-based analyzers are shown in Table 8–6.

Overall, the desk-top analyzers were again somewhat more variable than laboratory-based measurements for TC and TG. For TC, the among-laboratory

Table 8–4 ❖ Accuracy and Precision of Lipid and Lipoprotein Measurements Made in the Laboratory and With a Desk-top Analyzer

Analyte	*Pool*	*CDC Reference Value (mg/dL)*[2]	*Laboratory (mg/dL) Mean (SD)*	*CV (%)*	*Pool*	*DT 60 (mg/dL) Mean SD*	*CV (%)*
TC	Q17	166.8	165.2 (1.9)	1.2	I	158.2 (5.5)	3.5
	Q20	261.0	259.4 (4.0)	1.5	II	245.8 (10.2)	4.2
TG	Q18	103.4	103.5 (2.4)	2.3	I	132.9 (2.0)	1.5
	Q20	195.0	193.7 (4.5)	2.3	II	262.5 (23.3)	8.9
HDL-C	AQ13	51.9	49.7 (1.3)	2.6	I	29.4 (2.3)	7.9
	AQ14	43.2	44.3 (1.1)	2.5	II	41.9 (2.9)	6.9
LDL-C	Q17/AQ13	94.2[2]	94.8[2] (2.4)[3]	2.5[3]	I	102.4 (6.0)[3]	5.9[3]
	Q20/AQ13	170.1	171.0 (4.3)	2.5	II	151.3 (11.6)	7.7
	Q17/AQ14	102.9	100.2 (2.3)	2.2			
	Q20/AQ14	178.8	176.4 (4.2)	2.4			

[1]Data from the author's laboratory collected over a 1-y period. Desk-top analyzer data collected over 7-month period on controls provided by the manufacturer.
[2]To convert to mmol/L, divide TC or HDL-C by 38.7 and TG by 88.5.
[3]Reference values calculated from TC, TG, and HDL-C CDC reference values for the pools indicated using the Friedewald equation: LDL-C = TC – HDL-C – TG/5 for mg/dL. (Use TG/2.2 for mmol/L.) Laboratory and desk-top analyzer LDL-C means were similarly calculated using the respective means for the three analytes.
[4]Variances for LDL-C were calculated from the variances for the three analytes as follows: $LDL_{var} = TC_{var} + HDL_{var} + TG_{var}/25$. SDs and CVs were calculated from the LDL variances.

Table 8-5 ❖ Performance for Total Cholesterol (TC), HDL-Cholesterol (HDL-C) and Triglyceride (TG) Measurements with Two Desk-top Analyzer Systems[1]

			ABBOTT VISION		KODAK DT/DTII	
Analyte	*Pool*	*CDC confirming value (mg/dL)*[2]	*mean (mg/dL)*[2]	*Among-Laboratory CV (%)*	*mean (mg/dL)*[2]	*Among-Laboratory CV (%)*
TC[3,4,5]	LP-06	317	298	3.1	310	5.5
	LP-07	181	166	3.0	182	5.0
	LP-08	251	238	3.6	252	4.6
	LP-09	208	197	3.4	211	5.2
	LP-10	159	147	3.3	162	4.6
HDL-C[4,5]	LP-06	53.8	—	—	31.2	14.1
	LP-07	29.2	—	—	30.2	13.5
	LP-08	26.7	—	—	23.2	11.3
	LP-09	37.9	—	—	24.3	17.5
	LP-10	24.9	—	—	22.3	12.6
TG[3,4,5]	LP-06	202	228	5.5	200	4.8
	LP-07	79	89	7.8	85	5.3
	LP-08	161	175	8.1	153	4.5
	LP-09	152	159	5.0	155	4.0
	LP-10	105	111	6.1	112	4.1

[1]Data from 1993 CAP Surveys, Chemistry Series 1, Series 2 Surveys. College of American Pathologists, Chicago, IL.
[2]To convert to mmol/L, divide TC or HDL-C by 38.7 and TG by 88.5.
[3]Data for TC and TG have been rounded to nearest 1 mg/dL.
[4]Results were reported from 31–35 participants for TC and from 18–19 participants for TG.
[5]Results for TC, HDL-C, and TG were reported from 104–107, 37–39, and 67–69 participants, respectively.

CVs of 2.6–4.0% were obtained by the laboratory-based systems (Table 8–6) versus 3.0–5.5% for the desk-top analyzers (Table 8–5). For TG, among-laboratory CVs of 2.2–5.0% were obtained by the laboratory-based systems (except for one pool which had a CV of 9.7% with the Abbott spectrum), compared to 4.0–8.1% for the desk-top analyzers. In addition, the Abbott VISION desk-top analyzer gave negative biases of 5–8% for TC and positive biases of 5–13% for TG, compared to the CDC-confirming values for the survey pools. In comparison, with laboratory-based systems the largest biases given were 3% for TC and 7% for TG, except for a single pool (CDC confirming value 79 mg/dL) which had a 10% positive bias with all three laboratory systems shown in Table 8–6.

HDL-C measurements had among-laboratory CVs of 7–27% with the laboratory-based systems, versus 13–18% for the single desk-top analyzer system for

Table 8-6 ❖ Performance for Total Cholesterol (TC), HDL-Cholesterol (HDL-C), and Triglyceride (TG) Measurements with Three Laboratory-Based Systems[1]

Analyte	Pool	CDC Confirming Value (mg/dL)	ABBOTT SPECTRUM[2] mean (mg/dL)[5,6]	ABBOTT SPECTRUM[2] Among-Laboratory (CV (%))	KODAK EKTACHEM[3] mean (mg/dL)[5,6]	KODAK EKTACHEM[3] Among-Laboratory (CV (%))	HITACHI 704[4] mean (mg/dL)[5,6]	HITACHI 704[4] Among-Laboratory (CV (%))
TC	LP-06	317	309	3.5	307	3.5	315	2.9
	LP-07	181	179	3.9	180	3.8	183	3.1
	LP-08	251	251	3.2	248	3.6	253	2.8
	LP-09	208	205	3.4	209	3.8	209	2.6
	LP-10	159	159	3.6	160	4.0	161	2.9
HDL-C	LP-06	53.8	27.9	15.1	29.7	7.4	35.2	16.6
	LP-07	29.2	23.4	27.3	28.7	11.3	30.8	10.9
	LP-08	26.7	19.9	13.6	23.0	10.0	19.8	16.1
	LP-09	37.9	19.4	19.9	22.2	11.9	27.7	16.3
	LP-10	24.9	17.9	21.9	21.8	12.7	23.2	13.6
TG	LP-06	202	217	4.5	199	2.3	211	3.3
	LP-07	79	87	9.7	87	3.0	87	3.9
	LP-08	161	165	4.5	154	2.3	165	3.6
	LP-09	152	160	4.7	156	2.2	156	3.1
	LP-10	105	109	5.0	112	2.3	110	3.7

[1]Data from 1993 CAP Surveys, Chemistry Seris 1, Series 2, College of American Pathologists, Chicago, IL.
[2]Participating laboratories: TC, 413–415; HDL-C, 11–73; TG, 44–46.
[3]Participating laboratories: TC, 1024–1032; HDL-C, 626–631; TG, 1000–1009.
[4]Participating laboratories: TC, 101–104; HDL-C, 77–79; TG, 66–67.
[5]Data for TC and TG have been rounded to nearest 1 mg/dL.
[6]To convert to mmol/L, divide TC or HDL-C by 38.7 and TG by 88.5.

which data were reported. In addition, for two pools there were major discrepancies between both the laboratory-based and desk-top analyzer values reported and the CDC-confirming values for those pools. Since substantive "matrix effects" can occur with HDL-C measurement in lyophilized pools, these data probably underestimate the reliability of HDL-C measurements with both laboratory-based and desk-top analyzer systems. Thus, the survey pools themselves probably are not suitable for the assessment of HDL-C measurements. The fresh specimen data shown in Table 8-3 support this conclusion. Nevertheless, the data in Tables 8-5 and 8-6 do emphasize both the difficulty in assessing the reliability of HDL-C measurements and the need for survey pools that better emulate fresh specimens.

RECOMMENDATIONS

The use of desk-top analyzers for TC measurement was discussed in some detail at a workshop on TC screening sponsored by the NHLBI[32] and by the NCEP-LSP.[30] The following sections represent, for the most part, a summary of their recommendations. These recommendations can also be usefully applied to the measurement of TG and HDL-C.

Use an accurate system.

Select an instrument-reagent system for which the manufacturer has documented how the measurements were standardized and for which the measurements are traceable to the CDC reference values. This will help to ensure that at least, on average, the instrument is capable of providing accurate values.

Use a precise instrument.

Select a system that is precise; that is, repeat measurements made in the same sample should be reproducible. Within individual clinical laboratories, CVs of less than 3% can generally be obtained for TC, TG, and HDL-C. With desk-top analyzers, however, CVs are often about 5% or higher.

Use properly trained operators.

The measurements should be made by trained operators. Ideally, the operator should be a trained medical technician who is familiar with clinical chemical measurement and quality-control procedures. Before using the measurements for clinical purposes, the operator should be trained in the proper procedures for patient preparation, instrument operation, and regularly scheduled instrument maintenance according to the manufacturer's recommended procedures. Maintenance should be documented and records kept current.

Institute and use a quality-control system.

Establish a formal quality-control system using quality-control pools that are available in sufficient quantities to be analyzed for a minimum period of 6 months. Quality-control pools should cover the clinically significant range. For TC, two pools with TC concentrations of 180–200 mg/dL (4.66–5.18 mmol/L) and 240–270 mg/dL (6.22–6.99 mmol/L) should be used. For TG, use pools with concentrations of about 125–150 mg/dL (1.41–1.69 mmol/L) and 200–250 mg/dL (2.26–2.82 mmol/L). HDL pools should have concentrations of about 35 mg/dL and 60 mg/dL (0.90 and 1.55 mmol/L).

When first setting up the analytical system, these materials should be analyzed in duplicate on 5–10 different days during a 2-w period, and appropriate quality-control limits should be established for each. If assistance in calculating these limits is needed, a qualified local laboratory should be consulted.

New quality-control pools should be introduced before the current pools are exhausted and should similarly be analyzed to establish control limits for those pools before the original pools are discontinued. Quality-control records should be maintained and kept current.

In the laboratory, the results for patient samples are not used when the quality-control measurements are outside control limits; instead, the analyses are repeated. Such a procedure is not feasible with desk-top analysis because the results may have been reported before it was known that they were "out of control." For this reason, it is recommended that quality-control pools be analyzed at the beginning of the day to establish that the analyses are adequate, and again after every 20 samples (or every few hours if less than 20 samples are analyzed) to document that the instrument continues to function properly. If not, subsequent measurements should not be accepted, and appropriate corrective measures should be taken.

Use a system of split-sample analysis.

In some cases, measurements in quality-control pools can be influenced by factors arising from the way the pools are prepared, and quality-control pools may not accurately reflect the reliability of measurements in freshly drawn patient specimens. It is *not advisable* to rely solely on such pools for quality-control. Rather, the accuracy of the results should be monitored by comparing values generated by the desk-top analyzer to those in a qualified laboratory whose methods are ultimately traceable to reference values. This can be done in 5–10% of the samples by submitting aliquots of those samples to the laboratory. When desk-top analyzer measurements are made in capillary specimens obtained by finger stick, it is preferable that the split-sample comparison be performed in concurrently collected venous serum, since it may not be feasible to use split capillary specimens.

Follow proper patient preparation procedures.

Blood collection should be performed in a standard way to minimize changes that can lead to inaccurate measurements. For further details on patient preparation and specimen collection and handling, refer to Chapter 3.

SUMMARY

In general, desk-top analyzers tend to be fairly accurate in fresh specimens but less precise than laboratory-based analyses. These instruments can be useful for lipid screening or for following the progress of a patient being treated for hyperlipidemia once the lipoprotein abnormality has been diagnosed using the appropriate medical procedures and lipoprotein measurements. TC concentrations can be measured in samples from non-fasting patients, if necessary. HDL-C measured in fasting samples can be expected to be lower than in non-fasting samples by a few percentage points. TG should not be measured in non-fasting samples. The analyses are best performed in venous samples. If capillary samples are used, it is well to be aware that the values may not accurately reflect the patient's venous concentration, and in individual cases may differ substantially from venous values. Desk-top analyzers can be useful when they are operated by properly trained personnel, when adequate quality-control procedures are followed, and when the individuals charged with interpreting the results are aware of the limitations of the analyzers. ❖

Acknowledgments. This work was supported in part by NIH Grant (SCOR-A Grant No. HL 47212, NIH Contract No. HV 78102), and the Johns Hopkins Lipid Research Unit, Department of Pediatrics, The Johns Hopkins University School of Medicine. The author wishes to thank Kathleen Lovejoy for assembling and analyzing the quality-control data and Carol McGeeney for preparing the manuscript.

REFERENCES

1. Castelli WP, Garrison RJ, Wilson PWF, et al. Incidence of coronary heart disease and lipoprotein levels: the Framingham Study. JAMA 1986;256:2835–8.
2. Lipid Research Clinics Program. The Lipid Research Clinics Coronary Primary Prevention Trial results. I. Reduction in incidence of coronary heart disease. JAMA 1984;251:351–64.
3. Lipid Research Clinics Program. The Lipid Research Clinics Coronary Primary Prevention Trial results. II. The relation of reduction in incidence of coronary heart disease to cholesterol lowering. JAMA 1984;251:365–74.
4. Stamler J, Wentworth D, Neaton JD, for the MRFIT Research Group. Is the relationship between cholesterol and risk of premature death from coronary heart disease continuous and graded? Findings in 356,222 primary screenees of the Multiple Risk Factor Intervention Trial (MRFIT). JAMA 1986;256:2823–8.
5. Blankenhorn DH, Nessim SA, Johnson RL, et al. Beneficial effects of combined colestipol-niacin therapy on coronary atherosclerosis and coronary venous bypass grafts. JAMA 1987;257:3233–40.
6. Frick MH, Elo O, Haapa et al. Helsinki Heart Study: primary prevention trial with gemfibrozil in middle-aged men with dyslipidemia. N Engl J Med 1987;317:1237–45.
7. Brensike EJ, Levy RJ, Kelsey SF, et al. Effects of therapy with cholestyramine on progression of coronary arteriosclerosis: results of the NHLBI Type II Coronary Intervention Study. Circulation 1984;69:313–24.
8. Gordon T, Castelli WP, Hjortland MC, Kannel WB, Dawber TR. High density lipoprotein as a protective factor against coronary heart disease: the Framingham Study. Am J Med 1977;62:707–14.
9. Castelli WP, Doyle JT, Gordon T, et al. HDL cholesterol and other lipids in coronary heart disease: the co-operative lipoprotein phenotyping study. Circulation 1972;55:767–72.
10. Miller NE, Forde OH, Thelle DS, Mjos OD. The Tromso Heart Study. High density lipoprotein and coronary heart disease: a prospective case-control study. Lancet 1977;1:965–7.
11. Summary of the Second Report of the National Cholesterol Education Program (NCEP) Expert Panel on Detection, Evaluation and Treatment of High Blood Cholesterol in Adults (Adult Treatment Panel II). JAMA 1993;269:3015–23.

12. Report of the National Cholesterol Education Program Expert Panel on Detection Evaluation and Treatment of High Blood Cholesterol in Adults. Arch Intern Med 1988;148:36–69.
13. Allain CC, Poon LS, Chan CSG, Richmond W, Fu PC. Enzymatic determination of total serum cholesterol. Clin Chem 1974;20:470–6.
14. Demaker PNM, Boerma GJM, Baadenhuijsen H, et al. Evaluation of accuracy of 20 different test kits for the enzymatic determination of cholesterol. Clin Chem 1983;29:1916–22.
15. Bucolo G, David H. Quantitative determination of serum triglycerides by the use of enzymes. Clin Chem 1973;19:476–82.
16. Bachorik PS, Levy RI, Rifkind BM. Lipids and dyslipoproteinemia. In: Henry JB, ed. Clinical diagnosis and management by laboratory methods, 18th ed. Philadelphia: WB Saunders, 1991:188–214.
17. Bachorik PS, Rock R, Cloey T, et al. Cholesterol screening: comparative evaluation of on-site and laboratory-based measurements. Clin Chem 1990;36:255–60.
18. Sedor FA, Holleman CM, Heyden S, Schneider KA. Reflotron cholesterol measurement evaluated as a screening technique. Clin Chem 1988;34:2542–4.
19. Bachorik PS, Bradford RH, Cole T, et al. Accuracy and precision of analyses for total cholesterol as measured with the Reflotron cholesterol method. Clin Chem 1989;35:1734–9.
20. von Schenck H, Treichl L, Tilling B, Olsson AG. Laboratory and field evaluation of three desktop instruments for assay of cholesterol and triglyceride. Clin Chem 1987;33:1230–2.
21. Friedewald WT, Levy RI, Fredrickson DS. Estimation of the concentration of low density lipoprotein cholesterol in plasma without use of the preparative ultracentrifuge. Clin Chem 1972;18:499–502.
22. Koch TR, Mehtu U, Lee H, et al. Bias and precision of cholesterol analysis by physicians office analyzers. Clin Chem 1987;33:2262–7.
23. Sterling B, Kiang T, Subramanian K, et al. Simultaneous patient-side measurement of hemoglobin, glucose, and cholesterol in finger-stick blood. Clin Chem 1992;38:1658–64.
24. Kupke IR, Zeugner S, Gottschalk A, Kather B. Differences in lipid and lipoprotein concentrations of capillary and venous blood samples. Clin Chim Acta 1979;97:279–83.
25. Kaplan SA, Yuceoglu AM, Strauss J. Chemical microanalysis: analysis of capillary and venous blood. Pediatrics 1959;24:270–4.
26. Greenland P, Bowley NL, Melklejohn B, Doane KL, Sparks CE. Blood cholesterol concentration: fingerstick plasma vs. venous serum sampling. Clin Chem 1990;36:628–30.
27. Bachorik PS. Collection of blood samples for lipoprotein analysis. Clin Chem 1982;28:1375–8.
28. Laboratory Methods Committee, Lipid Research Clinics Program. Cholesterol and triglyceride concentrations in serum plasma pairs. Clin Chem 1977;26:60–3.
29. Cloey T, Bachorik PS, Becker D, et al. Re-evaluation of serum-plasma differences in total cholesterol concentration. JAMA 1990;263:2788–9.
30. Laboratory Standardization Panel. Recommendations for improving cholesterol measurement: a report from the Laboratory Standardization Panel of the National Cholesterol Education Program. NIH Publication No. 90-2964. Executive Summary, NIH Publication No. 90-2964A. Bethesda, MD: NIH, 1990.
31. Belsey R, Vandenbark M, Goitein RK, Baer DM. Evaluation of a laboratory system intended for use in physicians' offices. II. Reliability of results produced by health

care workers without formal or professional laboratory training. JAMA 1987; 258:357–63.

32. Recommendations regarding public screening for measuring blood cholesterol: a summary of a National Heart, Lung and Blood Institute workshop. NIH Publication No. 89-3045, Bethesda, MD: NIH, 1989.

CHAPTER 9

Clinical Significance and Measurement of Apolipoproteins AI and B

Evan A. Stein

INTRODUCTION AND CLINICAL IMPORTANCE

❖ Despite significant reductions in the incidence of atherosclerotic coronary heart disease (CHD) over the last three decades, it remains the major cause of mortality and morbidity in the developed world.[1] Positive results achieved in a number of major primary[2, 3] and secondary[4, 5] prevention trials that focused on reducing specific lipid fractions have made possible the development of "cut-points" and treatment guidelines for some lipid parameters, particularly low-density lipoprotein cholesterol (LDL-C). The concentrations outlined by the National Cholesterol Education Program (NCEP) Adult Treatment Panels I and II (ATP I[6] and ATP II[7]) have been accompanied by very specific guidelines for accuracy and precision in the measurements.[8] These dual developments in the area of clinical and laboratory lipidology as it pertains to the prevention of atherosclerosis have necessitated that similar constraints and standards be applied to all other lipid, lipoprotein, or apolipoprotein parameters in the clinical arena. Therefore, apolipoproteins AI and B must also meet these "new realities."

Apolipoprotein B-100 (apo B) currently appears to be the most clinically significant of the two parameters. It is an integral component of the four major atherogenic lipoproteins: very-low-density lipoprotein (VLDL), intermediate-density lipoprotein (IDL), low-density lipoprotein (LDL), and lipoprotein(a) [Lp(a)]. Apo B-100 should be distinguished from apo B-48, which is found only in lipoproteins of intestinal origin, such as chylomicrons and chylomicron remnants,[9] and is usually absent from the systemic circulation except in rare subjects who have Type I, III, or V hyperlipidemia.[10] Apo B's initial function in VLDL and IDL appears to be structural; however, with exposure of binding domains on LDL, it becomes responsible for interaction with high-affinity LDL or apo B receptors[11] on cell surfaces, which results in uptake and removal of LDL and its content from the circulation. For additional information on this issue, refer to Chapter 1.

Numerous cross-sectional epidemiological studies have shown a strong relationship between plasma apo B and coronary atherosclerosis.[12] In most case-controlled studies of subjects with and without CHD, apo B has been shown to be a better marker for atherosclerosis than the various plasma lipids.[12] More recently, pharmacological intervention in subjects with CHD and increased apo B demonstrated that reduction in this apolipoprotein results in fewer CHD-related events and regression of atherosclerosis.[13] Thus, apo B appears to have fulfilled all the requirements necessary to be designated "causative" in the atherosclerotic process.[14] The characteristics of the association of apo AI and B to CHD are given in Table 9-1.

Apo AI is the major protein constituent of lipoproteins in the high-density range ($d > 1.063$ kg/L). Its major function is apparently to act as a co-factor for the activation of lecithin-cholesterol acyltransferase (LCAT).[11] Recent studies have indicated that lipoprotein subspecies containing only apo AI (known as Lp AI particles) play a role in cholesterol efflux from cells,[15] while those containing apo AI and AII (Lp AI:AII particles) inhibit cholesterol efflux. The recent development of transgenic mice with human apo AI and AII genes has provided additional evidence of the potential importance of AI-only particles,[16] Apo AI may also be the ligand that binds to a proposed hepatic receptor for removal of high-density lipoprotein (HDL).[11]

In terms of being important in causation of atherosclerosis, apo AI, like HDL cholesterol (HDL-C), has not yet met all the criteria listed in Table 9-1. Especially absent is "clinical trial" evidence that increasing concentrations of apo AI alone reduces CHD mortality or morbidity. Thus, despite epidemiological and case-control studies that promote the clinical sensitivity and specificity of apo AI over other lipid fractions,[17] apo AI is still considered a "potential" risk factor for atherosclerosis, but cannot yet be designated as "causative." Apo AI may prove to be a contributing, co-existing, or even co-incidental factor for this multifactorial disease.

Table 9-1 ❖ Relationship of apo AI and B to Coronary Heart Disease

Characteristic of Association	*Apo AI*	*Apo B*
1. Strength	Medium	Medium
2. Independence	Medium	High
3. Biological consistency	Medium	High
4. Observational consistency	Medium	High
5. Measurement validity	Low	Low
6. Temporal relationship	Low	Medium
7. Interventional benefit	–	Low

Some investigators have attempted to derive a better index of atherosclerosis risk by dividing apo AI by apo B (AI/B ratio). Because of the pitfalls of merging these two not entirely independent parameters, especially when methodological problems exist, use of ratios is not recommended. This issue is discussed further at the end of this chapter.

EXISTING METHODS FOR MEASURING APO AI AND B

Unlike most other plasma proteins, apolipoproteins AI and B exist as part of complex macromolecules consisting of various lipids and apolipoproteins. Although only one apo B molecule exists within a lipoprotein complex, the associated lipids and other apolipoproteins vary considerably as the lipoprotein undergoes degradation from VLDL to LDL.[11] Even within the lipoprotein density classes, the composition of lipids/proteins is highly variable.[18] The conformational changes that occur in apo B during its transit through the systemic circulation alter its antigenic properties, thus creating the potential for antibody recognition in apo B quantitation procedures.[19] While this is generally not a problem with polyvalent antisera, it may present difficulties with assays based on monoclonal antibodies.[20] Furthermore, since the considerable variability in size and electrophoretic mobility of apo-B-containing lipoproteins can influence some assay systems,[21] such as radial immunodiffusion (RID) and electroimmunoassay (EIA), these factors need to be very carefully evaluated with multiple lipoprotein species. Assay systems that depend for an endpoint on light scattering or turbidity, such as nephelometric immunoassay (NIA) or turbidimetric immunoassay (TIA), may also be subject to direct interference by large lipoproteins such as chylomicrons or large VLDL when these are present in plasma in increased concentrations.[22]

While apo AI is found to be associated with a more limited number of lipoproteins, mainly in the HDL range (d = 1.063–1.21 kg/L), a large percentage of the protein is "buried" within the lipid component of the lipoprotein.[23] This can and does present problems for an assay system in which the detecting antibody is not targeted against an epitope or epitopes exposed on all apo-AI-containing particles at all times. Since apo AI is, for the most part, contained on smaller, denser lipoproteins, adverse affects on systems that depend on size, charge, or light scattering appear minimal.

Techniques for measuring both apolipoprotein AI and B are confined to immunological procedures that use antibodies directed at apo AI or B, including:

- Radio immunoassay (RIA): competitive or immuno-radiometric (IRMA) systems
- Enzyme-linked immunosorbent assay (ELISA): competitive or capture systems
- Fluorescence immunoassay (FIA): capture systems
- Radial immunodiffusion (RID)

- Nephelometric immunoassay (NIA)
- Turbidimetric immunoassay (TIA)
- Electroimmunoassay (EIA)

The potential advantages and disadvantages of each assay system are listed in Table 9-2. RIA and ELISA systems, being virtually free of matrix effects, tended to predominate in research and reference laboratories where assay systems were developed "from scratch."[24] These systems can recognize equally both purified apolipoprotein such as AI, or "narrow-cut" LDL containing only apo B, and the same apolipoproteins when they are contained in naturally occurring lipoprotein species. This enables standardization of these assay systems with "primary" standards and secondary reference material. These systems are also free of matrix effects such as that seen between fresh, frozen, or lyophilized sera in other systems.[25] However, improvements to NIA and TIA systems over the last few years have reduced matrix problems, resulting in assay systems of sufficient reliability for use in most clinical research situations. In the commercial arena, NIA and TIA systems predominate, most likely due to their large and rapid sample throughput, technical ease, and minimal handling requirements. In addition, with some systems, precision both within and between assays is better than that achieved in the more manual RIA and ELISA assays. The major drawback of NIA and TIA systems with apo B is that matrix effects are known to occur,[24, 25] thus preventing direct calibration with primary standards such as narrow-cut LDL for apo B. For apo B, even secondary reference material that has been frozen or lyophilized may not be suitable for calibrating systems that utilize immunoprecipitation endpoints.[25]

Over the last few years, a number of academic and research laboratories and industry-based scientists have devoted considerable time, effort, expense, and intellectual activity to overcoming the problems of apo AI and B standardization.[26] This effort was initiated in 1983 when the Centers for Disease Control and Prevention-International Union for Immunosystems (CDC-IUIS) surveyed laboratories performing apo AI and B assays and discovered that results obtained in these laboratories varied not so much because of assay technique, but because of method imprecision and differences in calibration.[27] Under an aggressive program organized by the International Federation of Clinical Chemistry (IFCC) subcommittee on apolipoproteins,[26] a comprehensive plan was put into action in 1988 to search for a suitable secondary serum reference material that could be used, without fear of matrix interactions, as a master calibrator for all current commercial assay procedures. Although commercial and governmental bodies around the world submitted numerous materials for evaluation, only a single stabilized liquid and a lyophilized material met all the predefined criteria for apo B and AI, respectively. These interim reference materials (IRM), designated SP3 for apo B and SP1 for apo AI, were sanctioned by the IFCC/World Health Organization (WHO) for utilization as worldwide calibrators.

Table 9-2 ❖ Advantages and Disadvantages of Current Assay Systems for Measurement of Apolipoprotein AI and B*

Method	*Advantages*	*Disadvantages*
RIA	Highly sensitive. No matrix effects. Uses polyvalent/monoclonals. Uses primary/secondary standards.	Automation difficult. Requires large dilutions. Moderately precise. Uses radioactivity. Short reagent stability.
ELISA	Highly sensitive. Polyvalent or monoclonals. Long shelf-life. Matrix insensitive. Uses primary/secondary standards.	Automation difficult. Moderately precise. Requires large dilutions.
NIA	Easily automated. Good precision.	Potential matrix interactions. Dedicated equipment. Specific antisera. Antisera "thirsty."
TIA	Easily automated. Uses common lab equipment.	Variable precision. Potential matrix interaction. Needs specific antisera. Antisera "thirsty."
RID	"Low-tech." Minimal special equipment. Small dilutions.	Slow, imprecise. Matrix interactions. Lipoprotein species variability. Needs specific antisera. Antisera "thirsty."
EIA	Small dilutions. Relatively "low-tech."	Slow, imprecise. Matrix interactions. Needs specific antisera. Antisera "thirsty."

*RIA = radio immunoassay, ELISA = enzyme-linked immunosorbent assay
NIA = nephelometric immunoasssay, TIA = turbidimetric immunoassay,
RID = radial immunodiffusion, EIA = electroimmunoassay

The remaining problem rests with assignment of "true" values to these materials. This would not present a significant problem if a widely accepted reference method, or even a selected method, for apo AI and/or B were available; however, this has not yet been accomplished. This final step has been delayed by continued dissention between the various scientific groups involved in the program over which assay systems to use for value assignment. Other problems that remain have to do with assigning values to the primary standards (purified apo AI and narrow-cut LDL). Despite these deficiencies, SP3 and SP1 were assigned values by a reference laboratory and were adopted by most manufacturers, who have begun to offer quantitation systems calibrated with, or traceable to, these IRMAs.[28]

REFERENCE METHOD

Currently no consensus-approved reference or definitive methods exist for apo AI or apo B measurement. A candidate reference method proposed by Albers, Lodge, and Curtiss[10] consisted of a capture ELISA assay utilizing two well-characterized monoclonal antibodies, MB 24 and MB 47, which are known to bind to apo B epitopes expressed on all apo-B-containing lipoproteins, including LDL and VLDL.[29] This method was based on a similar assay previously designed by Young et al.[29] in which MB 47 recognized only apo B-100 while MB 24 bound to both apo B-100 and B-48. In the configuration of the proposed reference method, however, only apo B-100 was quantitated. The assay therefore demonstrated the first requirement for consideration as a reference procedure: accuracy. It exhibited parallelism between isolated apo-B-containing lipoproteins, such as narrow-cut LDL and other apo-B-containing lipoproteins occurring in numerous different patient plasma samples. Delipidation of VLDL did not increase apo B epitope expression and binding.

The assay also demonstrated the second requirement for acceptance as a reference method: precision. This was especially the case with intra-assay reproducibility, with coefficients of variations (CV's) of 1.7–3.8% over a range of 95–146 mg/dL. The between-assay precision yielded CV's of 5.1–6.9% over the same concentrations, and was reasonably acceptable for a manual method. The potential to automate this assay makes it likely that precision could be significantly improved. No matrix effect was noted with this procedure, and fresh, frozen, and lyophilized material was measured accurately and consistently. Unfortunately, although this proposed reference method has been evaluated and validated in greater depth than any other proposed method, it has not become widely accepted or utilized. The method offers substantial benefits over RIA procedures and is more readily transferred to other laboratories, which would allow development of a reference laboratory network similar to that for measurements of total cholesterol.[30]

No similar assay system has yet been proposed for apo AI, although multiple or "pan" monoclonal antibodies which recognize apo AI on all AI-containing lipoproteins have also been described.[31] The CDC has developed a high-performance liquid chromatography–mass spectrometry (HPLC-MS) method for apo AI and continues to develop a reference RIA procedure for apo B. Problems still remain in assigning quantitative values to the primary standards for any reference method. Apo AI does not appear to present a substantial problem, as purified AI can be assigned protein mass using the traditional Lowry procedure or amino acid analysis. For apo B, however, the chromogenicity of the Lowry procedure differs from that of the albumin standard. In addition, it is suggested that the Lowry assay be carried out using sodium dodecylsulfate to reduce potential errors caused by protein aggregation. This has led to bias between the Lowry procedure and amino acid analysis in assignment of protein mass, and is still a source of error in assigning a value to a potential apo B calibrator.[10]

RECOMMENDED METHOD

Table 9–3 lists several theoretical and practical considerations that need to be addressed by clinical laboratories when selecting a method for measuring apo AI and B. Requirements for accuracy and precision will vary among laboratories, from the laboratory that services specialized lipid clinics, where highly abnormal samples are likely to be encountered, to the peripheral hospital or commercial laboratory that screens relatively "healthy" populations. Techniques with potential interference from severe lipidemia, such as RID, NIA or TIA, may suffice for the latter situation, where these types of samples are uncommon; however, in the former situation, alternative methods would also be needed. It is likely beyond the capacity of most clinical laboratories to develop the expertise and devote the resources needed to develop RIA or ELISA procedures, as this necessitates searching out and evaluating various antibodies, purification, labeling, and so on. For this reason, most laboratories rely on commercially available kits or analytical systems. Some commercial RIA procedures are available for both apo AI and B; however, the short shelf-life of the tracer used may prove unattractive to laboratories that do not perform these assays on a frequent basis. For laboratories that do not have a nephelometer and do not wish to use RIA, only three procedures remain: RID, EIA and TIA. Although electroimmunoassay also requires some specialized equipment, it does allow laboratories with fairly small, infrequent sample volumes to perform these assays in a reasonably efficient manner. Turbidimetry can be adapted to many analyzers, but currently only a limited number of "kit" systems are available.

NIA and TIA are the most widely used procedures in clinical laboratories in both the United States and Europe. These methods have the advantage of greater automation than any other procedure and allow for high sample throughput.

Nephelometry has been selected for description here as the recommended method. It does require specialized instrumentation, but of a type which is already in use in many clinical laboratories, especially those that perform other protein assays. Thus, it should be available in most larger hospital and commercial (private) laboratories where apo AI and B assays have been initially introduced. The

Table 9–3 ❖ Theoretical and Practical Considerations for Clinical Laboratories Selecting apo AI and B Methods

1. Availability of resources:	2. Potential sample throughput
Technical skills	3. Frequency of analysis
Time to invest	4. Clinical utility:
Availability of needed materials	Primary screening
Availability of equipment	Support of specialized clinical service
Financial issues	Clinical or epidemiological research

two major nephelometric systems used are rate nephelometry (Beckman Instruments, Palo Alto, CA) or two-point rate nephelometry (Behring Diagnostics, Sommerville, NJ). The latter system is described in the following sections.

Principle

When either apo AI or B in plasma or serum reacts with specific antisera directed against AI or B, insoluble complexes are formed. Infra-red light (wavelength 840 nm) is passed through the resultant suspension and scatters in all directions. The light that scatters in a forward direction is focused onto a detection system (photo-diode) by an optical lens system. The amount of light scattered and detected is proportional to the number of insoluble complexes formed, which in turn is dependent on the amount of apo AI or B present in the sample. Quantitation of the amount of apo AI or B is achieved by comparing the "signal" from unknown samples to values obtained from a set of standards or calibrators of known apo or B concentration.

Reagents

While it is theoretically possible to utilize reagents from other suppliers for both the Behring or Beckman nephelometric systems, considerable amount of time and effort are required to ensure complete harmonization of all components. Thus, in practice, it is usually more convenient and often more reliable to utilize the reagent and calibration systems supplied by the manufacturer, as they have been customized and optimized for that particular system. Furthermore, problems such as interference, precision, or inaccuracy (bias) in the system can be brought to the manufacturer's attention to be remedied.

Since apo AI and B can be analyzed together on the Behring Nephelometer 100 (BN-100), they are discussed simultaneously. The required reagents are listed below; all are nephelometric grade. Prior to use, reagents and samples should be warmed to room temperature.

- Antisera to both apo AI and apo B
- Apolipoprotein standard serum
- Apolipoprotein control serum
- Reaction buffer
- Diluent
- Supplement reagent/precipitation (for apo AI only)
- Washing Solution: Tween 20, four drops per liter of distilled water.

Procedure

Reagents are placed in correct positions according to color-coded directions on the analyzer. Because standard curve parameters for antisera are lot specific, these need to be re-entered into the instrument every time a new reagent lot is used. Selections are entered through a series of menus; each item is clearly explained in the BN-100 operating manual and the package insert accompanying the reagent. Samples are loaded into their respective positions on the sample rack and measurements are done automatically. The analytical parameters are shown in Table 9–4.

Sensitivity and Specificity

According to the manufacturer, the sensitivity for the assay is 18 mg/dL for apo AI and 15 mg/dL for apo B—well below the expected clinical range and sufficient even for quantitation of these apolipoproteins in various density fractions separated by ultracentrifugation.

The specificities for apo AI and B are totally dependent on the quality of the antisera. This is generally evaluated comprehensively by the manufacturer, but information as to cross-reactivity with other apolipoproteins (e.g., AII, AIV, E, C's) and other serum proteins should be available and carefully reviewed for each new lot of antisera. Information pertaining to the ability of the system to recognize apo AI or B on the different lipoprotein species (e.g., apo B in VLDL, IDL, and various LDL densities) should be carefully examined.

Precision

For apo AI at concentrations of 70 and 206 mg/dL, the manufacturer indicates within-run CV's of 4.64% and 1.85% and between-run CV's of 7.3% and 2.25% respectively. Studies in our laboratory have found slightly better precision, with within-run CV's of 2.4% and 1.8% at concentrations of 80 and 126 mg/dL respec-

Table 9–4 ❖ Analytical Parameters for Analysis of Apo AI and Apo B Using the Behring Nephelometer 100

Parameter	*Apo AI*	*Apo B*
Sample volume (μL)	10	20
Sample dilution	1:20	1:20
Antisera volume (μL)	40	40
Supplemental reagent (μL)	10	–
Reaction Buffer a. to sample (μL) b. to antisera (μL)	 80 80	 80 80
Standard Curve dilutions (6 point)	1:5/1:10/1:20 1:40/1:80/1:160	1:5/1:10/1:20 1:40/1:80/1:160
Working Range (mg/dL)	18–580	15–480

tively ($N = 8$) and between-run CV's of 3.7%, 3.9%, and 3.7% at concentrations of 85, 129, and 141 mg/dL respectively ($N = 10$). For apo B, the manufacturer claims within-run CV's of 0.77% and 1.06% at concentrations of 49 and 243 mg/dL respectively, and between-run CV's of 3.66% and 2.56% at the same concentrations. Independent evaluation in our laboratory found a within-run CV of 2.2% at 88 mg/dL ($N = 8$) and between-run CV's of 7.9%, 5.9% and 7.1% at concentrations of 92, 159, and 208 mg/dL respectively ($N = 10$).

Comparison to Other Procedures

The manufacturer has compared the procedure to its own RID assay with the following results: apo AI by NIA = 0.91 RID + 1.93 mg/dL ($N = 109$) with correlation coefficient $r = 0.963$; apo B by NIA = 1.02 RID – 3.4 mg/dL ($N = 49$) with correlation coefficient $r = 0.987$.

In an independent evaluation conducted in our laboratory, the BN-100 system was compared to an identical BN-100 system in a second laboratory, the Institute Pasteur in Lille, France. To assess the comparability of the same systems in different laboratories, 50 frozen patient samples were analyzed on BN-100 systems in the two laboratories. These instruments were not in any way specially calibrated, but were operated according to the manufacturer's instructions using the manufacturer's reagent systems and calibrators. Comparisons were as follows: Apo AI by Lab 1 = 0.845 Lab 2 + 19 mg/dL, $r = 0.960$; Apo B by Lab 1 = 1.01 Lab 2 + 12 mg/dL, $r = 0.850$.

Interferences

Because lipemic samples may cause positive interference with the apo B procedure, the manufacturer's suggestion to pretreat such samples by ultracentrifugation prior to analysis is not recommended, as this would remove substantial amounts of apo B in VLDL or IDL and thus give erroneously low results. Sample pretreatment with lipase or detergents may be preferable, provided it can be clearly demonstrated that this would not result in exposure of additional antigenic sites, which may result in falsely increased results or reduce secondary interactions needed to produce the light-scattering effect.

Studies have indicated interferences (matrix effects) with samples for apo B that have been previously frozen or lyophilized; these effects are minimal or absent for apo AI.[25] (For additional information on matrix effects, refer to Chapter 10.)

Finally, this system requires calibration with a serum-based material, as reactivity with purified apo AI and isolated narrow-cut LDL for apo B does not produce parallel curves to fresh serum samples.

CALIBRATION AND STANDARDIZATION

By 1993, most manufacturers had switched value assignment to the IFCC/WHO interim reference material described previously.[27]

QUALITY ASSURANCE

Although manufacturers supply material for quality control purposes, it is recommended that all laboratories analyze stable and suitable human-based material at at least two concentrations (high and low) for both apo AI and B. In automated systems such as NIA, samples should be analyzed once for every 10 patient samples, alternating high and low pools. In batch-type procedures (RID, EIA, RIA, and ELISA), both pools should be analyzed at least once in every run. Each laboratory needs to establish its own operational limits based on repeated analysis of its quality assurance material, and then follow the standard criteria it uses for other analytes.

External quality assurance is virtually absent, although the College of American Pathologists (CAP) has been collecting data in recent surveys.[32] Unfortunately, many of the survey pools were not entirely human based, contained altered lipoproteins, and had significant matrix interactions, which prevented obtaining useful comparative data. It is anticipated that the work done by the IFCC and industry to develop matrix-free calibrators will eventually lead to a suitable material for use in blinded external quality assurance programs.

INTERPRETATION OF RESULTS

A number of practical difficulties continue to hinder the transfer of apo AI and B from the research to the clinical laboratory:

1. The lack of standardization.
2. The lack of reference methods.
3. The change from reference ranges based on population distributions to cutpoints based on relative atherosclerotic risk and likelihood of benefit from dietary or pharmacological intervention.

At the present time, no large comprehensive, statistically and demographically sound population study has employed a fully standardized method as a reference base for apo B or AI. Numerous small studies, often based on blood samples obtained from subjects being evaluated at medical facilities, have been conducted.[33] However, in most of these studies, standardized or validated methods were not used, calibration was not traceable to a primary standard, and blood collection was not obtained in a defined manner. While apo AI and B are being measured as part of a number of large, well-defined population groups such as the National Health and Nutrition Examination Survey III (NHANES III), the methodology used has potential matrix interactions and is not directly traceable to the currently proposed IFCC/WHO calibrators.

As with total and LDL cholesterol, there is a definite increase in apo B with age, and males tend to have higher values than females.[34] Based on relatively small numbers of generally younger subjects (age 20–50 y) using immunological methods not requiring precipitation endpoints (e.g., RIA), apo B averages 80–90 mg/dL. In one recent intervention study,[13] subjects were selected for entry based on an apo B above 125 mg/dL—about the 90th percentile for a group of healthy men in Seattle, where the study was conducted.[13] Reduction of apo B in two groups of men using a combination of lipid- and apo-B-lowering drugs resulted in fewer cardiovascular events and more reversal of angiographically demonstrable coronary atherosclerosis than was observed in a group of men on a placebo, in whom no lipid or apo B reduction was occurred.[13] This study, which was the first to select subjects based on apo B, potentially establishes a therapeutic cutpoint for apo B of 125 mg/dL. To utilize this cutpoint, however, it is essential that other laboratories using different methods produce results that are both traceable and standardized (i.e., exhibit accuracy) to the procedure used in the intervention study.

Methods using immunological procedures with precipitation endpoints (e.g., RID, EIA, NIA, or TIA) have generally yielded higher values in population studies;[33] however, this may reflect calibration/standardization differences rather than inherent method bias.

In terms of a developed and clinically usable data base, the position of apo AI is similar to that of apo B. The changes with age are not as clearly elucidated. However, in terms of sex differences, females have substantially higher apo AI concentrations than males,[33] similar to those seen for HDL-C. The difference in values for apo AI between methods that involve immunoprecipitation endpoints and those that do not are small and are not problematic as with apo B. For adults, average apo AI concentrations are in the order of 125–135 mg/dL for males and 135–145 mg/dL for females. The relationship of apo AI to atherosclerotic risk has not been sufficiently developed to establish specific cutpoints.

A number of investigators and laboratories have suggested using the apo AI/B ratio for designating risk for atherosclerosis or instituting treatment. Not only are ratios inherently flawed for diagnosing or treating specific disorders of lipid, lipoprotein, or apolipoprotein metabolism, but, given the lack of standardization of either apo AI or B, the poor between-run precision of some assay systems, and the absence of soundly based reference ranges, use of such ratios as an index of atherosclerosis risk should be avoided. ❖

REFERENCES

1. Havlik RJ, Feinleib M. Proceeding of the conference on the decline of coronary heart disease mortality (NIH publication no. 79-610). Bethesda, MD: National Heart, Lung, and Blood Institute, 1979.
2. Lipid Research Clinics Program. The Lipid Research Clinics Coronary Primary Prevention Trials results. I. Reduction in incidence of coronary heart disease. JAMA 1984:251:351–64.
3. Frick MH, Elo O, Haapa K, et al. Helsinki Heart Study. Primary-prevention trial with gemfibrozil in middle-aged men with dyslipidemia. N Engl J Med 1987;317:1237–45.
4. Blankenhorn DH, Nessim SA, Johnson RL, Sanmarco ME, Azen SP, Cashin-Hemphill. Beneficial effects of combined colestipol-niacin therapy on coronary atherosclerosis and coronary venous bypass grafts. JAMA 1987;257:3233–40.
5. Canner PL, Berg KG, Wenger NK, et al. Fifteen years mortality in coronary drug project patients: long-term benefit with niacin. J Am Coll Cardiol 1986;8:1245–55.

6. Report of the National Cholesterol Education Program Expert Panel on Detection, Evaluation, and Treatment of High Blood Cholesterol in Adults. Arch Intern Med 1988;148:36–69.
7. Expert Panel on Detection, Evaluation, and Treatment of High Blood Cholesterol in Adults. Summary of the Second Report of the National Cholesterol Education Program (NCEP) Expert Panel on Detection, Evaluation, and Treatment of High Blood Cholesterol in Adults (Adult Treatment Panel II). JAMA 1993;269:3015–23.
8. Naito HK. Current status of blood cholesterol measurement in clinical laboratories in the United States: a report from the laboratory standardization panel of the National Cholesterol Education Program. Clin Chem 1988;34:193–201.
9. Kane J, Hardman DA, Paulus HE. Heterogeneity of apolipoprotein B: isolation of a new species from human chylomicrons. Proc Natl Acad Sci USA 1980;77:2465–9.
10. Albers JJ, Lodge MS, Curtiss LK. Evaluation of a monoclonal antibody based enzyme-linked immunosorbent assay as a candidate reference method for the measurement of apoprotein B-100. J Lipid Res 1989;30:1445–58.
11. Brewer HB, Gregg RE, Hoeg JM, Fojo SS. Apolipoproteins and lipoproteins in human plasma: an overview. Clin Chem 1988;34:B4–B8.
12. Albers JJ, Brunzell JD, Knopp RH. Apoprotein measurements and their clinical application in cholesterol screening. Clin Lab Med 1989;9:137–52.
13. Brown G, Albers JJ, Fisher LD, et al. Regression of coronary artery disease as a result of intensive lipid-lowering therapy in men with high levels of apolipoprotein B. N Engl J Med 1990;323:1289–98.
14. Heiss G, Tyroler HA: Are apolipoproteins useful for evaluating ischemic heart disease? A brief overview of the literature. In: Lippel K, ed. Proceedings of the workshop on apolipoprotein quantification. Bethesda, MD: National Institutes of Health, 1983:7–24.
15. Barbaras R, Puchois P, Ailhaud G. Cholesterol efflux from cultured adipose cells is mediated by Lp A-I particles but not by Lp A-II particles. Biochem Biophys Res Commun 1987;142:63–69.
16. Schultz JR, Verstuyft JG, Gong EL, Nichols AV, Rubin EM. Protein composition determines the anti-atherogenic properties of HDL in transgenic mice. Nature 1993;365:762–4.
17. Maciejko JJ, Homes DR, Konke BA, Zinsmeister AR, Dinh DM, Mao SJT. Apolipoprotein A-I as a marker of angiographically assessed coronary artery disease. N Engl J Med 1983;309:385–9.
18. Sniderman A, Shapiro S, Marpole D, Skinner B, Teng B, Kwiterovich PO. Association of coronary atherosclerosis with hyperbetalipoproteinemia [increased protein but normal cholesterol levels in human plasma low density (beta) lipoproteins]. Proc Natl Acad Sci USA 1980;77:604–8.
19. Schonfeld GW, Patch BP, Witzum JL, Weidman W. Lipolysis produces changes in the immunoreactivity and cell reactivity of very low density lipoproteins. J Clin Invest 1979;64:1288–97.
20. Tikkanen MJ, Cole TG, Hahm KS, Krul ES, Schonfeld G. Expression of apolipoprotein B epitopes in very low density lipoprotein subfractions: studies with monoclonal antibodies. Arteriosclerosis 1984;4:138–46.
21. Lutalo-Bosa AJ, Adolphson JL, Albers JJ. Evaluation of the measurement of B protein of plasma low density lipoproteins by radial immunodiffusion. J Lipid Res 1985;26:995–1001.
22. Labeur C, Shepherd J, Rosseneu M. Immunological assays of apolipoproteins in plasma: methods and instrumentation. Clin Chem 1990;4:591–7.
23. Albers JJ, Cheung MC. Immunoassay of apoprotein A-I and A-II: effect of different polyclonal antisera. Lippel K, ed. In: Lippel K, ed. Proceedings of the work-

shop on apolipoprotein quantification. Bethesda, MD: National Institutes of Health, 1983:349–53.

24. Smith SJ, Henderson LO, Hannon WH, Cooper GR. Effects of analytical method and lyophilized sera on measurements of apolipoproteins A-I and B: an international survey. Clin Chem 1990;36;290–6.
25. Marcovina S, Adolphson JL, Parlavecchia M, Albers J. Effects of lyophilization of serum on the measurement of apolipoproteins A-I and B. Clin Chem 1990;36: 366–9.
26. Marcovina S, Albers J. Apolipoprotein assays: standardization and quality control. Scand J Clin Lab Invest 1990;50:58–65.
27. Henderson O, Hannon W, Smith J, Cooper G. An international collaborative study on standardization of apolipoproteins A-I and B. Part II. Evaluation of contributions of antisera to among-laboratory variance components. Clin Chem 1987;33: 2250–6.
28. Marcovina SM, Albers JJ. International Federation of Clinical Chemistry study on the standardization of apolipoproteins A-I and B. Curr Opin Lipidol 1991;2:355–61.
29. Young SG, Smith RS, Hogle DM, Curtiss LK, Witzum JL. Two new monoclonal antibody-based enzyme-linked assays of apolipoprotein B. Clin Chem 1986;32: 1484–90.
30. Cooper G, Myers G. Reference system for cholesterol measurements. The Scand J Clin Lab Invest 1990;50:27–31.
31. Hogle DM, Smith RS, Curtiss LK. Quantitation of plasma apolipoprotein AI using two monoclonal antibodies in an enzyme-linked immunosorbant assay. J Lipid Res 1988;29:1221–9.
32. Naito HK. Reliability of lipid, lipoprotein, and apolipoprotein measurements. Clin Chem 1988;34:58.
33. Assmann G. Lipid metabolism and atherosclerosis. In: FK Schattauer, ed. Lipid metabolism and atherosclerosis. Stuttgart, Germany: 1982:155–60.
34. Alaupovic P, McConathy WJ, Fesmire J, Tavella M, Bard JM. Profiles of apolipoproteins and apolipoprotein B-containing lipoprotein particles in dyslipoproteinemias. Clin Chem 1988;34:B13–B27.

CHAPTER 10

Matrix Effects in the Measurement and Standardization of Lipids and Lipoproteins

W. Greg Miller

INTRODUCTION

The goal of clinical laboratory testing is to produce reliable information about an individual patient's clinical condition. Standardization of test method results is critically important for lipid testing that involves using epidemiologically derived national guidelines to interpret laboratory results. Standardized test results require a calibration process that produces equivalent results irrespective of the method or location performing the assay. The most practical way to achieve standardized lab results is to have reference materials that can be used by all methods to transfer and validate accuracy from a definitive or reference method to routine field methods. Unfortunately, the reference materials in general use have not been satisfactory because of differences in chemical reactivity between these processed materials and freshly collected specimens from individual patients.

Koch et al. demonstrated this limitation in reference materials in 1988.[1] They documented that the accuracy of four routine total cholesterol (TC) methods for freshly collected non-frozen individual patient's sera had an average bias of < 2.7% compared to the Abell-Kendall reference method. Also assayed with each of the four methods were National Institute for Standards and Technology (NIST) Standard Reference Material (SRM) 909 and three College of American Pathologists (CAP) Lipid Reference Materials—all processed, freeze-dried serum materials intended for calibration of TC. When the results for the reference materials were compared to the Abell-Kendall assigned target values, two of the routine methods had insignificant apparent biases ranging from −1.1% to +2.3%, while the other two routine methods had significant apparent biases ranging from

–9.8% to –4.8%. These altered chemical reactivity properties of processed reference materials are referred to as "matrix effects."

Current routine methods for lipid analysis are almost exclusively based on enzymatic or immunological reactions. Because pure TC or triglyceride (TG) molecules are only soluble in organic solvents, the primary standard materials are not suitable for calibration of routine assay methods. Lipoproteins can be purified, but their tertiary and quaternary structures are usually altered in the process, so preparation of primary aqueous standards with these compounds has not been practical. Consequently, secondary standards are commonly used for calibration of routine laboratory methods. The secondary standards are pooled serum (usually human) based materials that have been modified by various processing steps including addition of concentrated lipoprotein complexes with or without removal of various constituents. The materials are either frozen or, more commonly, freeze-dried, which induces additional modification and alteration of physical and chemical properties.

The concentration of an analyte in a secondary standard is not quantitatively determined by the preparation method and must be assigned by an assay process. Proficiency testing materials used for accuracy assessment are manufactured in essentially the same manner. Because of the manufacturing process, serum-based secondary standards or proficiency testing materials commonly have altered chemical reactivity properties and hence exhibit a matrix bias compared to native clinical specimens.

Laboratory test methods must be designed to have appropriate analytical attributes to ensure that test results for patient specimens provide clinically reliable information. One important analytical attribute is recovery of the correct amount of analyte without interference from other components in the specimen. An additional analytical attribute required by current clinical laboratory regulations is that routine method performance be validated on a regular basis by successful analysis of proficiency testing materials. Thus, routine methods must be able to accurately quantitate an analyte in clinical specimens and in commercially available proficiency testing materials.

The term "commutability" was used in 1973 by Fasce et al.[2] to refer to the ability of a processed material to show inter-assay properties comparable to authentic clinical specimens. Ideally, processed materials used for routine calibration and for standardization and verification of accuracy among laboratories would be commutable with freshly collected serum specimens. Unfortunately, processed materials frequently have assay properties that are different from those of fresh patient specimens. If an analysis system is sensitive to the difference between an analyte's reactive properties in a processed material and in fresh human specimens, the change in measured response to that analyte is referred to as a method/material-specific matrix interaction or matrix effect.[3] The matrix effect produces a matrix-induced bias in the analyte result for that processed material, which precludes using that result to evaluate accuracy of results for patient specimens. Matrix effects can be unique to a single lot of processed material and a single lot of reagents used in an analytical system. It is more common that matrix effects are consistently observed between an analytical system and a processed serum-based material, although the magnitude of the matrix bias may be unique to each lot of materials or reagents.

The matrix of a specimen or material has had various definitions in different contexts. The National Committee for Clinical Laboratory Standards (NCCLS) has defined matrix as the "milieu of the analyte"[4] and is considering a more specific definition in the context reviewed here.[5] The American Society for Testing and Materials defines matrix as "the principle element or elements in a sample" and defines an interference as "an effect due to the presence of a constituent or characteristic that influences the measurement of another constituent or characteristic."[6] Rej has provided a comprehensive definition of matrix as "the entire milieu or environment in which an analyte resides and all components (other than the analyte) or attributes."[7]

With these broad definitions, a matrix interference or effect could be ascribed to any processing induced or physiologically derived component of the material. Rej cautions that "matrix effects are often ascribed to interferences due to unknown or uncharacterized substance(s) or factor(s) in the biological specimen."[7] He prefers to consider matrix interferences as the result of lack of analytical specificity for the analyte, but notes that at some point a processed material may "differ so fundamentally from the normally encountered clinical specimen that the specificity of even the most robust method is unfairly challenged".[7]

In the context of this chapter, matrix effects are considered to be analytical interferences caused by alterations in the human serum material induced by preparation and processing of that material. Processing includes human and nonhuman additives to the material as well as any physical or chemical changes that occur as a result of the preparation or manufacturing procedure. Matrix effects in this context do not include interferences from physiological or drug-induced metabolic substances present in normal or abnormal concentrations. However, such physiological interferences are significant issues and represent analytical challenges that must be solved by acceptable field methods. Matrix effects may be caused by undefined substances in the material, but this phenomenon should not be used to arbitrarily dismiss inadequate methodological specificity for the analyte of interest.

Although commutability between pooled-serum materials and fresh individual specimens is desirable, it is not necessary for all aspects of calibration. Secondary standard calibration materials used by a manufacturer solely to calibrate a unique analytical system do not need to be commutable with any other analytical system nor with patient specimens. In this case, any matrix bias in the calibration materials can be accommodated by assignment of the nominal analyte value in such a way that the analytical system's response is set to produce accurate results for individual patient specimens. The process for value assignment of the secondary standard and verification of accurate results for those specimens must be linked to a validated reference system such as the National Reference System for the Clinical Laboratory (NRSCL) of the NCCLS (see Chapter 11).

PREPARATION OF COMMERCIAL MATERIALS FOR CALIBRATION AND PROFICIENCY TESTING

Serum-based reference materials undergo a substantial number of processes during their manufacture that can change the matrix from that of native serum.[8-11]

The starting material is commonly plasma collected at a commercial donor center. The plasma may be stored in liquid form or, more typically, frozen for several months prior to use. Off-the-clot serum can be used to obtain the base material, but it is considerably more expensive than plasma due to the time delay for clotting and its special handling requirements. The plasma is defibrinated and converted to serum by addition of calcium and possibly clotting agents. The recovered serum may undergo various additional processes such as filtration, dialysis, ion exchange, reverse osmosis, charcoal adsorption, or lipid stripping[12] to reduce endogenous constituents to low levels.

Various spiking materials are added back to the base serum pool to produce the desired composition and concentrations of analytes. Other additives may also be added to prevent microbial growth, to enhance product stability, or to enhance compatibility with an analytical system. In the case of lipid products, the common spiking materials are low-density lipoprotein (LDL) rich and high-density lipoprotein (HDL) rich fractions from human serum obtained by various modifications of the ethanol and pH precipitation procedures originally described by Cohn.[9, 13] TG-rich additives are typically obtained from hen egg yolk.[10, 11] Ultracentrifugation has also been used to obtain concentrated lipoprotein fractions from serum when smaller quantities of additive are required.[14]

Once the pool is prepared and analytes adjusted to desired concentrations, the material is filtered to 0.22 μm to remove bacteria, mixed to homogeneity, and dispensed into vials. The vials can be frozen or freeze-dried for storage and distribution. Freezing below −50° C is necessary to fully crystallize water[15] and provide maximum stability of serum materials. Freeze-drying is most commonly used because the materials can be stored, packaged, and shipped (usually at ambient temperatures) considerably more economically than when frozen.

However, the freeze-drying process has been shown to alter the physical properties of serum materials. Rej[7, 16] reported that lyophilization of an aliquot of serum produced moderate changes in osmolality, surface tension, and viscosity, and produced large changes in pH, number and size of particles, and absorbance. It is well accepted that freeze-drying produces denaturation of lipoprotein particles, which makes them difficult to resolubilize in water and increases turbidity in reconstituted specimens. Kroll et al.[17] showed that freeze-drying two serum pools decreased the recovery of TC for four routine enzymatic methods compared to freezing at −20° C or −70° C . Marcovina et al.[18] showed a systematic decrease in recovery of apolipoprotein B by seven immunological methods after lyophilization of six human serum pools. Kroll and Chesler[19] further reported that five commercial and two in-house lyophilized pooled serum materials, including NIST SRM 909, had decreased recovery of TC with the duPont aca due to a decreased reaction rate for the enzymatic hydrolysis of TC esters in these materials. The reaction time allowed by the analyzer was adequate for fresh human specimens but too short for complete hydrolysis of the matrix-altered lipoproteins in the lyophilized materials. Noel et al.[20] and Wiebe and Bernert[21] reported that cholesteryl esterases used in some enzymatic reagents did not completely hydrolyze cholesterol arachidonate present in some processed serum materials.

IDENTIFICATION OF MATRIX LIMITATIONS IN MATERIALS

Ultimately, commutability of a material and method must be evaluated by comparison of a material's results to those for individual patient specimens. Each individual's serum has variability in many components, including the distribution of lipoproteins. This variability becomes greater when sera from diseased persons are included to evaluate the full analytical range for a method. Thus, a group of freshly collected patient sera will have physiological differences in composition, and these sera may exhibit some variability in chemical reactivity when assayed by a routine method with inadequate chemical specificity. When comparing the analytical performance of clinical specimens between routine and reference methods, any observations of "outliers" must be evaluated carefully, as they may represent methodological limitations of the routine method when used with patient specimens. Because of inherent physiological variability and the analytical imprecision of measurement, an adequate number of representative freshly collected patient specimens is essential for evaluation of a method's performance.

Any inference about the presence of matrix bias in a processed material must be based on a statistically sound evaluation protocol. The CAP has developed a protocol to detect the presence of matrix effects in processed materials when a reliable accuracy comparison method is available.[22] Briefly, the protocol selects 20 or more freshly collected individual human specimens and assays them along with the processed materials under evaluation using the test method and a suitable comparison method. The comparison method should be the most robust method available, ideally a NRSCL-credentialed reference method, such as Abell-Kendall for TC, or the best-characterized consensus method available for the analyte, such as a Centers for Disease Control and Prevention (CDC) standardized laboratory method for HDL-cholesterol (HDL-C) or TG.

Regression analysis on the patients' results is used to establish a 95% prediction interval, and the processed materials are evaluated for conformance to that interval. This protocol identifies commutability of a material but does not specifically quantitate the magnitude of any matrix bias that may be present. When used with a chemically specific reference method, this procedure does validate the accuracy of a routine method for patient specimens and determines whether processed materials can be used for assessment of the method's accuracy. When used with a non-credentialed comparison method, this procedure can only confirm commutability of a material between the two methods. The NCCLS is currently developing a similar protocol for determining the presence of matrix effects in materials.

Rej et al.[7] have described use of correspondence analysis[23] as a statistical tool to evaluate the commutability of processed materials with authentic clinical specimens. This technique does not require an accuracy-based reference method and does not quantitate matrix bias or calibration bias. The approach is very useful for evaluating commutability among materials by simultaneously comparing reactivity characteristics among several materials, including authentic clinical specimens, assayed by several analytical methods. This procedure uses a multivariate statistical analysis and produces a two-dimensional projection of the most significant analytical components. Materials and methods that have commutable reactivity characteristics cluster in the same region as patient specimens, while materials and methods with matrix interactions are located elsewhere.

EXAMPLES OF MATRIX LIMITATIONS

A CAP-sponsored conference in 1992 titled "Matrix Effects and Accuracy Assessment in Clinical Chemistry"[24] reviewed the state of the art in understanding and controlling matrix effects and developed recommendations for future improvement of accuracy transfer and assessment. At this conference, Naito et al.[25] demonstrated the possible errors in interpretation of proficiency testing results for TC with processed serum materials. They described the Veterans Administration National Center for Laboratory Accuracy Standardization (VANCLAS) program for TC. Target values for the materials used were set by the CDC using the Abell-Kendall method. This standardization program included 174 VA medical centers and outpatient clinics, each of which received six fresh, non-frozen, non-supplemented, pooled human serum samples and five freeze-dried, human Cohn fraction supplemented, pooled human serum based materials as used in the 1990 CAP proficiency testing program. The fresh, non-frozen serum pools were assumed to be free of any matrix interactions with the routine methods used by the participants.

Based on the fresh serum results, 112 laboratories (63%) met the ± 3% analytical bias specifications of the National Cholesterol Education Program (NCEP), and their results were used to evaluate matrix biases in the freeze-dried processed materials. In the laboratories that met the NCEP bias goal, six of eight routine methods had interactions with the matrix-modified processed materials which resulted in matrix-induced apparent biases that ranged from −8.9% to +4.4%. Methods showing no significant matrix bias with the proficiency testing materials were the Hitachi/Boehringer Mannheim (Boehringer Mannheim Corporation, Indianapolis, IN) and Abbott Spectrum EPX (Abbott Diagnostics, Abbott Park, IL). Methods showing an apparent negative bias with proficiency testing materials were the duPont Dimension (E. I. duPont de Nemours and Company, Wilmington, DE), Beckman CX4,5,7 (Beckman Instruments, Brea, CA), IL Monarch (Instrumentation Laboratories, Lexington, MA), and Baxter Paramax (Baxter Diagnostics, McGaw Park, IL). Methods showing an apparent positive matrix bias with the proficiency testing materials were the Kodak Ektachem (Eastman Kodak Company, Rochester, NY) and Technicon SMAC and RA (Miles, Tarrytown, NY). Figure 10–1, reprinted from Natio et al.,[25] illustrates the insidious nature of matrix bias using the duPont Dimension as an example. In this case, although results for patient specimens were correct, the proficiency testing results incorrectly indicated an apparent bias in results. The results from proficiency testing materials could not be used to judge the performance of a laboratory for patient specimens.

At the CAP conference, Ross et al.[26] reported results from an extensive evaluation of matrix effects in 37 routine method peer groups in the 1989 CAP proficiency testing program. A sample of 915 participants out of approximately 5,000 assayed a single, freshly collected, off-the-clot, non-supplemented, twice-

Figure 10–1 ❖ *Example of Matrix Effects between duPont Dimension and CAP Proficiency Testing Materials*

The ideal line is the line of identity. The patient-regression line represents mean values of triplicate measurements on each of the five fresh human specimens plotted against CDC reference-method values (duplicate measurements). The CAP-regression line represents triplicate measurements on each of the five 1990 CAP Comprehensive Chemistry Survey specimens plotted against the CDC reference-method values. Data also indicate the laboratory's overall precision, mean bias on measurement of fresh human materials, and mean bias on measurement of CAP Survey materials. CV = Coefficient of variation. *Source:* Naito et al.[25] © Amer. Medical Assn. Reprinted by permission.

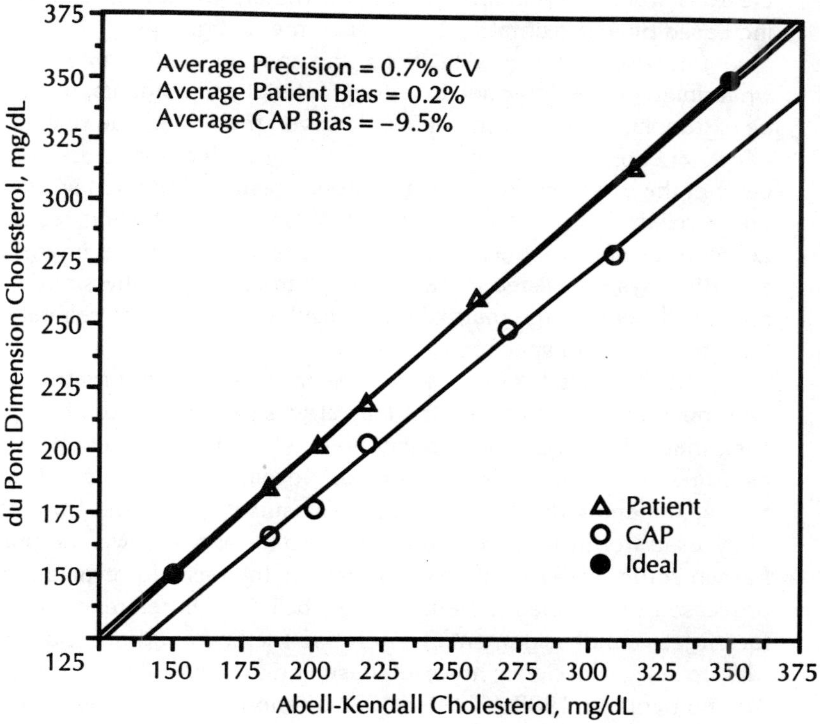

(To convert mg/dL to mmol/L, multiply by 0.0259.)

frozen, pooled serum specimen as well as the two freeze-dried, human Cohn fraction supplemented, pooled human serum based materials usually used in the proficiency testing program. The fresh-frozen material was assumed to be free of matrix interactions with the various methods. Target values were assigned to the fresh-frozen serum and the freeze-dried survey materials by both the NIST using the definitive isotope dilution mass spectrometry method and the CDC using the Abell-Kendall reference method. Method calibration bias was determined as the difference between the reported result and the target value for the fresh-frozen specimen. Method matrix bias was determined as the mean difference between the reported result for one freeze-dried survey material and its target value minus the calibration bias. The freeze-dried proficiency testing material chosen to determine method matrix bias had a reference method TC of 204 mg/dL (5.29 mmol/L), which closely matched the fresh-frozen serum value of 198 mg/dL (5.11 mmol/L).

The participants were divided into 37 unique method groups based on instrumentation and reagent source. Table 10–1, which is adapted from Ross et al.,[26] indicates that 70% of the method groups exhibited a significant matrix bias with the freeze-dried survey material. The data in Table 10–1 used the NIST definitive method as the basis for accuracy evaluation. At that time, the definitive method results were on average 1.6% lower than the CDC reference method.[27] Thus, if the reference method were used for evaluation, the calibration bias in Table 10–1 would be decreased by approximately 1.6 and the matrix bias would be increased by approximately 1.6 for each method group.

Because matrix modifications and specific method interactions are somewhat unique to each production lot of material and reagents, these results cannot be extrapolated to other materials. However, they clearly demonstrate that matrix effects are common and that raw proficiency testing results are not a reliable indicator of the true accuracy of a laboratory's results for patient specimens. Inappropriate reliance on and interpretation of proficiency testing results can have a detrimental effect on the accuracy of results for patient specimens. Several of the analytical systems listed in Table 10–1 had an apparently small survey bias but had a sufficiently large matrix bias to result in a significant calibration bias for the fresh-frozen serum specimen.

The CAP has taken an active role in understanding matrix bias limitations, and much of the data on matrix bias effects has come from that program. However, matrix bias is a general phenomenon that has been reported for many processed reference materials. Lasky et al.[28] evaluated the matrix bias of proficiency testing materials used in eight different survey programs from four countries. They used frozen human serum pools, some of which were supplemented with human serum cholesterol concentrates, as the basis for evaluation of the highly processed freeze-dried materials. The Abell-Kendall reference method was used for target value assignment. Ten routine methods had acceptable performance with frozen serum pools; however, based on the Clinical Laboratory Improvement Amendments of 1988 (CLIA) ± 10% criterion, nine methods would have failed proficiency testing with one or more materials.

Franzini and Luraschi[29] evaluated the commutability of 24 lyophilized commercial control materials from 11 manufacturers with 107 individual patients' sera assayed by two methods using Boehringer Mannheim enzymatic reagents. They showed that if the lyophilized materials had been used as calibrators, seven of the materials had a matrix interaction which would have introduced a mean bias of > 3% into the patients' results.

Frozen serum pools are commonly considered to be less prone to matrix modifications than the typical freeze-dried commercial serum based materials. However, frozen pools can represent a wide range of preparation protocols and have been reported to exhibit method- and material-specific matrix biases.

Table 10–1 ❖ Instrument/Reagent System Specific Mean Survey Bias, Calibration Bias, and Matrix Bias versus the Definitive Method for one 1989 CAP Cholesterol Proficiency Testing Material[a]

Instrument/Reagent System, Manufacturer	*Survey Bias (%)*	*Calibration Bias (%)*	*Matrix Bias (%)*
Am Monitor Perspective/Am Monitor	1.6	8.2	–6.5
Olympus Demand/Technicon	–1.5	7.2	–8.7
Roche Cobas Mira/Roche	2.2	6.6	–4.4
Kone Progress/Kone	1.8	6.3	–4.5
Am Monitor KDA/Am Monitor	1.7	6.3	–4.7
Baker Encore/Baker	–1.9	5.8	–7.7
Roche Cobas/Roche	0.5	5.7	–5.2
Abbott TDX/Abbott	–4.7	4.8	–9.5
Baker Centrifichem/Baker	–0.1	4.1	–4.2
duPont Dimension/duPont	–0.6	4.1	–4.6
Gilford Impact 400/Ciba	3.6	3.9	*
Boehringer Mannheim 736, 737/BMD	1.3	3.6	–2.3
Baxter Paramax/Paramax	1.9	3.4	–1.5
Technicon SMAC/Technicon	2.3	3.2	*
Instrumentation Lab Monarch/Inst Lab	–3.0	2.8	–5.8
Olympus 5000/Olympus	1.5	2.6	–1.1
Technicon 12-60/Technicon	–0.6	2.1[†]	*
Technicon RA-1000/Technicon	0.5	2.0	–1.5
Abbott Spectrum/Abbott	–0.4	1.9	–2.2
Olympus Demand/Olympus	0.7	1.7[†]	–1.0
Ciba 550 Express/Ciba	–0.9	1.5[†]	*
Technicon RA-1000/Sigma	–1.4	1.5[†]	*
Technicon Chem 1/Technicon	1.1	1.0[†]	*
Am Monitor Parallel/Am Monitor	–2.9	0.7[†]	–3.6
duPont *aca*/duPont	–11.0	0.6[†]	–11.6
Boehringer Mannheim 8700/BMD	0.1	0.5	*
Boehringer Mannheim 704, 705/BMD	0.0	0.2[†]	*
Am Monitor Parallel/Behring	–0.2	0.1[†]	*
Boehringer Mannheim 717/BMD	–0.3	–0.3[†]	*
Ektachem DT 60/Kodak	0.9	–0.9[†]	–1.7

(Continues on next page)

Table 10-1 ❖ *(Continued)*

Instrument/Reagent System, Manufacturer	*Survey Bias (%)*	*Calibration Bias (%)*	*Matrix Bias (%)*
Coulter Dacos/Coulter	–3.1	–1.3	–1.8
Beckman Synchron CX4,5/Beckman	–5.1	–1.5[†]	–3.6
Abbott VP/Abbott	–5.8	–1.5[†]	–4.4
Beckman Astra 4,8 Ideal/Beckman	–12.3	–1.5[†]	–10.9
Instrumentation Lab Multistat III/Beckman	–11.9	–1.9[†]	–10.1
Ektachem 400, 700/Kodak	–1.1	–3.2	2.1
Electronuclear Gemini/Electro	–6.6	–3.9[†]	*

*Matrix bias not significant, P <0.05.
[†]Calibration bias not significant, P < 0.05.
[a]Adapted from Table 3 in Ross et al. Arch Pathol Lab Med 1993;117:393–400.

Waymack et al.[30] evaluated eight frozen human serum pools used in the CDC Lipid Standardization Program. Some of the pools were supplemented with human Cohn fraction serum lipid concentrates. Fresh, non-frozen individual patients' sera with Abell-Kendall value assignment were used to establish assay performance. Five of the eight frozen pools had a matrix bias that was not a function of the reagent used but of unique instrument assay parameters.

Myers et al.[31] reported that three CDC Cohn fraction cholesterol supplemented, frozen human serum pools showed significant matrix bias with two of 16 routine methods tested, and three commercial liquid stabilized human serum pools showed matrix bias with nine of the 16 methods. For this study, non-supplemented, fresh-frozen serum pools were used as the controls for no matrix effects.

Miller et al.[32] evaluated matrix bias in pooled serum materials. Two non-frozen pools were prepared from two sets of eight freshly collected, off-the-clot individual patients' sera and tested with four routine methods. These non-frozen serum pools had no matrix effects as assessed by TC results which were the same as the mean of the individual sera making up the pools. Four frozen serum pools were prepared from a different set of off-the-clot, non-supplemented human serum which was pooled in 2–3 liter bottles, frozen in bulk, thawed, remixed, aliquoted, and re-frozen. One to four of these frozen human serum pools showed significant matrix bias as compared to non-frozen individual donor sera for each of four routine methods.

Tetrault et al.[14] prepared three frozen human serum pools from off-the-clot serum which was frozen, thawed, pooled, aliquoted, and re-frozen. One of the pools was supplemented with an ultracentrifugally concentrated human serum LDL fraction from outdated blood bank plasma. Matrix bias was determined as the difference in apparent bias indicated by the mean of 16 fresh non-frozen patient sera versus the frozen pooled materials. All concentration values were assigned by the CDC reference methods for TC and HDL-C. One or more of the frozen human serum pools showed matrix bias for four of seven routine methods for TC and for six of seven routine methods for HDL-C.

Holani and Miller[33] used the CAP matrix effect evaluation protocol[22] to test five CAP processed freeze-dried serum pools and nine CDC supplemented frozen serum pools using three commercial reagents for TG measurement. Twenty-four freshly collected individual sera were used as controls for no matrix bias. All specimens were assayed by the CDC chromotropic acid TG reference method and by each glycerol blank corrected commercial reagent system adapted to a Cobas-BIO analyzer (Roche Instruments, Nutley, NJ). Results of two of the commercial reagents for both CAP and CDC proficiency testing materials were commutable with fresh sera. Results of the third reagent system were commutable with the CDC frozen pools but had a significant matrix bias for four out of five CAP freeze-dried materials.

Substantial progress has been made toward identification of commutable materials for standardization of apolipoprotein AI (apo AI) and apolipoprotein B (apo B). All routine methods are based on immunoassay, and there are no established reference methods for either of these analytes. Consequently, efforts have focused on identification of reference materials that give equivalent results with the variety of methods in common use.

Smith et al.[34] reported an international investigation of four non-supplemented lyophilized human serum pools by eight method principles in 140 laboratories. They found that some lyophilized materials had equivalent mean values for apo AI among the method groups, but all materials had significantly different apparent mean values for apo B assays.

Marcovina et al.[18] prepared three non-supplemented human serum pools for apo AI and six such pools for apo B. Half of the vials were stored frozen at −78° C and the other half were lyophilized. Each material was assayed by five method principles with a common frozen serum pool used to standardize calibration of each method. Apo AI values were the same for frozen and lyophilized pools. Apo B values for the lyophilized pools differed from those of the frozen materials by −26% to +4% depending on the assay method.

Marcovina et al.[35] reported results from an investigation of reference materials for calibration of apolipoproteins AI and B which was conducted by the International Federation of Clinical Chemistry Committee on Apolipoprotein Standardization. The inter-method commutability of 15 candidate reference materials for apo AI and 11 materials for apo B were evaluated by 26 and 28 methods, respectively, in 28 laboratories in nine countries. Three non-supplemented, fresh-frozen human serum pools were used to standardize calibration between the laboratories. Ten fresh-frozen individual donor sera were used to confirm standardization with an inter-method coefficient of variance (CV) of 5% for apo AI and 6.3% for apo B. For the candidate materials, inter-method CVs for apo AI ranged from 4.4% to 80.6% with four of 15 materials having an inter-method CV $< 5\%$. For apo B, inter-method CVs ranged from 6.0% to 35.1% with three of 11 materials $\leq 7\%$. Only three materials for each apolipoprotein had no turbidity and nearly normal lipoprotein patterns on electrophoresis. The most suitable candidate reference materials for apo AI were lyophilized and those for apo B were liquid stabilized.

Subsequent work by Albers et al.[36] utilized linearity and parallelism evaluation to identify one material for apo AI (inter-method CV = 3%) and one for apo B (inter-method CV = 6%) which were sufficiently commutable for use as calibration reference materials.

Marcovina et al.[37] reported use of the apo AI reference material to standardize calibration of 30 routine methods involving 23 manufacturers. Assays of 50 fresh-frozen individual donor sera by each participant resulted in an inter-laboratory CV for each sample ranging from 2.1% to 5.6%. This material has been approved as Apolipoprotein A-I International Reference Material by the World Health Organization. Credentialing of the apo B reference material awaits publication.

PRACTICAL IMPACT OF MATRIX LIMITATIONS ON CALIBRATION AND PROFICIENCY TESTING

Accuracy Verification

There are no currently available reference materials that can be used as universal calibrators or to verify calibration of all analytical systems for TC, HDL-C, or TG. The National Cholesterol Reference Method Network, described in Chapter 11, provides a mechanism for manufacturers and individual laboratories to establish and/or verify calibration for TC based on split assay of fresh sera. The Reference Method Network will soon have available reference equivalent methods for HDL-C and TG. Analytical system and reagent manufacturers use the Reference Method Network to establish the calibration traceability of their specific methodology to the accuracy base of the national reference methods. The calibration material provided by an analytical system manufacturer is uniquely formulated to provide correct calibration for patient specimens assayed by that specific method. Thus, one cannot use calibration materials from one manufacturer with a reagent system from a different manufacturer unless that specific combination has been validated using the National Reference Laboratory protocol. The validation can be performed by a material vendor or by the individual laboratory.

TC calibration can be accomplished by manufacturers as described above. However, this calibration must be transferred to hundreds of laboratory locations, which is a substantial analytical challenge.[38, 39] Manufacturing of equipment and reagents is performed to tolerances defined to allow the total error of the final analytical system to meet medical requirements at an acceptable cost. Manufacturing tolerances result in some degree of variability among individual instruments, lots of reagents, and lots of calibrators. Some variables, such as shipping and storage conditions, operating environment conditions, and equipment maintenance, are beyond the control of manufacturers yet can affect the accuracy of results and must be accommodated in the total analytical system design.

The success of TC accuracy transfer to routine laboratories has been investigated. Naito et al.[25] used six non-supplemented, fresh, non-frozen serum pools to evaluate TC performance in 174 Veterans Administration Hospitals and reported that TC calibration bias was ≤ 3% of the CDC reference method for 63% of the laboratories. Ross et al.[26] used a non-supplemented, fresh-frozen serum pool to evaluate TC performance in the CAP proficiency testing program and found that TC mean calibration bias was ≤ 3% of the CDC reference method for 65% of 37 methods.

Although TC is well standardized by most manufacturers, because of the number of variables unique to each testing location, each laboratory must validate calibration on a regular basis. HDL-C and TG are not yet standardized using the National Reference Laboratory approach, and calibration must be established and verified by each individual laboratory. LDL-C is commonly calculated from TC, HDL-C, and TG using the Friedewald equation.[40] The new direct assays for LDL-C will require calibration verification analogous to those for HDL-C.

Because of the matrix limitations of processed materials, primary accuracy verification for all these analytes will continue to be accomplished by splitting fresh, non-frozen individual patients' sera with a National Reference Laboratory for TC and with a CDC standardized laboratory for HDL-C, LDL-C, and TG. Processed materials with National Reference System assigned values can be used for accuracy verification in special cases when an analytical method has been shown to be free of matrix bias for that specific material.

Monitoring Relative Accuracy within a Laboratory

Accuracy verification using sera split with a certified reference laboratory is time consuming and expensive and requires several weeks to obtain results. Thus, it is recommended that each laboratory identify stable reference materials which are value assigned by the individual laboratory at the same time that absolute accuracy is verified with the patient comparison technique. Figure 10–2 illustrates this procedure for monitoring relative accuracy. The target value assignment for the reference material compensates for any matrix bias that may be present with the particular lot of reagents in use.

These in-house value-assigned materials can be assayed periodically to verify that accuracy has been maintained relative to the time when the absolute accuracy was validated. In many cases, a well-designed process-control (quality-control) system will provide this function. The key operational requirement for such materials is that they not have reactive characteristics that change with new lots of reagents or other components of the analytical system. The materials selected can exhibit matrix effects with the analytical system as long as the matrix interaction is constant and predictable between reagent lots or other assay system variables that the in-house reference material must control. Such properties will permit these materials to be used to validate accuracy following changes in lots of reagent or manufacturer's calibrators or after a major instrument component is changed.

Some materials that can be useful for monitoring relative accuracy within an individual laboratory are listed in Table 10–2 in approximate order of matrix modification during preparation. Non-supplemented, fresh-frozen serum pools can be prepared in-house or can be obtained from Pacific Biometrics Research Foundation (1100 Eastlake Avenue, Seattle, WA 98109) and from Solomon Park Research Laboratories (12815 N.E. 124th Street, Suite 1, Kirkland, WA 98034). The CAP (325 Waukegan Road, Northfield, IL 60093-2750) excess survey materi-

Figure 10–2 ❖ Schematic of Calibration Adjustment based on Individual Patients' Sera and Matrix Compensated Value Assignment of a Reference Material for Subsequent Verification of Field (Routine) Method Accuracy

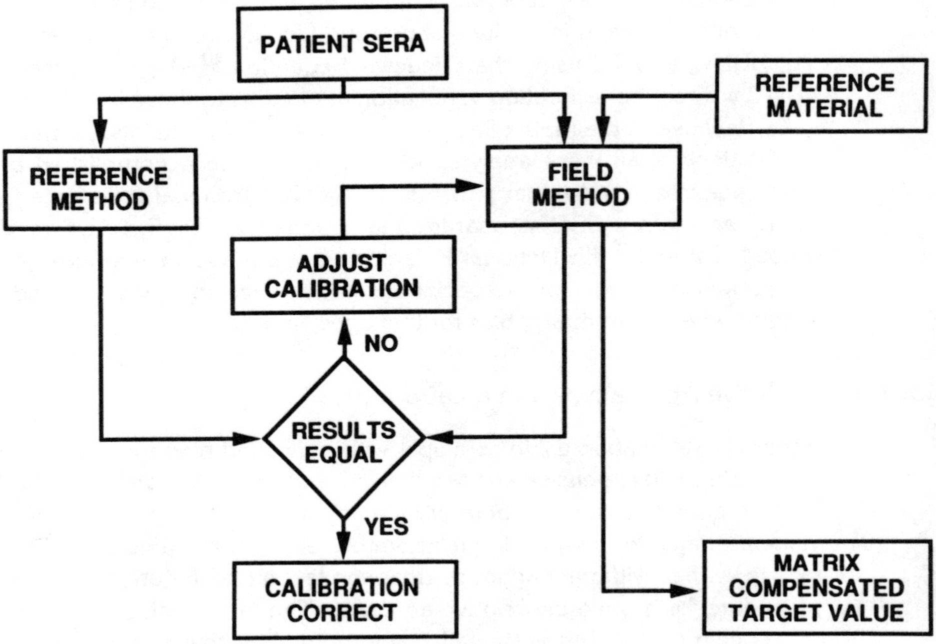

als are supplemented, freeze-dried pooled serum materials which have been assayed by thousands of laboratories and have statistically reliable peer-group consensus mean values. The peer-group value assignment of the CAP materials circumvents the matrix issue but does not provide a direct verification of accuracy for an individual peer group.

The CAP is actively developing lipid materials with improved commutability with fresh human serum. NIST SRM 909a (United States Department of Com-

Table 10–2 ❖ Materials Useful to Monitor Relative Accuracy
● Frozen (–70° C) aliquots of a single donor serum.
● Non-supplemented frozen (–70° C) serum pools.
● Supplemented frozen (–70° C) serum pools.
● College of American Pathologists survey validated materials.
● NIST SRM 909a or 909b, SRM 1951, SRM 1952a.
● Commercial lipid control materials.

merce, National Institute of Standards and Technology, Gaithersburg, MD 20899) is a two-concentration set of supplemented, freeze-dried pooled serum materials that have definitive and reference method values assigned for TC. NIST SRM 1951 (currently sold out) is a set of three non-supplemented, frozen human serum pools with definitive and reference method TC values and reference method TG values. NIST 1952a is a set of three supplemented, freeze-dried human serum pools with definitive method TC values. These CAP- and NIST-processed serum based materials do exhibit matrix bias with various methods and must be used in conjunction with appropriate accuracy validation based on fresh sera.

The in-house value assignment of these reference materials must be done during a period when the analytical system is working well as evidenced from daily process quality control, and when the method's analytical calibration is verified by a split patient correlation with a certified reference laboratory. The materials selected must be assayed a sufficient number of times to have a statistically useful confidence interval (CI) for the mean value and to represent expected analytical imprecision over an extended period of time. Multiple analytical runs on multiple days must be used, ideally with more than one lot of reagents and several calibration events. As an example, Table 10-3 shows the CI for the mean of various numbers of replicates (N) for a mean TC of 200 mg/dL (5.17 mmol/L) with a method CV of 2.5%. The CI indicates the concentration range within which the mean is likely to be found with a defined probability. For example, with 5 replicates, under the conditions above, the mean of 200 mg/dL (5.17 mmol/L) has a 0.95 probability of being in the interval 194–206 mg/dL (5.02–5.32 mmol/L) and a 0.90 probability of being in the interval 195–205 mg/dL (5.05–5.30 mmol/L). The practical implication is that adequate replication is necessary when assigning a target value to a reference material. A correlate is that a more reliable calibration of a method will be achieved when the standard materials are assayed in replicate.

Apolipoprotein assays present a different challenge to control for matrix interactions between different lots of reagents and calibrators and inter-instrument analytical variability. As discussed earlier, there are substantial method/material matrix interactions with apolipoprotein assays. However, there are no credentialed reference methods with well-established performance characteristics. Thus, the method of accuracy transfer or validation based on splitting fresh human sera is not possible. Standardization of accuracy is being accomplished by sharing commutable serum reference materials validated by the International Federation of Clinical Chemistry among analytical method manufacturers. Thus, the state of the art depends on the continued commutability of the reference materials and the ability of manufacturers to transfer that accuracy to the secondary calibration materials provided with the assay systems. It would be prudent for individual laboratories to maintain in-house value-assigned frozen serum aliquots as described above for use in validating relative accuracy when introducing new lots of reagents or manufacturer's calibrators or after major maintenance or repair to the analytical equipment.

Table 10-3 ❖ Confidence Intervals (CI) for Various Numbers of Replicates (N)[a]

	For 200 mg/dL (5.17 mmol/L and method CV = 2.5%	
N	*95% CI*	*90% CI*
2	± 45 mg/dL (1.17 mmol/L)	± 22 mg/dL (0.57 mmol/L)
3	± 12 mg/dL (0.31 mmol/L)	± 8.4 mg/dL (0.22 mmol/L)
5	± 5.7 mg/dL (1.15 mmol/L)	± 4.8 mg/dL (0.12 mmol/L)
10	± 3.6 mg/dL (0.09 mmol/L)	± 2.9 mg/dL (0.08 mmol/L)
20	± 2.3 mg/dL (0.06 mmol/L)	± 1.9 mg/dL (0.05 mmol/L)

[a]Calculated as: $CI = \pm (t_{N-1, (1-a)/2}) * (s^2/N)^{1/2}$ where t is the Student's t value for $N - 1$ degrees of freedom and (1 – a)/2 significance level, and s is the standard deviation for the assay.

Proficiency Testing

Proficiency testing (PT) materials in common use exhibit sufficiently widespread matrix interactions that PT results cannot be used to evaluate a laboratory's ability to produce accurate lipid results for patient specimens. Ross et al.[26] reported that 70% of the enzymatic methods in use for TC exhibited non-specific matrix interaction with one or more materials used in the 1989 CAP survey program. Almost all providers of PT in the United States are currently evaluating laboratory lipid assays by comparison of results to the peer-group mean. Peer-group evaluation compares an individual laboratory result to the mean of all laboratories using the same method. Variability within a peer group can be caused by an individual laboratory's skill at applying the method and by calibration variability in the manufacturer's accuracy transfer process. Passing peer-group evaluation should not be affected by the presence of matrix effects since the method/material-specific interactions are a property of the assay system and should be the same for all participants. However, there can be variable matrix bias within a homogeneous peer group if manufacturing variability between lots produces changes in the matrix interaction with a PT material.

Some of the newer compact-analyzer methods that use dry-format reagents and membrane and filtration technology may be particularly sensitive to lot-specific material interactions. Variable matrix bias can also be present within a peer group composed of hybrid analytical systems which, for example, use reagents or calibration materials from different vendors. Another limitation of peer-group evaluation is that a statistically valid mean cannot be derived when methods are subdivided into very small groups or when new methods are employed by very few users. In these cases, either no evaluation is possible or evaluation can be performed versus some other target value, which can introduce a significant matrix bias.

A peer group evaluation protocol allows individual laboratories to ascertain whether they are applying a particular methodology in the same manner as other laboratories. However, the inability to compare results between methods precludes determining whether the accuracy necessary for use of the NCEP clinical decision guidelines is being achieved successfully in a routine laboratory. Recent literature reports[25, 26] suggest that most routine laboratories can provide satisfactory lipid results by using methods that have been validated by the manufacturer through the National Reference Method Network program. However, a substantial number of laboratories have not achieved the performance goals recommended by the NCEP. Laboratories that desire to maintain the highest levels of accuracy will need to validate performance in-house by splitting individual patient specimens with a certified reference laboratory. When necessary, the manufacturer's calibration will need to be adjusted to achieve accurate results for patient sera.

Laboratories that find it necessary to adjust the manufacturer's calibration settings will need to report PT results in an appropriate way to permit useful evaluation. The method/material matrix interaction is a function of the methodology and is unaffected by calibration adjustments. However, the calibration offset will invalidate comparison of PT results to the peer group mean. One way to address this issue is to report the method as "Other," reflecting that the method has been modified in-house. This may not be satisfactory if the proficiency testing provider evaluates "Other" method groups against a reference method assigned target value, the all-method mean, or some other target value derived from the participants' results. In these cases, the potential matrix-induced biases with the proficiency testing materials will make reliable evaluation impossible. The "Other" category may be appropriate for these methods if they have been optimized to be sufficiently robust to eliminate matrix interactions with the PT materials and if evaluation is made against an accurate target value. Another option is to report proficiency testing results after removing adjustment factors to revert to the manufacturer's original calibration settings. In this way, PT results can be compared to an appropriate peer-group target value to evaluate the laboratory's ability to correctly use the analytical system. It is also necessary to provide separate documentation of the laboratory's calibration and accuracy verification process based on splitting patient specimens with a certified reference laboratory.

The CLIA regulations state, "The laboratory must examine or test, as applicable, the proficiency testing samples it receives from the proficiency testing program in the same manner as it tests patient specimens."[41] One interpretation of this regulation suggests that the same calibration settings should be used for proficiency testing samples as are used for patient testing. If this interpretation is followed, calibration adjustment to produce accurate results for individual patients cannot be performed when necessary. The laboratory will pass PT based on peer group evaluation but will have no independent verification of accuracy for individual patient specimens.

Another interpretation of the CLIA regulations requires use of calibration settings appropriate for the type of specimen being tested. This interpretation allows calibration adjustment when necessary to produce accurate results for patient specimens and allows readjustment to manufacturer's settings for PT specimens. A competent laboratory will assay all specimens with method calibrations appropriate to those specimens. Clinical specimens with unique fluid matrix properties, such as pleural fluid or spinal fluid, are frequently assayed using the same methods as serum, but with calibration settings adjusted to obtain results that can be properly interpreted. The requirement to test proficiency testing samples "in the same manner" as patient specimens means that any special requirements of the PT specimen must be properly accommodated in the testing process. Good laboratory practice requires use of calibration settings appropriate for a PT specimen with its unique matrix requirements in order to obtain a result that can be correctly interpreted. Readjustment of PT results to original manufacturer's calibration settings is the only technique that permits appropriate evaluation within a peer group.

The specialty program offered by Pacific Biometrics Research Foundation offers an option for PT of lipid methods. This program uses freshly collected, non-supplemented, non-frozen pooled human serum specimens that are likely to have minimal method/material matrix interactions. Target values are assigned by certified reference methods. This approach offers the possibility for reliable evaluation of accuracy. Detailed evaluation of the effectiveness of this PT program awaits publication.

FUTURE DIRECTIONS IN CONTROLLING MATRIX LIMITATIONS

Current laboratory practice requires large quantities of secondary standards to perform daily or periodic method calibration in routine laboratories, as well as large quantities of proficiency testing materials to independently verify accuracy in individual laboratories. For example, a single mailing of five 5 mL vials to 10,000 laboratories requires 250 liters of pooled serum. Constraints of storage and shipping stability and costs make freeze-dried materials the only currently practical solution for widely used reference materials. Commutability of reference materials with patient specimens is desirable, and substantial effort is needed to develop materials with optimal properties and minimal residual matrix biases. Realistically, some degree of matrix modification of serum reference materials is unavoidable. An essential property of the manufacturing process must be to produce a material with constant and predictable matrix interactions with most or all methods in routine use.

In addition to processed reference materials with consistently predictable matrix modifications, analytical methods must be manufactured with robust assay characteristics to permit use of commercially available reference materials. Good laboratory practice requires standardized accurate results to meet national recommendations for medical diagnosis and therapeutic intervention for lipid disorders. Current regulations require that accuracy be independently verified at each laboratory testing location using a PT program. Consequently, all analytical methods must be designed to recover analyte from the usual patient specimens as well as from currently available reference materials. New measurement technology must address the PT requirement as an integral part of the assay system. Robust assay characteristics must be uniformly manufactured into subsequent lots of reagents, calibrators, and other analytical components of the measurement system.

Peer-group evaluation does not assess accuracy. Thus, improved technology and new evaluation procedures are necessary to meet clinical requirements.

It would be possible to evaluate accuracy in the presence of matrix bias if the method/material-specific matrix bias could be measured and were constant and predictable among all laboratories using a method. Longitudinal evaluation would be possible if subsequent lots of materials and method components maintained predictable method/material matrix interactions. Ross et al.[26] tested this concept for evaluation of TC proficiency testing in the 1989 CAP program discussed earlier. A matrix bias correction factor was calculated from the measured calibration and matrix biases and applied to the survey results for three mailings which shared the same freeze-dried material. The aggregate results indicated that after correction for matrix bias, 41% of the 37 method groups had a calibration bias that exceeded the NCEP ± 3% limit as compared to the CDC reference method. However, most methods had such excellent precision that 92–93% of methods were within a ± 8.9% total error range based on current NCEP laboratory performance goals.[42] If the proficiency testing results were not corrected for matrix bias, only 85–86% of results would be within the acceptable total error range.

Matrix bias correction factors are expensive to determine because of the requirement for a commutable serum material assayed by a statistically reliable sample of routine method groups and credentialed reference methods. Thus, only periodic re-validation of the correction factors would be economically feasible. Successful application of a matrix bias correction technique requires development of proficiency testing materials and assay systems with method/material matrix interactions that are predictable between lots of materials and lots of method reagents/components over extended time periods and among all method locations. In addition, a suitable number of credentialed reference methods and a network of reference laboratories in the clinical chemistry community are needed to provide the accuracy base to verify routine methods. The correction factor approach offers promise as an effective tool to allow proficiency testing to evaluate accuracy for routine laboratory methods and to allow proficiency testing results to be used for standardization of patients' results among various methods. The CAP is further developing the matrix bias correction tool in the 1994 survey program and extending its application to other lipid and common analytes.

The NCCLS Subcommittee on Cholesterol Materials is developing specifications for preparation of a fresh-frozen serum pool with the goal of minimizing matrix modifications. The protocol will test the pool and the individual serum units that make up the pool to determine whether pooling and freezing can induce method/material matrix interactions. The objective of the development protocol is to determine manufacturing specifications for a highly commutable reference material. Frozen serum pools are probably too expensive to manufacture and distribute to meet the need for routine calibration or proficiency testing products. However, such materials would be very effective to use for periodic accuracy verification, for accuracy transfer at the manufacturing level, to establish calibration for a reference method network, or to provide a material to establish matrix correction factors for other, more economical mass-produced reference materials.

The matrix modifications present in current processed serum materials require expensive and time-consuming techniques to provide reliable accuracy transfer and validation for routine lipid testing. Current matrix limitations in proficiency testing materials preclude using these programs to evaluate accuracy for routine methods. Reference materials must have minimal but constant and predictable matrix modifications. All assay methods must be chemically robust to recover analyte from clinical specimens and from available reference materials. Credentialed reference methods must be available to the clinical chemistry community to provide an accuracy base for calibration and PT. Cooperative development among providers of reference materials, analytical methods, and PT programs will reduce the limitations of method/material matrix interactions on accuracy and standardization in clinical laboratory measurements.

REFERENCES

1. Koch DD, Hassemer DJ, Wiebe DA, Laessig RH. Testing cholesterol accuracy. Performance of several common laboratory instruments. JAMA 1988;260:2552–7.
2. Fasce CF, Rej R, Copeland WH, Vanderlinde, RE. A discussion of enzyme reference materials: applications and specifications. Clin Chem 1973;19:5–9.
3. Miller, WG, Kaufman H. College of American Pathologists Conference XXIII. Introduction: matrix effects and accuracy assessment in clinical chemistry: Arch Pathol Lab Med 1993;117:343–4.
4. National Committee for Clinical Laboratory Standards. Nomenclature and definitions for use in NRSCL and other NCCLS documents: proposed guidelines, 2d ed. NCCLS Publication NRSCL8-P2. NCCLS, 771 E. Lancaster Ave., Villanova, PA 19085, 1993.
5. National Committee for Clinical Laboratory Standards. Identification and quantification of matrix effects. NCCLS Publication EP14-P. NCCLS, 771 E. Lancaster Ave., Villanova, PA 19085, in preparation.
6. ASTM Committee on Terminology. Compilation of ASTM standard definitions. Philadelphia, PA: American Society for Testing and Materials, 1990.
7. Rej R. Accurate enzyme activity measurements: two decades of development in the commutability of enzyme quality control materials. Arch Pathol Lab Med 1993;117:352–64.
8. Posner A. Problems in formulating method insensitive proficiency testing materials. Arch Pathol Lab Med 1993;117:422–4.
9. Kuchmak M, Hazlehurst JS, Olansky AS, Taylor L. Reference sera with graded levels of high density lipoprotein TC. Clin Chim Acta 1984;144:237–43.
10. Williams JH, Taylor L, Kuchmak M, Witter RF. Preparation of hypercholesterolemic and/or hypertriglyceridemic sera for lipid determinations. Clin Chim Acta 1970; 28:247–53.
11. Kuchmak M, Taylor L, Williams JH. Preparation of reference sera with desired levels of TC and TG. Clin Chim Acta 1981;114:127–35.
12. Kuchmak M, Taylor L, Olansky AS. Low lipid level reference sera with human serum matrix. Clin Chim Acta 1981;116:125–30.
13. Cohn EJ, Strong LE, Hughes WL Jr, et al. Preparation and properties of serum and plasma proteins IV: a system for the separation into fractions of the protein and lipoprotein components of biological tissues and fluids. J Am Chem Soc 1946;68:459–75.
14. Tetrault GA, Miller WG, Chinchilli VM, and others. Regional interlaboratory standardization of determination of cholesterol, high-density lipoprotein cholesterol, and triglycerides. Clin Chem 1990;36:145–9.

15. Van Brunt N, Egensperger H, Greenberg N, Halstead R, Cassano T, Doptis P. Serum based quality control fluids and calibrators: is there an advantage of lyophilization over vacuum drying? [abstract]. Clin Chem 1989;35:1101.
16. Rej R, Jenny RW, Bretaudiere JP. Quality control in clinical chemistry: characterization of reference materials. Talanta 1984;31:851–62.
17. Kroll MH, Chesler R, Elin RJ. Effect of lyophilization of results of five enzymatic methods for cholesterol. Clin Chem 1989;35:1523–6.
18. Marcovina SM, Adolphson JL, Parlavecchia M, Albers JJ. Effects of lyophilization of serum on the measurement of apolipoproteins A-1 and B. Clin Chem 1990; 36:366–9.
19. Kroll MH, Chesler R. Effect of serum lyophilization on the rate constants of enzymatic methods for measuring cholesterol. Clin Chem 1990;36:534–7.
20. Noel SP, Dupras R, Filion AM. The activity of cholesterol ester hydrolase in the enzymatic determination of cholesterol: comparison of five enzymes obtained commercially. Analytical Biochemistry 1983;129:464–71.
21. Wiebe DA, Bernert JT. Influence of incomplete cholesteryl ester hydrolysis on enzymic measurements of cholesterol. Clin Chem 1984;30:352–6.
22. Eckfeldt JH, Copeland KR. Accuracy verification and identification of matrix effects: the College of American Pathologists' protocol. Arch Pathol Lab Med 1993;117:381–6.
23. Greenacre M. Correspondence analysis in medical research. Stat Meth Med Res 1992;1:97–117.
24. Miller, WG, Kaufman H, McLendon WW. College of American Pathologists Conference XXIII: Matrix effects and accuracy assessment in clinical chemistry. Arch Pathol Lab Med 1993;117(4):343–436.
25. Naito HK, Yun-Sik K, Hartfiel JL, et al. Matrix effects on proficiency testing materials: Impact on accuracy of cholesterol measurement in laboratories in the nation's largest hospital system. Arch Pathol Lab Med 1993;117:345–51.
26. Ross JW, Myers GL, Gilmore BF, Cooper GR, Naito HR, Eckfeldt J. Matrix effects and the accuracy of cholesterol analysis. Arch Pathol Lab Med 1993;117:393–400.
27. Ellerby P, Myers GL, Cooper GR, et al. A comparison of results for cholesterol in human serum obtained by the reference method and by the definitive method of the National Reference System for Cholesterol. Clin Chem 1990;36:370–5.
28. Lasky FD, Powers DM, Hassemer DJ, Wiebe DA. Quality of fluids used in external quality control programs affects the reliable assessment of accuracy of routine methods, as documented by cholesterol. Quality Control in the Clinical Laboratory '91. Proceedings of the 7th International Symposium on Quality Control; Tokyo, June 15–16, 1991;199–208.
29. Franzini C, Luraschi P. Commutability of control materials in cholesterol measurement. Scand J Clin Lab Invest 1993;53:51–55.
30. Waymack PP, Miller WG, Myers GL. Assay instrument dependent matrix effects in standardization of cholesterol measurements. Clin Chem 1993;39:2058–62.
31. Myers GL, Schap D, Smith SJ, et al. College of American Pathologists–Centers for disease control collaborative study for evaluating reference materials for total serum cholesterol measurements. Arch Pathol Lab Med 1990;114:1199–1205.
32. Miller WG, Brown B, Dalby T. Cholesterol matrix effects in fresh non-frozen and frozen serum pools [abstract]. Clin Chem 1991;37:920.
33. Holani KK, Miller WG. Robustness of triglyceride reagents for matrix effects of proficiency testing materials [abstract]. Clin Chem 1993;in press.
34. Smith SJ, Henderson LO, Hannon WH, Cooper GR. Effects of analytical method and lyophilized sera on measurements of apolipoproteins A-1 and B: an international survey. Clin Chem 1990;36:290–6.

35. Marcovina SM, Albers JJ, Dati F, Ledue TB, Ritchie RF. International federation of clinical chemistry standardization project for measurements of apolipoproteins A-1 and B. Clin Chem 1991;37:1676–82.
36. Albers JJ, Marcovina SM, Kennedy H. International Federation of Clinical Chemistry standardization project for measurements of apolipoproteins A-1 and B. II. Evaluation and selection of candidate reference materials. Clin Chem 1992; 38:658–62.
37. Marcovina SM, Albers JJ, Henderson LO, Hannon WH. International Federation of Clinical Chemistry standardization project for measurements of apolipoproteins A-1 and B. III. Comparability of apolipoprotein A-1 values by use of international reference material. Clin Chem 1993;39:773–81.
38. Lasky FD. Achieving accuracy for routine clinical chemistry methods by using patient specimen correlations to assign calibrator values: a means of managing matrix effects. Arch Pathol Lab Med 1993;117:412–19.
39. Gochman N, Hall S. Manufacturer's approach to transferring accuracy to multiple field installations. Arch Pathol Lab Med 1993;117:420–1.
40. Friedewald WT, Levy RI, Frederickson DS. Estimation of the concentration of low-density lipoprotein cholesterol in plasma, without use of the preparation ultracentrifuge. Clin Chem 1972;18:499–502.
41. Federal Register, Vol. 57, No. 40 (Feb. 28, 1992), p. 7146, section 493.801, part b. Washington DC: US Government Printing Office.
42. Recommendations for improving cholesterol measurement: a report from the Laboratory Standardization Panel of the National Cholesterol Education Program. (NIH Publication No. 90-2964.) Washington, DC: U.S. Dept of Health and Human Services, Public Health Service, 1990.

CHAPTER 11

Standardization of Lipid and Lipoprotein Measurements

Gary L. Myers, Gerald R. Cooper, L. Omar Henderson,
David J. Hassemer, and Mary M. Kimberly

BACKGROUND

❖ The National Cholesterol Education Program (NCEP) has mounted a concerted national effort to identify and treat every American adult who has high blood cholesterol and therefore has increased risk of coronary heart disease (CHD). The association between increased total cholesterol (TC), due to increased concentrations of low-density lipoprotein cholesterol (LDL-C), and the risk of premature CHD has been well documented.[1, 2] In addition, high-density lipoprotein cholesterol (HDL-C) concentrations are known to be negatively associated with CHD risk.[3, 4] The association of triglycerides (TG) with the risk of CHD continues to be somewhat controversial; TG are generally considered to be less predictive of CHD risk than TC.[4–7] The importance of TG rests in the estimation of LDL-C by the Friedewald equation.[8] Data from many epidemiological and clinical studies suggest that apolipoprotein A-I (apo AI) and apo B-100 (apo B) may have promise as predictors of CHD risk and may eventually serve as additional markers of CHD risk.[9]

In its first report, the Adult Treatment Panel (ATP) of the NCEP established uniform decision points for identifying individuals who have increased risk of CHD.[2] The ATP's guidelines outlined a systematic approach to treatment of increased blood cholesterol which based initial case findings on TC concentrations and cholesterol-lowering therapy on LDL-C as the primary decision point.

In 1993, the second report issued by the Adult Treatment Panel (ATP II) reaffirmed that lowering TC and LDL-C concentrations reduces the risk of CHD;[10] (see Chapter 2 for additional information). However, ATP II contains some new issues that distinguish it from ATP I. For the clinical laboratory, the most significant difference is the increased emphasis on HDL-C as an independent risk factor for CHD. Specifically, ATP II recommends the addition of HDL-C to initial TC testing, so that whenever serum TC is measured for assessing CHD risk, HDL-C should also be measured.

All this attention to TC and the lipoproteins has focused unprecedented national attention on clinical laboratories and the need to provide reliable measurements of TC, TG, HDL-C, and LDL-C. Given the central role of cholesterol measurement in the assessment and management of CHD risk, the NCEP established the Laboratory Standardization Panel on Blood Cholesterol Measurement to assess the state of reliability of cholesterol measurement in clinical laboratories and to recommend means to improve the precision and accuracy of cholesterol testing. In 1988, the Laboratory Standardization Panel[11] recommended that cholesterol measurements be standardized so that values are traceable to the Centers for Disease Control and Prevention (CDC) cholesterol reference method (using a modification of the Abell-Kendall method),[12, 13] or to the National Institute of Standards and Technology (NIST) definitive method for cholesterol (using isotope dilution–mass spectrometry).[14] An NCEP Lipoprotein Measurement Working Group is currently preparing for publication guidelines for improving the measurement of HDL-C, LDL-C, and TG. (Additional information on these guidelines is given Chapter 8.)

Since laboratories employ a variety of methodologies, it is not possible or practical to require that all laboratories use the same methods, calibrators, and controls. The key point to emphasize here is that the basis of standardization should be *traceability*: all results should be tied to a defined common accuracy base. Standardization requires that a reproducible analytical base be established to serve as a common denominator that permits laboratories to compare results and monitor analytical performance to ensure continuing accurate and precise results. The major components of this benchmark accuracy base comprise a hierarchy of methods and materials and include a definitive method, primary reference materials, a reference method, and secondary reference materials.

THE ACCURACY BASE

For each lipid, lipoprotein, and apolipoprotein, an analytical method must be widely accepted as the basis for universal reference. While it is generally accepted that a definitive method establishes the true value for an analyte, definitive techniques are not widely available to fulfill all of the tasks necessary to provide a point of reference for effective laboratory standardization. As a result of such limitations, a definitive method is primarily used to validate a more widely available reference method designed for broader use and application.

If reference methods are to serve as the reference point for laboratory standardization, the problem then becomes how to interface the accuracy base with the laboratory community; that is, the base must be transferable on a broad scale. The most rapid and economical means of broadening the transfer of this accuracy base and of multiplying the impact of laborious reference-method

Table 11-1 ❖ Lipid and Lipoprotein Reference Systems*

	Definitive Method	*1° Reference Material*	*Reference Method*	*2° Reference Material*
Cholesterol (NRS/CHOL)	ID-MS (NIST)	NIST SRM 911b Pure cholesterol	Abell-Kendall (CDC)	CDC Frozen Pools NIST SRM 909
HDL-C	Not Available	Not Available	UC/Heparin-Mn^{2+}-Abell-Kendall (CDC). Recommended by NCEP.	CDC Frozen Pools
Triglyceride	ID-MS (NIST) (Candidate)	NIST SRM 1595 tripalmitin	Methylene chloride-silicic acid-chromotropic acid (CDC). Recommended by NCEP.	CDC Frozen Pools
LDL-C	Not Available	Not Available	Beta-quantification (CDC). Recommended by NCEP.	CDC Frozen Pools
APO AI	HPLC-MS (CDC) (primary standard only) (Candidate)	BCR-CRM 393 Purified APO AI	Not Available	WHO Reference Reagent SP1-01 (for manufacturers). Labelled by CDC-RIA comparison method.
APO B	Not Available	1.030–1.050 d UC purified LDL	Not Available	WHO Reference Reagent SP3-07 (for manufacturers). Labelled by NWLRL-immuno-nephelometry comparison method.

*CDC = Centers for Disease Control, NCEP = National Cholesterol Education Program, WHO = World Health Organization, NWLRL = Northwest Lipid Research Laboratories

analyses has traditionally been through the use of high-quality secondary reference materials. Table 11-1 summarizes the available components that make up reference systems for the various lipids and lipoproteins.

However, this traditional approach to standardization by assigning target values to reference materials by definitive and/or reference methods has been complicated by the occurrence of fluid matrix effects in some analytical systems. This phenomenon is discussed later in this chapter and is presented in greater detail in Chapter 10.

Cholesterol

The standard for accuracy in blood cholesterol measurements is the National Reference System for Cholesterol (NRS/CHOL) established by the National Committee for Clinical Laboratory Standards (NCCLS).[15] The NRS/CHOL is a unique voluntary consensus standard endorsed by professional, industrial, and government organizations. It is made up of the NIST definitive method, the CDC reference method, and a certified pure cholesterol standard, NIST SRM 911b. Both the definitive and reference methods are calibrated using the NIST SRM 911b pure cholesterol material. The two methods are monitored on an ongoing basis, and comparisons have demonstrated a small but persistent positive bias in the reference method of +1.6% compared to the definitive method.[16] The results of further investigations suggest that more than half of the difference in cholesterol values determined by the two methods is from small contributions from cholesterol precursor sterols and phytosterols, which are measured by the reference method.[17] Since the observed bias is consistent and quite small, the reference method provides a practical basis for standardization of routine methods used in the clinical environment, whereas the definitive method would be impractical for use in this manner.

HDL-C

The accuracy base for HDL-C is not as well defined and complete as that for TC. A significant problem is that the lipoproteins, as a heterogeneous mixture of lipids and proteins (including HDL-C), are not rigidly defined. As a result, significant overlap can exist in the physical properties of the major lipoprotein classes. Therefore, developing a reference system for HDL-C will require that it be precisely defined. The current operational definition of HDL-C is based on physical separation, such as ultracentrifugation ($d > 1.063$ Kg/L) or chemical precipitation.

A definitive method for HDL-C does not yet exist. The reference point for HDL-C measurement, as recommended by the NCEP Lipoprotein Measurement Working Group, is the CDC reference method, a multi-step procedure involving ultracentrifugation, precipitation, and cholesterol analysis.[18] The method combines removal of very-low-density lipoprotein (VLDL) by a beta-quantification ultracentrifugation procedure, isolation of HDL-C by precipitation of LDL-C from the beta-quantification bottom fraction (d = 1.006 Kg/L) by 46 mmol/L heparin–manganese (Mn^{+2}), and cholesterol analysis of the HDL-C supernate by the NRS/CHOL cholesterol reference method. The major problems with the HDL-C reference method are that it is technically demanding, requires a large sample volume (5 mL), has low throughput, and needs an expensive ultracentrifuge, which may precludes its widespread use for routine standardization of HDL-C measurements. However, since this HDL-C method has served as the reference method for the epidemiologic and clinical investigations by the CDC and National Heart, Lung and Blood Institute (NHLBI), from which CHD risk estimations and population distributions have been derived, there is considerable justification for continuing this procedure as a national reference method. The CDC is now working with the reference laboratories participating in the Cholesterol Reference Method Laboratory Network to evaluate and adopt a "designated comparison method" for routine use in establishing traceability to the reference method.

LDL-Cholesterol

Although the primary decision factor for patient follow-up and treatment is based on the measurement of LDL-C, no formal effort has been made to develop, validate, and approve a reference method and system for standardizing its measurement. The process of establishing an accuracy target for LDL-C is complicated due to its heterogeneity and the difficulty of precisely defining the fraction of interest.

The CDC has adopted as a reference procedure for LDL-C a variation of the multi-step beta-quantification procedure used by the Lipid Research Clinics[18] which combines separation by ultracentrifugation and chemical precipitation. In the CDC procedure, heparin-manganese precipitation is performed on the d >1.006 Kg/L serum fraction rather than on whole plasma. The CDC reference value for LDL-C is calculated as the difference in cholesterol between the measured HDL-C and the cholesterol recovered in the d >1.006 Kg/L fraction obtained by ultracentrifugation.

TG

A reference method for TG was established at CDC in 1963 on the basis of the method of Carlson,[19, 20] and the techniques of Van Handel and Zilversmit[21] and Lofland[22] and has been used as the CDC's accuracy base for standardization programs. The method is complicated, requiring extraction with silicic acid and chloroform for removal of free glycerol and other interferences, filtration to remove the silicic acid particles, hydrolysis with alcoholic potassium hydroxide, and color development by metaperiodate-arsenite-chromotropic acid reagent. In 1993 the CDC reference method was modified to replace chloroform with methylene chloride in order to eliminate the use of chloroform in the extraction step and to eliminate the need for filtration to remove the silicic acid particles. CDC is now collaborating with the Canadian Reference Foundation (Vancouver, British Columbia) to further improve the TG method by combining the hydrolysis and extraction steps of the CDC method with color development by enzymatic endpoint.

NIST has recently developed a candidate definitive method for TG utilizing isotope dilution–mass spectrometry.[23] The NIST method is capable of giving results for total glycerides and for pure TG only. TG results are calculated in terms of triolein. CDC and NIST are now evaluating the relationship between the definitive and reference methods.

The CDC uses a triolein/tripalmitin (2:1 ratio) standard to reflect the average unsaturated/saturated TG composition in human serum. A tripalmitin standard (99.5% pure) is available from NIST, but because it is insoluble in aqueous medium, it is unsuitable for enzymatic methods.

Apo AI and B

Ideally, gravimetrically weighed pure apolipoproteins should be used as primary standards in apolipoprotein analysis. However, as a matter of practicality, purified apolipoprotein may not be suitable for some immunological techniques, and the physical and chemical properties of purified apo B do not lend themselves to immunochemical analysis. Purified apo AI can be used as a primary standard in some immunochemical techniques, and an ultracentrifugally prepared LDL fraction (d = 1.030–1.050 Kg/L) that does not include intermediate-density lipoprotein or lipoprotein(a) is currently the best choice for use as a surrogate standard for apo B. A purified apo AI has been developed by the European Bureau of Reference for use as a primary standard.[24] Reference methods for the immunochemical analysis of apo AI and B have not been universally accepted in the scientific community. Also, determination of the protein content of LDL presents significant problems[25] which are under investigation but have not been resolved.

The preferred approach to establishing an accuracy base is to use an accepted and well-documented reference method or definitive method with primary standards to set apolipoprotein mass units on a secondary reference material (SRM) in a suitable matrix base. These SRMs can then be used to set apolipoprotein mass units on calibration materials used by manufacturers of apolipoprotein measurement kits. This process is not possible at present because there is no universally accepted reference method for analysis of apo AI or B.[26]

Reference and definitive methods for proteins depend on highly accurate protein determinations of primary standards for accuracy of labeled values. CDC has investigated and established an enzymatic digestion, liquid chromatography, isotope dilution–mass spectrometry method to assign mass values to an apo AI primary standard for use as a model for apolipoprotein primary standards.[27]

The World Health Organization (WHO) has accepted two serum matrix-based materials as reference reagents: SP1-01, a lyophilized material for apo AI, and SP3-07, a stabilized liquid preparation for apo B. At present, these WHO Reference Reagents are distributed only to manufacturers of reagents and equipment used in apolipoprotein analysis. Mass units have been assigned using standardized immunoassay techniques and purified apo AI and LDL (d = 1.030–1.050 Kg/L) as primary calibrators.[28]

NATIONAL PERFORMANCE CRITERIA

The role of cholesterol and the lipoproteins in the assessment of CHD risk has accentuated the need for more precise and accurate laboratory measurements.[29] In order to properly assess the quality of analytical performance in individual laboratories, meaningful criteria to evaluate the reproducibility of measurements and the deviation from a target value are needed. CDC was the first organization to establish specific performance criteria for (TC, TG, and HDL-C) designed to improve the performance of specialized lipid laboratories participating in the CDC-NHLBI Lipid Standardization Program.[30]

In 1988, the NCEP Laboratory Standardization Panel recommended laboratory goals for both bias and precision of TC measurement.[11] Specifically, the panel recommended that, as a national goal, clinical laboratories should achieve a bias of ± 3% of the CDC reference method and an overall precision consistent with a coefficient of variation (CV) of 3% or less.[11] For a single analysis, the allowable total error (imprecision plus bias) would be ± 8.9% using L = 0.05 in a two-tailed test.[31] This was one of the most important recommendations from the Laboratory Standardization Panel in that it established for the first time specific national performance criteria by which clinical laboratories can judge the reliability of cholesterol assays.

Similar nationally accepted performance criteria for TG, HDL-C, LDL-C, and apolipoprotein measurement by routine laboratories have not officially been established. However, the NCEP Lipoprotein Measurement Working Group has evaluated the performance necessary for reliable TG, HDL-C, and LDL-C measurements and has recommended to the NCEP Coordinating Committee national performance criteria for these analytes. The forthcoming NCEP performance goals for TG, HDL-C, and LDL-C will be presented in terms of allowable total error; in other words, method bias and method precision will be considered in terms of total error and must satisfy the goals for total error.

Presented in Table 11-2 are the performance criteria for specialized lipid laboratories participating in the CDC-NHLBI Lipid Standardization Program, and the NCEP performance goals for clinical laboratories. In addition, under the Clinical Laboratory Improvement Amendments (CLIA) of 1988, the federal government has established national proficiency testing criteria for a large number of clinical analytes. The CLIA '88 evaluation criteria (total error) for TC, TG, and HDL-C are also presented in Table 11-2.

STANDARDIZATION ISSUES

When attempting to standardize laboratory measurements, one must consider factors that contribute to inaccuracy in the measurement process and affect the reliability of test results. Traditionally, major emphasis has been on analytical factors—factors that compromise accuracy during measurement. However, as efforts to control analytical imprecision and inaccuracy progress, concern has increased significantly about the effect on test results of preanalytical sources of variation—factors that operate before, during, and after blood collection and storage.[32] It is important, therefore, to carefully consider both preanalytical sources of variation and laboratory sources of error.

Preanalytical Consideration

Since the purpose of lipid and lipoprotein measurement is to determine the status of a patient's metabolic parameters, physicians need to understand and control, as much as possible, factors that can transiently alter a patient's usual lipid and lipoprotein concentrations.[33] Preanalytical factors include behavioral factors, the clinical state of the patient, the conditions of patient preparation, and the collection, handling, storage, and shipment of patient samples.[32] Preanalytical factors are covered in detail in Chapter 3; hence, only a brief overview is given here.

Considerable variation in average intra-individual biological variation has been reported in the literature. In order to obtain the best estimates of the average intra-individual biological variability (expressed as CV_b) in the concentrations in

Table 11–2 ❖ National Performance Criteria for Lipid and Lipoprotein Measurement[a]

Analyte	*CDC Standardization Criteria*			*NCEP Performance Criteria*[b]			*CLIA Evaluation Criteria*[c]
	Concentration Range	Bias	Imprecision	Bias	Imprecision	Total Error	Total Error
Low TC	< 50 mg/dL 50-99.9 mg/dL	≤ 2.0 mg/dL ≤ 3.0 mg/dL	SD[d] ≤ 2.0 mg/dL SD ≤ 3.0 mg/dL				
TC	100–149 mg/dL ≥ 150 mg/dL	≤ 0.03 RV[e] ≤ 0.03 RV	SD ≤ 4.0 mg/dL CV[f] ≤ 3.0%	≤ 3% RV	CV ≤ 3%	≤ 9%	± 10%
TG (mmol/L)	0.00–0.99 1.00–1.99 2.00–2.49 ≥ 2.50	≤ 0.10 ≤ 0.11 ≤ 0.12 0.05 RV	SD ≤ 0.08 SD ≤ 0.09 SD ≤ 0.11 CV ≤ 5.0%	≤ 5% RV	CV ≤ 5%	≤ 15%	± 25%
HDL-C	< 40 mg/dL 40–59.9 mg/dL ≥ 60 mg/dL	≤ 0.10 RV ≤ 0.10 RV ≤ 0.10 RV	SD ≤ 2.5 mg/dL[g] SD ≤ 3.0 mg/dL SD ≤ 3.5 mg/dL	≤ 10% RV *1998:* ≤ 5% RV	SD ≤ 2.5 at < 42 mg/dL. CV ≤ 6% at ≥ 42 mg/dL. 1988: SD ≤ 1.7 at < 42 mg/dL. CV ≤ 4% at ≥ 42 mg/dL.	≤ 23% *1998:* ≤ 13%	± 30%
LDL-C				≤ 4% RV	CV ≤ 4%	≤ 12%	

[a]TC = Total Cholesterol, TG = Triglyceride, HDL-C = High-Density Lipoprotein Cholesterol, LDL-C = Low-Density Lipoprotein Cholesterol. To convert cholesterol values from mg/dL to mmol/L, multiply by 0.0259. To convert TG values from mmol/L to mg/dL, divide by 0.0113.

[b]Performance criteria for TC recommended by the NCEP Laboratory Standardization Panel in: Clin Chem 1988;34:193–201. Performance criteria for TG, HDL-C, and LDL-C are forthcoming from the NCEP Lipoprotein Measurement Working Group and will be cast primarily in terms of total error. In addition, interim goals for HDL-C will be recommended, with tighter criteria to take effect in 1998.

[c]Federal Register, February 28, 1992.

[d]SD = standard deviation.

[e]RV = reference value assigned by CDC reference methods

[f]CV = coefficient of variation

[g]criteria for HDL-C in the CDC program. Most commonly used lipoprotein separation techniques are biased relative to the CDC reference method for HDL-C. Most biases are negative and become more negative at higher concentrations. The CDC criteria for systematic error allow the use of methods which have negative bias at concentrations of HDL-C above 50 mg/dL (1.30 mmol/L).

serum of TC, TG, HDL-C, and LDL-C, results from 30 studies published from 1970 to 1992 were evaluated by meta-analysis.[34]

A method based on relative range was developed to estimate the effect of biological variation on the total variation of the mean serum TC result.[33] The method can also serve as a guide for determining the number of specimens required from an individual to minimize the effect of biological variation on the individual's TC result.[33] The relative range method has been extended (using the meta-analysis results for CV_b and the forthcoming NCEP performance goals) to estimate and establish goals for minimizing biological variation for TG, HDL-C, and LDL-C mean serum concentrations.[35] The in-depth discussion of preanalytical variation presented in Chapter 3 includes detailed recommendations issued in 1990 by the NCEP Laboratory Standardization Panel pertaining to factors that must be considered for minimizing or controlling the influence of preanalytical factors.[31] The NCEP report does not cover apolipoproteins specifically, and although no thorough temporal studies on variation in apo AI or apo B have been conducted, similar precautions should be taken to help minimize preanalytical sources of variation in apolipoprotein measurements.[36]

Matrix Effects of Reference Materials

The traditional approach for standardization has been to use reference materials and calibrators that have been value-assigned by the reference or definitive methods. This approach is based on the assumption that reference materials mimic fresh patient specimens when assayed. However, recent studies have indicated that significant discrepancies in the measurement of cholesterol exist when some analytical systems are compared to the reference method on lyophilized and frozen processed serum materials.[37-45] As a result of their preparation, the processed materials may undergo changes that affect their assay characteristics. These matrix effects complicate standardization because different methods may produce different results for the same reference material and may yield erroneous conclusions about the performance of an analytical system. In addition to the matrix interactions affecting the cholesterol analysis, any changes in the specimen that affect the separation step may also give rise to matrix effects in the measurement of lipoproteins, including HDL-C. Matrix effects in the measurement of TG have also been reported.[46]

Until reference materials free of matrix effects are available, an interim approach that involves implementation of the NCEP Laboratory Standardization Panel's recommendations is to directly compare the analytical system with the cholesterol reference method through the use of split fresh patient specimens. The use of fresh-frozen serum may alleviate the problems associated with matrix interactions, but even the process of freezing appears to affect some analytical methods.[47] Establishing traceability by comparing split fresh patient specimens is valid for both manufacturers of diagnostic systems and for clinical laboratories seeking to verify the performance of a cholesterol method. Since the same type of matrix limitations exist for lipoproteins and TG, the cholesterol experience is a good model for standardization.

Purified apo AI is usually acceptable, but apo B will not work well as a primary standard in all immunological methods of analysis. Freshly collected serum is the specimen of choice, but practical considerations of distribution and stability preclude using fresh materials. Lyophilized serum-based reference materials with assigned concentrations of apo AI seem to work well in most immunological methods;[48] however, when certain methods, including radial immunodiffusion and immunoturbidimetry, are used, systematic differences in apo B measurements occur.[49, 50] Fresh-frozen serum seems to perform well in most immunological methods if the protein epitope bound by the antibody is not sensitive to freeze/thaw cycles. Stabilized liquid preparations of whole serum appear to work for apo B analysis in most immunological techniques.[49, 50]

There are inherent problems in maintaining stability of large quantities of frozen or stabilized liquid apolipoprotein reference materials.[50] For a given method, reagent system, or combination of instrument system and reagents, the primary responsibility for establishing and maintaining an accuracy base for mass units rests with the manufacturer, who should calibrate each lot of secondary calibrators to the appropriate WHO Reference Reagents.

TG Measurement and Free Glycerol

The TG measurement poses a unique matrix-effect type of problem resulting from the contribution of free glycerol. Controversy over whether to blank TG analyses for free glycerol continues.[51–53] A major justification for accurate TG measurements is its inclusion in the Friedewald equation used to estimate LDL-C.[8] Thus, not correcting for free glycerol may result in erroneous estimation of LDL-C. In a study of both inpatients and outpatients at a university hospital, the need for free glycerol blanking was evaluated, and, except in those situations in which abnormally increased glycerol is expected, no significant consequence resulted for the average individual when a glycerol blank was not used.[52] Serum glycerol blanking does, however, become extremely important when attempting to standardize TG measurements. Glycerol content in commercially prepared materials, such as survey samples, controls, and calibrators, can be quite variable, depending on the matrices of the source material and the fortification to increase lipid concentrations.[46] This situation may result in confusion in accuracy transfer, making the standardization of TG measurements across the nation's clinical laboratories very difficult and the assessment of the accuracy of results impossible. It therefore becomes very important that laboratories evaluate each lot of materials for free glycerol.

In the preparation of reference materials for TG standardization, the CDC routinely screens serum samples for free glycerol concentrations using an enzymatic method. In an effort to establish the target glycerol content of reference pools, the CDC developed an isotope dilution–gas chromatography–mass spectrometry method for the analysis of serum free glycerol.[54] Comparison of glycerol concentrations for several reference serum pools determined by using different enzymatic kits sometimes produced substantially different results.[54] However, all the methods evaluated provided similar results when pure aqueous standards were tested and when standard addition studies were conducted.[54] Thus, potential matrix effects with different enzymatic free glycerol assays must also be considered.

In addition to the issue of free glycerol blanking, enzymatic methods to determine TG may also be affected by the turbidity of the specimen.[55] In a recent study, effects of specimen turbidity and glycerol blanking on nine enzymatic TG methods were compared.[56] The results showed variability among the methods evaluated ranging from 6.2% to −15.6% of the measured result.[56] It was found that methods using a bichromatic blank better minimized the interference from turbidity than did a reagent or serum blank, and therefore a bichromatic glycerol-blanked method was recommended to obtain a more accurate TG result for turbid specimens.[56]

RESOURCES AVAILABLE FOR STANDARDIZATION

The NCEP's Laboratory Standardization Panel stated that to ensure reliable cholesterol results, the measurement of cholesterol in all clinical laboratories must be traceable to a national accuracy base such as the NRS/CHOL.[11] Achieving traceability to the NRS/CHOL or the accuracy bases for TG, the lipoproteins, and the apolipoproteins will require a unified national effort involving manufacturers, government agencies, and the clinical professions. The programs and resources available to assist clinical laboratories in standardizing the measurement of lipids and lipoproteins are summarized below. While this discussion may not be exhaustive, it does represent the major efforts toward standardization.

CDC-NHLBI Lipid Standardization Program

Since 1957, the CDC has been developing and evaluating the resources needed to standardize the analytical measurement of cholesterol and other metabolically related lipids, lipoproteins, and apolipoproteins. Through standardization programs, the CDC has offered these resources nationally and internationally to epidemiologic, lipid methodology research, proficiency testing programs, and manufacturers' laboratories.[30] Standardization efforts have focused on:

1. epidemiologic laboratories, to ensure comparability of population studies
2. lipid methodology research laboratories, to help in the development of accurate methods
3. reference laboratories, to provide targets for proficiency testing programs
4. laboratories of manufacturers of cholesterol diagnostic products, through which standardization assistance can be provided to all clinical laboratories.

These standardization resources were developed because many investigators interested in the relationship between serum lipid concentrations and cardiovascular disease were concerned about the reliability of lipid analyses within laboratories and the comparability of results among laboratories. CDC annually standardizes laboratories that support over 100 studies funded by the National Institutes of Health that investigate CHD and other related factors, including diabetes, nutrition, genetics, and health issues of women and minorities.

The CDC-NHLBI Lipid Standardization Program has three major phases. Part I is the preliminary evaluation phase in which CDC reference materials and fresh patient specimens are analyzed concurrently. During this phase, a split-sample comparison with a laboratory participating in the Cholesterol Reference Method Laboratory Network is made to verify the performance of CDC materials versus patient samples when the participant's method is used. If analytical criteria are met by the participant and significant matrix effects on CDC pools are not seen, performance is assumed to be sufficient to ensure the quality of the relationship determined in the participant's laboratory between CDC's reference values and the participant's values. Part II is the principal phase in standardization and confirms the level of accuracy and precision found in Part I as well as the laboratory's ability to achieve and maintain acceptable performance. To ensure representative performance on the part of the analyst, the CDC samples are provided to the participant without concentration values. Part III monitors the ability of a participant laboratory to maintain acceptable performance for 12 months or longer. In this phase, CDC reference materials are distributed at quarterly intervals.

Criteria for acceptable performance in the CDC program were discussed earlier and summarized in Table 11–2. CDC standardization performance evaluations of individual laboratory results are based on fixed limits combined with target values based on the CDC reference method. In this way, standardization permits evaluation of all components of method error.

For more information, contact the Special Activities Branch, Division of Environmental Health Laboratory Sciences, Centers for Disease Control and Prevention, 4770 Buford Highway, NE (F25), Atlanta, GA 30341-3724, Telephone 404-488-4126, Fax 404-488-4192.

Apolipoprotein Standardization Programs of the CDC and International Federation of Clinical Chemistry (IFCC)

In 1981 the CDC formed an apolipoprotein working group to work on standardizing apolipoprotein measurements. In 1983 a large pool of lyophilized human serum (CDC-1883) was prepared and used in several international surveys to determine the sources of variability in apolipoprotein measurements as well as the utility of this lyophilized material as a reference material for apolipoprotein measurement.[57–59] In 1985 a group of expert laboratories measured CDC-1883 using their own analytical techniques and calibrators. Preliminary-consensus mass units for apo AI and B were assigned to CDC-1883. The accelerated thermal stability and temporal stability over 5 years have shown that this material is suitable as a stable point of reference for apo AI and B measurement, but CDC-1883 demonstrates a matrix bias when used in certain immunological techniques to measure apo B. CDC-1883 has a labeled value of 124 mg/dL for apo AI and is traceable to the WHO Reference Reagent for apo AI (SP1-01).

CDC is now collaborating with WHO and IFCC to distribute the WHO Reference Reagents to manufacturers of apolipoprotein products. At present, these WHO Reference Reagents are available only to manufacturers of apolipoprotein diagnostic products. Under the auspices of the IFCC, the Northwest Lipid Research Laboratories are offering three frozen serum pools with assigned values for apo AI and B (traceable to the WHO Reference Reagents) to clinical laboratories for calibration or confirmation of accuracy of their calibrators.

THE CHOLESTEROL REFERENCE METHOD LABORATORY NETWORK

The problems associated with matrix effects on the accuracy of cholesterol measurement were mentioned earlier in this chapter and are discussed in detail in Chapter 10. The recommended approach to assess how well a laboratory method correlates with the reference method is to directly compare the two methods using split fresh patient specimens.

To facilitate access to the NRS/CHOL by manufacturers and clinical laboratories, the CDC established a national network of laboratories that use the cholesterol reference method.[30] A list of participating Network laboratories and contact persons is presented in Table 11–3. The Network laboratories perform the Abell-Kendall procedure exactly as it is performed at CDC and are standardized to CDC through participation in the CDC-NHLBI Lipid Standardization Program. The performance of the cholesterol reference method in each Network laboratory is monitored monthly. These monthly surveys involve either the analysis of CDC serum reference pools or in-house fresh-frozen, single-donor samples. Under this program, the Network laboratories must maintain a CV of less than 2% and a bias to CDC of less than 1%. Over the period of one year, the within-lab CV ranged from 0.5% to 1.1% with an average of 0.8%.[60] During the same period, the average bias of Network laboratories ranged from –0.8% to 0.4% with an overall average of –0.3%.[60]

In addition to the nine U.S. Network laboratories, six international laboratories participate in the Network and serve as a point of reference for their individual countries. A list of international members of the Cholesterol Reference Method Laboratory Network is presented in Table 11–4. These laboratories are standardized using the same criteria as the U.S. Network laboratories. Thus, the reference methods used in these laboratories are traceable to the NRS/CHOL. The programs offered by the international participants vary depending on the needs of the individual countries. The international laboratories are also capable of providing certification services to manufacturers and clinical laboratories.

The strategy for transferring the accuracy base for cholesterol to manufacturers and clinical laboratories via the Cholesterol Reference Method Laboratory Network is shown in Figure 11–1.

Table 11–3 ❖ U.S. Laboratories Participating in the Cholesterol Reference Method Laboratory Network

State Laboratory of Hygiene
University of Wisconsin
Center for Health Sciences
465 Henry Mall
Madison, WI 53706
David Hassemer, M.S.
Phone 608-833-1770, Ext. 102
Fax 608-833-2803
BMH@STOVALL.SLH.WISC.EDU

Pennsylvania State Department of Health
Bureau of Laboratories
Division of Chemistry and Toxicology
Post Office Box 500
Exton, PA 19341-0500
Irene Daza
Phone 215-363-8500
Fax 215-436-3346

University of Minnesota Hospital and Clinic
Department of Laboratory Medicine
and Pathology
Box 198 UMHC
420 Delaware Street
Minneapolis, MN 55455-0392
John H. Eckfeldt, M.D., Ph.D.
Phone 612-626-3176
Fax 612-625-6994
ECKFE001@STAFF.TC.UMN.EDU

Northwest Lipid Research Laboratories
Core Laboratory
2121 N. 35th Street
Seattle, WA 98103
Santica Marcovina, Ph.D.
Phone 206-685-3331
Fax 206-685-3279

Wadsworth Center for Laboratories and Research
New York State Department of Health
Empire State Plaza
Albany, NY 12201
Robert Rej, Ph.D.
Phone 518-473-0117
Fax 518-474-7992
BOBREJ@RACHEL.ALBANY.EDU

Washington University School of Medicine
Lipid Research Center
4566 Scott Avenue
St.Louis,MO 63110
Thomas G. Cole, Ph.D.
Phone 314-362-3522
Fax 314-362-7657

USDA Human Nutrition Research Center on Aging
Tufts University
711 Washington Street, Room 501
Boston, MA 02111
Judith R. McNamara, M.T. (ASCP)
Phone 617-556-3104
Fax 617-556-3103
MCNAMARA_LI@HNRC.TUFTS.EDU

Pacific Biometrics Research Foundation
1100 Eastlake Avenue East
Seattle, WA 98109
G. Russell Warnick, M.S., M.B.A.
Elizabeth Teng Leary, Ph.D.
Phone 206-233-9151
Fax 206-233-0198

The Cleveland Clinic Foundation
Department of Biochemistry, L-11
9500 Euclid Avenue
Cleveland, OH 44195
Joan A. Waletzky
Phone 216-444-8301
Fax 216-444-4414

Table 11-4 ❖ International Laboratories Participating in the Cholesterol Reference Method Laboratory Network

Rotterdam University Hospital "Dijkzigt" Department of Clinical Chemistry Lipid Reference Laboratory 3015 GD Rotterdam The Netherlands Christa M. Boersma-Cobbaert, Ph.D. Phone 011-31-10-4633493 Fax 011-31-10-4367894	**Canadian Reference Foundation** 307-2083 Alma Street Vancouver, British Columbia V6R 4N6 Canada David W. Seccombe, M.D., Ph.D. 73361.1047@COMPU.SERVE COM Jennifer Hamilton, Ph.D. Phone 604-222-1879 Fax 604-222-0134 72772.1340@COMPU.SERVE.COM
Center for Adult Diseases, Osaka Department of Epidemiology and Mass Examination for CVD 3 Nakamichi 1-chome Higashinari-ku Osaka 537 Japan Masakazu Nakamura, Ph.D. Minora Lida, M.D. Phone 011-81-6-972-1181, Ext. 2211 Fax 011-81-6-972-7749	UNDERGOING STANDARDIZATION: **H.S. Raffaele** Laboratorio Analisi Cliniche Via Olgettina 60 20132 Milano Italy Ferruccio Ceriotti, Ph.D. Phone 011-39-2-2643-2315 (or 2313) Fax 011-39-2-2643-2640 CERIOTF%IMIHSRA.BITNET@VM.CNUCE.CNR.IT
Institute of Biochemistry Department of Pathological Biochemistry Glasgow Royal Infirmary 4th Floor Alexandra Parade Glasgow G31 2ER Scotland Great Britain Chris J. Packard, Ph.D. Phone 011-44-41-552-3535 Fax 011-44-41-553-2558	**Centre de Medicine Preventive** Laboratoire de Biologie Clinique 2, avenue du Doyen J. Parisot 54500 Vandoeuvre-les-Nancy France Prof. Gerard Siest Josiane Steinmetz, Ph.D. Phone 011-33-83-444415 Fax 011-33-83-440361

Certification of Instrument Systems, Reagents, and Reference Materials for the Measurement of Cholesterol

Manufacturers of instruments and reference materials are key to achieving standardization of cholesterol measurements in clinical laboratories. Manufacturers should perform comparison analyses with the reference method using fresh patient specimens in order to document the performance of their own complete instrument systems, reagent applications on other instrument systems, reference materials, and calibrators. To document performance and traceability to the NRS/CHOL, the Network has adopted a methods evaluation protocol developed by NCCLS entitled "Method Comparison and Bias Estimation Using Patient Samples."[61] The comparison protocol requires analyzing 40–50 patient specimens

*Figure 11–1 ❖ Strategy for Transferring the Cholesterol Accuracy Base to Manufacturers and Clinical Laboratories**

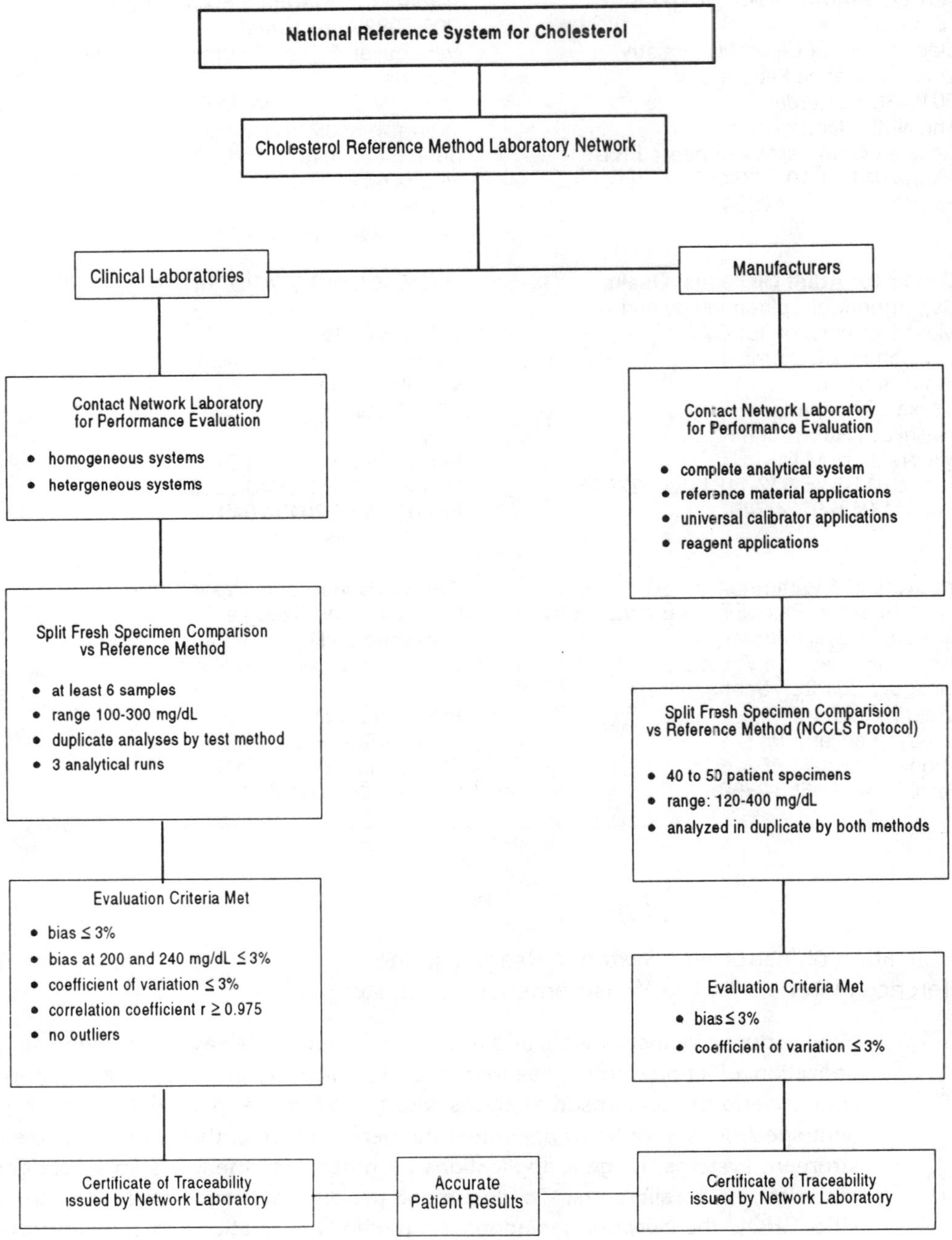

*To convert cholesterol values from mg/dL to mmol/L, multiply by 0.0259.

throughout the range of interest, in duplicate by each method. Once evaluation of the test method has been completed and certification criteria have been met, a dated certificate is issued to the manufacturer stating that the analytical system, reference material, or reagent application was traceable to the NRS/CHOL under the conditions tested. Manufacturers are also encouraged to recertify their systems whenever a new reagent or calibrator lot is prepared.

The list of analytical systems, reagents, and calibrators that have undergone evaluation over the last 15 months and have met the NCEP criteria appears annually in *Clinical Chemistry News*, published by the American Association for Clinical Chemistry.

For more information about how to participate in the manufacturer certification program, contact any one of the Network laboratories listed in Tables 11–3 and 11–4 or contact the Special Activities Branch, Centers for Disease Control and Prevention, 4770 Buford Highway NE, Atlanta, GA 30341-3724, Telephone 404-488-4126.

Certification of Clinical Laboratories

Clinical laboratories will use either of two types of analytical systems, homogeneous or heterogeneous. In a homogeneous analytical system, the instrument, reagent, and calibrator are supplied by a single manufacturer, whereas in a heterogeneous system, each may come from different sources. For laboratories using a homogeneous system, documentation of performance and traceability to the reference method can be obtained from the manufacturer. However, the manufacturer's demonstration of traceability to the NRS/CHOL in itself does not guarantee accuracy in the hands of every eventual user. For this reason, it is recommended that laboratories using certified systems verify their system's performance by performing a direct comparison with a Network laboratory using fresh patient specimens. Laboratories using heterogeneous analytical systems must assume primary responsibility for documenting performance and establishing traceability to the NRS/CHOL. In many cases, the test system used by a laboratory will be unique to that laboratory, thus requiring individual documentation of accuracy.

For practical reasons, certification of a laboratory's traceability to the NRS/CHOL is not as extensive as that required for a manufacturer. However, laboratory certification is still based on a direct comparison with the reference method using fresh patient specimens. The current protocol requires that the laboratory analyze at least six patient specimens in duplicate in each of three runs. The TC values should cover a range that includes the NCEP recommended cholesterol decision points of 200 and 240 mg/dL (5.18 and 6.22 mmol/L).[2] Acceptable performance, indicating traceability to the NRS/CHOL, is documented by a certificate which is valid for 6 months from the date of comparison.[62] The minimum cost to the clinical laboratory for this evaluation is $180.

CDC has implemented a Voice Information System to provide information about the clinical laboratory certification program. This system provides a general explanation of the Network and information about how clinical laboratories can verify accuracy in cholesterol testing. It also offers physicians a regional listing of labs that have met the national performance goals to which they can send their patients' samples. Listed are clinical laboratories that have successfully participated in the certification program within the last 12 months and the dates they received a Certificate of Traceability.

For more information about how to participate in the clinical laboratory certification program, contact any one of the Network laboratories listed in Tables 11–3 and 11–4 or telephone the CDC Voice Information System at 404-332-2592.

Reference Services for HDL-C and TG

Since matrix limitations also exist for HDL-C and TG, the mechanism for transferring the accepted accuracy base will also involve comparisons using fresh specimens. Manufacturers can establish traceability to the reference method and receive a certificate that indicates the total error, bias, and coefficient of variation observed with the analytical system.

HDL-C

A reference system equivalent to the NRS/CHOL does not exist for HDL-C. The NCEP Lipoprotein Measurement Working Group has recommended the method used at CDC as the accepted accuracy base for HDL-C. This is the ultracentrifugation with heparin-manganese precipitation method described previously in this chapter and in Chapter 6. Although the method can be transferred, it is too labor intensive and expensive to utilize in the Network for providing reference analyses for split-sample comparisons. As an alternative, the Network laboratories have chosen to utilize a "designated comparison method" (titled based on NCCLS nomenclature) for establishing traceability to the CDC reference method for HDL-C.[63] The designated comparison method will employ direct precipitation of the apo-B-containing lipoproteins, followed by determination using the Abell-Kendall reference method for cholesterol. Only samples with TG concentrations < 200 mg/dL (1.26 mmol/L) can be used with this protocol. At the time of this writing, the Network is evaluating dextran sulfate of 50 kDa with magnesium as the candidate designated comparison method.

TG

A complete reference system equivalent to the NRS/CHOL does not exist for TG. A candidate definitive method utilizing isotope dilution–mass spectrometry has been developed at the NIST.[23] The NCEP Lipoprotein Measurement Working Group will recommend the CDC reference method (a chromotropic acid method) as the accepted accuracy base for TG. For similar reasons as apply to the HDL-C method, it is considered impractical to use the CDC's TG reference method as it

Table 11-5 ❖ CLIA-Approved Proficiency Testing Programs

Program	Sample Type	Telephone
Accutest	Lyophilized	800-356-6788
American Academy of Family Physicians	Lyophilized	800-274-2237
American Academy of Pediatrics	Lyophilized	708-228-5005
American Association of Bioanalysts	Lyophilized	800-234-5315
American Society of Internal Medicine	Lyophilized	800-338-2746
College of American Pathologists (CAP)	Lyophilized	800-323-4040
EXCEL/CAP	Lyophilized	800-323-4040
Idaho Bureau of Labs	Lyophilized	208-334-2235
New Jersey Health Department	Lyophilized	609-530-6172
New York State Health Department	Frozen liquid	518-474-8683
Pacific Biometrics Research Foundation	Fresh serum	206-233-9151
Puerto Rico Health Department	Lyophilized	809-764-6945
Wisconsin State Laboratory of Hygiene	Lyophilized	800-462-5261

Source: Health Care Financing Administration Division of Laboratory Standards & Performance/OSC, letter dated January 26, 1994

is now performed for providing TG reference services in the Network. The CDC and the Canadian Reference Foundation are collaborating to modify the reference method by combining the same extraction and hydrolysis steps as the current CDC reference method, followed by an enzymatic endpoint for determining TG. If successful, this method will have the advantages of ease in transfer and operation.

NATIONAL PROFICIENCY SURVEYS

The Clinical Laboratory Improvement Amendments (CLIA) passed by Congress in 1988 established standards designed to improve the quality of clinical laboratory testing in U.S. laboratories that conduct testing on human specimens for health assessment or for the diagnosis, prevention, or treatment of disease. CLIA '88 mandates proficiency testing as a means to externally evaluate the quality of a laboratory's performance. Each participating laboratory is challenged in three testing events annually. In each testing event, five unknown samples for each analyte or test are provided. Listed in Table 11-5 are the proficiency testing programs that have been approved by the Health Care Financing Administration.

All of the programs offer surveys that include TC, TG, and HDL-C, with the exception of the American Academy of Pediatrics, which provides TC only. Since all of the programs, except the Pacific Biometrics Research Foundation and the New York State Health Department, employ processed survey materials, assessing accuracy for most systems by comparison of survey results to the reference or definitive method values is not possible due to matrix effects. For this reason, participant performance is now evaluated using the peer group mean as the target value. The Pacific Biometrics Research Foundation program uses fresh human serum pools and the New York State Health Department uses frozen serum pools, thus eliminating matrix effects inherent in processed materials and permitting assessment of testing accuracy versus the reference method.

OTHER PROGRAMS

The University of Washington's Reference Lipoprotein Analysis Basic Survey

The Northwest Lipid Research Laboratories at the University of Washington, Seattle, conducts a quarterly survey called the Reference Lipoprotein Analysis Basic Survey (ReLABS). The goals of the program are to achieve and maintain the NCEP performance guidelines for cholesterol analysis and to standardize HDL-C, TG, LDL-C, and apo AI and B measurements through routine monitoring and traceability testing of laboratory performance using fresh patient specimens. Participating manufacturers and clinical laboratories receive quarterly statistical reports and troubleshooting assistance. Additionally, traceability of cholesterol testing to the National Reference System for Cholesterol is assessed every 6 months. A Certificate of Traceability is issued to laboratories that meet the established criteria for bias and precision in cholesterol testing. Traceability testing for HDL-C and TG analysis will be added to the program when these services are provided by the Cholesterol Reference Method Laboratory Network.

For more information, contact Northwest Lipid Research Laboratories, 2121 N. 35th Street, Seattle WA 98103, Telephone 206-685-3317.

Canadian Reference Foundation's LIPID^{PLUS} and Offsite^{PLUS}

The Canadian Reference Foundation (CRF) is a non-profit society which serves as the support center for standardization in Canada. The CRF operates a blinded external proficiency testing program which uses fresh human serum. With each LIPID^{PLUS} challenge, the laboratory receives by overnight courier three concentrations of fresh human serum on ice. Target values are assigned by the NRS/CHOL reference method and other standardized lipid and lipoprotein methods. CRF also administers a program called Offsite^{PLUS}, which is a blinded external proficiency testing program for cholesterol testing by finger-stick method using compact analyzers, offered in association with wellness programs.

For more information, contact The Canadian Reference Foundation, 307-2083 Alma Street, Vancouver, British Columbia V6R 4N6, Canada, Telephone 604-222-1879.

MANUFACTURERS' PROGRAMS

Apolipoproteins

The IFCC Committee on Apolipoproteins and manufacturers of apolipoprotein diagnostic kits began a collaborative program in 1988 to produce and evaluate SRMs for standardizing apolipoprotein measurements. In early 1989, more than 20 manufacturers and several reference laboratories evaluated 26 candidate SRMs and measured 10 frozen sera with assigned values for apo AI and B. A lyophilized serum for apo AI (SP1-01) and liquid stabilized serum for apo B (SP3-07) were selected as reference materials on the basis of homogeneity, stability, reproducibility, and linearity upon dilution. Upon submission to WHO, both were accepted as WHO Reference Reagents for apo AI and B, respectively. These reference reagents are available for use only by manufacturers in standardizing apolipoprotein diagnostic products.

For more information, contact Northwest Lipid Research Laboratories, Seattle, WA 98103, Telephone 206-685-3331.

NCCLS Subcommittee on Apolipoproteins

The Subcommittee on Apolipoproteins of the NCCLS has prepared a proposed guideline for apolipoprotein assays (I/LA 15-P). For information contact: NCCLS, Villanova, PA 19085, Telephone 215-525-2435.

SOURCES OF STANDARDS AND REFERENCE MATERIALS

Although matrix effects complicate the standardization of lipid and lipoprotein measurements, reference materials still play a significant role in assessing the performance of test systems. The matrix sensitivity of an analytical system (either homogeneous or heterogeneous) to a specific reference material should be established by concurrent analysis with fresh patient specimens. If an evaluated instrument system has been shown to be matrix insensitive or if a consistent matrix discrepancy can be documented for a particular material, then that material could be used with reasonable confidence to assess performance of the system.

Available standards, reference materials, and calibrators are listed in Tables 11-6 through 11-8. While these lists are certainly not comprehensive, they are representative of the resources currently available to the clinical laboratory. ❖

Note: Use of trade names in this chapter is for identification only and does not constitute endorsement by the Public Health Service or the U.S. Department of Health and Human Services.

Table 11-6 ❖ Primary Standards

Organization	*Telephone*	*Standard*
	PUBLIC SOURCE	
Natl. Institute for Standards & Technology	301-975-6776	Cholesterol (99.8%): SRM 911b Tripalmitin (99.5%): SRM 1595
	COMMERCIAL SOURCES	
Aldrich Chemical Co.	800-558-9160	Cholesterol (99+%) Glycerol (99.5%) Tripalmitin (99%)
Alfa Aesar (Morton Thiokol Inc.)	800-343-0660	Glycerol (99.5+%)
American Biorganics, Inc.	716-283-1434	Cholesterol (99%) Tripalmitin (99+%) Triolein (99%)
Boehringer Mannheim Biochemicals	800-428-5030	Cholesterol (Precical) Triolein (99%) Glycerol (99.5%)
Calbiochem Biochemicals	800-854-3417	Cholesterol (99+%)
Eastman Kodak Co. (Distributed by Fisher Scientific)	800-766-7000	Cholesterol (99.8%) Glycerol (99.5%) Tripalmitin
Gallard-Schlesinger	516-333-5600	Glycerol (99+%)
ICN Biomedicals, Inc.	714-545-0113	Cholesterol (99+%) Glycerol (99.5%) Triolein (95+%)
J.T. Baker Chemical Co.	800-582-2537	Cholesterol "Baker" Glycerol (99.5%)
Sigma Chemical Co.	800-325-3010	Cholesterol (> 99%) Glycerol (99%) Triolein (99%) Tripalmitin (99%)

Table 11-7 ❖ Secondary Reference Materials		
Organization	*Telephone*	*Reference Material(s)*
	PUBLIC SOURCES	
National Institute for Standards and Technology	301-975-6776	Cholesterol SRM 909c
Centers for Disease Control and Prevention	404-488-4126	Frozen serum pools prepared from human based material. Target values assigned by the CDC reference methods for cholesterol, TG, and HDL-C. Available, on a limited basis, through the CDC-NHLBI Lipid Standardization Program.
College of American Pathologists	708-446-8800	Reference Material Serum Lipids: RM016 Tri-level human based lyophilized reference materials. Assigned values for TC by NIST definitive method and CDC reference method Assigned values for TG and HDL-C assigned by CDC reference methods
	COMMERCIAL SOURCES	
Abbott Diagnostics	800-527-1869	Cholesterol calibrator for TDx Cholesterol and HDL-C standards for VP Cholesterol, HDL-C, and TG calibrator for Spectrum Cholesterol, TG, and HDL-C calibrator for Vision
Baxter Diagnostics, Inc.	305-633-6461	Cholesterol standard: 100 mg/dL (2.59 mmol/L) DADE Lipid-Trol: assigned for cholesterol and TG DADE Choles-Trol: assigned for cholesterol
Beckman Instruments, Inc.	800-526-3821	Cholesterol reference serum: 50, 200 mg/dL (1.30, 51.8 mmol/L) Elevated Lipid Control: cholesterol, TG, and HDL Triglyceride calibrator: 250 mg/dL (2.83 mmol/L) Decision Liquid Comprehensive Controls: 3 concentrations, value assigned for cholesterol and TG For Astra System: Calibrator E: TG Cal 1, Cal 2: multi-analyte For Synchron Systems CX4, CX5: HDL calibrator Multi-calibrator: cholesterol Multi-level controls: 3 concentrations, value assigned Triglyceride controls: 3 concentrations, value assigned

Continued

Table 11-7 ❖ Continued

Organization	Telephone	Reference Material(s)
Boehringer Mannheim	800-426-2336	Precical: serum cholesterol calibrator Precinorm L: cholesterol, TG, HDL-C calibrator Preciset cholesterol: aqueous Precimat glycerol: aqueous, 200 mg/dL (21.72 mmol/L) Precimat cholesterol: aqueous, 50 mg/dL (1.30 mmol/L) Precitrol A control serum Precitrol N control serum
CASCO Standards	800-367-4010	Cholesterol standards: serum, 5 concentrations 100–500 mg/dL (2.59–12.95 mmol/L) Triglyceride standards: aqueous glycerol, 5 concentrations 50–1000 mg/dL (5.43–108.6 mmol/L)
Diagnostic Chemicals	800-325-2436	DC-Trol multianalyte: 2 concentrations DC-Cal multianalyte calibrator HDL control: 2 concentrations
Dupont	800-435-7222	Cholesterol calibrator: 3 concentrations HDL calibrator: 3 concentrations Triglyceride calibrator: aqueous glycerol
Gilford Systems, Ciba-Corning Diagnostics	800-255-2121	Multicalibrator: cholesterol and TG HDL calibrator: 50, 100 mg/dL (1.3, 2.59 mmol/L)
Miles, Inc., Diagnostic Division	914-524-2265	Test Point 1 control: cholesterol, TG, HDL-C Test Point 2 control: cholesterol, TG, HDL-C Set Point Cal 1, Cal 2: for SMAC system Set Point Universal: Axon, DAX, Chem 1, RA systems
New England Reagent Laboratory	800-556-7575	NERL 1305: aqueous cholesterol standards NERL 1307: aqueous TG standards
Olympus Corporation	800-223-0125	Multianalyte calibrator: cholesterol and TG HDL standard
Pacific Biometrics Research Foundation	206-233-9151	Fresh frozen human serum pools with target values for lipids/lipoproteins assigned by accurate reference methods
Quantimetrix	800-624-8380	Normal lipid Control: assigned for cholesterol and TG Abnormal lipid control: assigned for cholesterol and TG HDL plus control: cholesterol, TG, HDL-C Lipoprotein control for Lipoprint system

Continued

Table 11–7 ❖ Continued

Organization	*Telephone*	*Reference Material(s)*
Reagents Applications, Inc.	800-438-6100	Cholesterol standards: 50, 100, 200, 300, 400 mg/dL (1.30, 2.59, 5.18, 7.77, 10.36 mmol/L) Triglyceride standards: glycerol, 100, 300, 500 mg/dL (10.86, 32.58, 54.30 mmol/L)
Roche Diagnostics Systems	800-526-1247	Roche calibrator (Cholesterol, TG, HDL-C)
Seradyn, Inc.	800-428-4007	Lipid control level 1 and 2 (cholesterol, TG, HDL-C, and LDL-C)
Sigma Chemical Co.	800-325-3010	Aqueous cholesterol standards: 100, 200, 400 mg/dL (2.59, 5.18, 10.36 mmol/L) Cholesterol stock standard: 500 mg/dL (12.95 mmol/L) HDL-C calibrator: 50 mg/dL (1.30 mmol/L) Cholesterol calibrator: 200 mg/dL (5.18 mmol/L) Triolein triglyceride standards
Solomon Park Research Laboratories	206-821-7005	Frozen fresh pooled human serum materials Cholesterol: value assigned by NRS/CHOL reference method HDL: value assigned by CDC standardized method Triglyceride: value assigned by CDC standardized method

Table 11–8 ❖ Sources of Reference Materials for Apolipoproteins*

Centers for Disease Control and Prevention apolipoprotein repository
WHO Reference Reagent SP1-01 for apo AI
WHO Reference Reagent SP3-07 for apo B

*Available to manufacturers only. Request from: Northwest Lipid Research Laboratories, Seattle, WA. Telephone 206-685-3317.

REFERENCES

1. Lipid Research Clinics Program. The Lipid Research Clinics Primary Prevention Trial results. I: Reduction in incidence of coronary heart disease. II: The relationship of reduction in incidence of coronary heart disease to cholesterol lowering. JAMA 1984;251:351–75.
2. The Expert Panel. Report of the National Cholesterol Education Program Expert Panel on Detection, Evaluation, and Treatment of High Blood Cholesterol in adults. Arch Intern Med. 1988;148:36–9.
3. Gordon DJ, Rifkind BM. High density lipoprotein: the clinical implications of recent studies. New England Journal of Medicine 1989;148:1311–16.
4. NIH Consensus Conference. Triglyceride, high-density lipoprotein, and coronary heart disease. JAMA 1993;269:505–10.
5. Consensus Development Conference. Treatment of hypertriglyceridemia. JAMA 1984;251:1196–1200.
6. Stein EA, Steiner PM. Triglyceride measurement and its relationship to heart disease. Clinics in Laboratory Medicine 1989; 9(1):169–85.
7. Austin MA. Plasma triglyceride and coronary heart disease. Arteriosclerosis and Thrombosis 1991;11:2–15.
8. Friedewald WT, Levy RI, Fredrickson DS. Estimation of the concentration of low-density lipoprotein cholesterol in plasma, without use of the preparative ultracentrifuge. Clin Chem 1972;18:499–502.
9. Albers JJ, Brunzell JD, Knopp RH. Apoprotein measurements and their clinical application. Clin Lab Med 1989;9:137–52.
10. The Expert Panel. Summary of the Second Report of the National Cholesterol Education Program (NCEP) Expert Panel on Detection, Evaluation, and Treatment of High Blood Cholesterol in Adults (Adult Treatment Panel II). JAMA 1993;269:3015–23.
11. Current status of blood cholesterol measurement in clinical laboratories in the United States: A report from the laboratory standardization panel of the National Cholesterol Education Program. Clin Chem 1988;34:193–201.
12. Duncan IW, Mather A, Cooper GR. The procedure for the proposed cholesterol reference method. Atlanta, Ga: Centers for Disease Control, 1988.
13. Cooper GR, Smith SJ, Duncan IW, et al. Interlaboratory testing of the transferability of a candidate reference method for total cholesterol in serum. Clin Chem 1986;32:921–9.
14. Cohen A, Hertz HS, Mendel J, et al. Total serum cholesterol by isotope dilution–mass spectrometry: a candidate definitive method. Clin Chem 1980;26:854–60.
15. Vanderlinde RE, Bowers GN, Jr, Schaffer R, Edwards GC. The National Reference System for Cholesterol. Clin Lab Med 1989;9:89–104.
16. Ellerbe P, Myers GL, Cooper GR, et al. A comparison of results for cholesterol in human serum obtained by the reference and by the definitive method of the National Reference System for Cholesterol. Clin Chem 1990;36:370–5.
17. Bernert JT, Jr, Akins JR, Cooper GR, et al. Factors influencing the accuracy of the National Reference System Total Cholesterol Reference Method. Clin Chem 1991;37:2053–61.
18. Hainline A, Jr, Karon J, Lippel K (eds.). Manual of laboratory operations: Lipid and lipoprotein analysis (2nd ed). [HEW Pub. No. (NIH) 75-628 (rev.). U.S. Government Printing Office Publication No. 1982-361-132:678.] Bethesda, MD: National Heart, Lung and Blood Institute, Lipid Research Clinics Program.
19. Carlson LA, Wadstrom LB. Determination of glycerides in blood serum. Clin Chim Acta 1959;4:197–205.
20. Carlson LA. Determination of serum triglycerides. J Ather Res 1963;3:334–6.

21. Van Handel E, Zilversmit DB. Micromethod for the direct determination of triglycerides in serum. J Lab Clin Med 1957;50:152–7.
22. Lofland HB, Jr. A semiautomated procedure for the determination of triglycerides in serum. Anal Biochem 1964;9:393–400.
23. Ellerbe P, Sniegoski LT, Welch MJ. Determination of total glycerides and triglyceride in human serum by isotope dilution–mass spectrometry as a candidate definitive method. (submitted to Clin Chem).
24. Shepherd J, Rosseneu M, Vercaemst R, Colinet E, Profilis C. Purification and certification of human apolipoprotein A-I and A-II reference materials (CRM 393 and 394). Community Bureau of Reference, 1991. (BCR Information series No. CD-NA-13393-EN-C, ISBN 92-826-2402-1.) Luxembourg: Commission of the European Communities, 1991, 78 pp.
25. Henderson LO, Powell MK, Smith SJ, et al. Impact of protein measurements on standardization of assays of apolipoprotein AI and B. Clin Chem 1990;36:1911–7.
26. Albers JJ, Marcovina SM. Standardization of apolipoprotein B and AI measurements. Clin Chem 1989;35:1357–61.
27. Barr JR, Maggio VL, Alexander LR, et al. Quantification of apolipoprotein A-I by isotope dilution FAB MS. Proceedings of the 41st American Society of Mass Spectrometry Conference on Mass Spectrometry and Allied Topics, San Francisco, CA, May 31, 1993.
28. Marcovina SM, Albers JJ, Henderson LO, Hannon WH. International Federation of Clinical Chemistry standardization project for measurement of apolipoproteins. III: Comparability of apo A-I values by the use of common reference material. Clin Chem 1993;39:773–81.
29. Naito HK. The need for accurate total cholesterol measurement. Clin Lab Med 1989;9:37–60.
30. Myers GL, Cooper GR, Winn CL, Smith SJ. The Center for Disease Control, National Heart, Lung and Blood Institute Lipid Standardization Program: An approach to accurate and precise lipid measurements. Clin Lab Med 1989;9:105–35.
31. Recommendations for improving cholesterol measurement: A report from the Laboratory Standardization Panel of the National Cholesterol Education Program. NIH Publication No: 90-2964. Bethesda, MD: National Institutes of Health, February 1990.
32. Cooper GR, Myers GL, Smith SJ, Sampson EJ. Standardization of lipid, lipoprotein, and apolipoprotein measurements. Clin Chem 1988;34:B95–B105.
33. Cooper GR, Myers GL, Smith SJ, Schlant RC. Blood lipid measurements. Variations and practical utility. JAMA 1992;267:1652–60.
34. Smith SJ, Cooper GR, Myers GL, Sampson EJ. Biological variability in the concentration of serum lipids: sources of variation among results from published studies and composite predicted values. Clin Chem 1993;39:1012–22.
35. Cooper GR, Smith SJ, Myers GL, et al. Estimating and minimizing effects of biologic sources of variation by relative range when measuring the mean of serum lipids and lipoproteins. Clin Chem 1994;40:227–32.
36. Cooper GR, Hannon WH, Henderson LO, Smith SJ. Apolipoprotein measurements: Pre-analytical issues and standardization. Fourth Asian-Pacific Congress of Clinical Biochemistry Proceedings. Hong Kong: Gardiner-Caldwell Communications, 1990:392-7.
37. Greenberg N, Li ZM, Bower GN. National Reference System for Cholesterol (NRS-CHOL): Problems with transfer of accuracy with matrix materials. Clin Chem 1988;34:1230–1.
38. Kroll MH, Chesler R, Elin RJ. Effect of the lyophilization on results of five enzymatic methods for cholesterol. Clin Chem 1989;35:1523–6.

39. Kroll MH, Chesler R. The effect of surfactant on the enzymatic determination of cholesterol with lyophilized materials. [Abstract] Clin Chem 1990;36:960.
40. Lasky FD, Powers DM, Hassemer DJ, Wiebe DA. Quality of fluids used in external QC programs affects the reliable assessment of accuracy of routine methods, as documented by cholesterol. In: Kawai T, Ohba Y, Kanno T, Kawano K, Ueda K, Tatsumi E, eds. Quality Control in the Clinical Laboratory '91. Princeton, NJ: Excerpta Medica 1992:199–208.
41. Koch DD, Hassemer DJ, Wiebe DA, et al. Testing cholesterol accuracy performance of several common laboratory instruments. JAMA 1988;260:2252–7.
42. Myers GL, Schap FD, Smith SJ, et al. CAP-CDC collaborative study for evaluating reference materials for total serum cholesterol measurements. Arch Pathol Lab Med 1990;114:1199–205.
43. Naito HK, Kwak YS, Hartfiel JL, et al. Matrix effects on proficiency testing materials: Impact on accuracy of cholesterol measurement in laboratories in the nation's largest hospital system. Arch Pathol Lab Med 1993;117:345–51.
44. Ross JW, Myers GL, Gilmore BF, et al. Matrix effects and the accuracy of cholesterol analysis. Arch Pathol Lab Med 1993;117:393–400.
45. Waymack PP, Miller WG, Myers GL. Assay instrument-dependent matrix effects in standardization of cholesterol measurements. Clin Chem 1993;39:2058–62.
46. Holani KK, Miller WG, Waymack PP. Robustness of three triglyceride reagents for matrix effects of proficiency testing materials. Clin Chem 1993;39:1126.
47. Miller WG, Levine J, Santulli M, et al. Effect of freezings on cholesterol in individual sera. [Abstract] Clin Chem 1990;36:965.
48. Smith SJ, Henderson LO, Hannon WH, et al. Effects of analytical method and lyophilized sera on measurements of apolipoproteins AI and B: An international survey. Clin Chem 1990;36:290–6.
49. Marcovina SM, Albers JJ, Kennedy H, et al. International Federation of Clinical Chemistry standardization project for measurement of apolipoproteins AI and B. IV: Comparability of apo B values using international reference materials. Clin Chem 1994;(in press).
50. Mei JV, Powell MK, Henderson LO, et al. Method-dependent variation in the stability of apolipoprotein B in a stabilized liquid reference material. Clin Chem 1994;(in press).
51. Rautela GS, Stater S, Arvon DA. Assessment of the need for triglyceride blank measurements. Clin Chem 1973;19:1193–5.
52. Jessen R, Cass C, Eckfeldt J. Do enzymatic analyses of serum triglycerides really need blanking for free glycerol? Clin Chem 1990;36:1372–5.
53. Cole T. Glycerol blanking in triglyceride assays: Is it necessary? Clin Chem 1990;36:1267–8.
54. Bernert JT, Jr, Bell CJ, McGuffey JE. Determination of free glycerol in human serum reference materials by isotope-dilution gas chromatography-mass spectrometry. J Chromatography 1992;578:1–7.
55. Klotzsch SG, McNamara JR. Triglyceride measurements: a review of methods and interferences [Review]. Clin Chem 1990;36:1605–13.
56. Sampson M, Ruddel M, Elin RJ. Effects of specimen turbidity and glycerol concentration on nine enzymatic methods for triglyceride determination. Clin Chem 1994;40:221–6.
57. Cooper GR, Smith SJ, Wiebe DA, et al. International survey of apolipoproteins AI and B measurements (1983–84). Clin Chem 1985;31:223–8.
58. Smith SJ, Henderson LO, Cooper GR, Hannon WH. An evaluation of the trends in analytical performance of international apolipoprotein AI and B assays. Clin Chem 1988;34:1644–6.

59. Smith SJ, Cooper GR, Henderson LO, Hannon WH. Apolipoprotein Standardization Collaborating Group: An international collaborative study on standardization of apolipoproteins AI and B. Part I: Evaluation of a lyophilized candidate reference and calibration material. Clin Chem 1987;33:2240–9.
60. Kimberly MM, Myers GL, Howard SM, Waymack PP. Performance of laboratories participating in the Cholesterol Reference Method Laboratory Network (CRMLN). [Abstract] Clin Chem 1992;38:1061.
61. National Committee for Clinical Laboratory Standards. Method comparison and bias estimation using patient samples: tentative guideline. (NCCLS document EP9-T, ISBN 1-56238-189-X). Villanova, PA: NCCLS, 1993.
62. Centers for Disease Control. Clinical laboratory measurements traceable to the National Reference System for Cholesterol. Morbidity and Mortality Weekly Report. U.S. Department of Health and Human Services, Public Health Service. 1994;43:149–50.
63. National Committee for Clinical Laboratory Standards. Development of designated comparison methods for analytes in the clinical laboratory: proposed guideline (2nd ed.). (NCCLS document NRSCL6-P2, ISBN 1-56238-227-6). Villanova, PA: NCCLS, 1993.

CHAPTER 12

Measurement and Clinical Significance of High-Density Lipoprotein Cholesterol Subclasses

G. Russell Warnick

CLINICAL SIGNIFICANCE

❖ The inverse relationship between high-density lipoprotein cholesterol (HDL-C) values and risk of coronary heart disease (CHD) is well established in epidemiological studies.[1-7] A consensus panel sponsored by the National Institutes of Health (NIH) and the Adult Treatment Panel of the National Cholesterol Education Program (NCEP) confirmed the protective role of high-density lipoprotein (HDL) and recommended its measurement in making treatment decisions.[8-11] Chapter 6 provides detailed information about the clinical value of HDL-C and recommendations for its measurement.

In fact, the predictive association between HDL and CHD risk is much more complicated because HDL is actually a heterogeneous mixture of lipoprotein particles differing in size, density, and lipid and apolipoprotein composition as well as function. In 1954, DeLalla, Elliot, and Gofman,[12] pioneers in establishing the implications of the lipoproteins in CHD risk, first reported the heterogeneous nature of HDL based on observations using the analytical ultracentrifuge. DeLalla and colleagues observed two major subclasses, designated as HDL_2 and HDL_3, as well as a relatively minor subclass designated as HDL_1. They also reported substantial gender differences, with females having significantly higher HDL_2 values than males, but relatively similar HDL_3 values. In 1966, these same researchers reported results from a prospective study indicating that subjects with CHD had lower HDL_2 values but similar HDL_3 values relative to subjects without CHD.[13]

These early reports on the association between CHD risk and HDL and its subclasses were largely ignored for many years but have been confirmed by more recent studies using a variety of measurement techniques and study designs with various populations.[14-22] For example, in a study of coronary angiography, Miller and colleagues[14] demonstrated that the extent of atherosclerosis is associated with decreased HDL_2-C rather than with HDL_3-C values. Ballantyne et al.[15] reported that HDL_2-C but not HDL_3-C was reduced in survivors of myocardial infarction relative to control subjects. In a more recent retrospective study,[21] concentrations of the larger HDL_2 particles were shown to be most strongly associated with extent and progression of CHD. However, results from a case-control design within the Physician's Health Study[22] indicated the strongest association between the smaller HDL_3 particles and CHD risk. The inference from these studies is that measurement of the HDL subclasses might improve the prediction of risk for CHD.

HDL is in fact more heterogeneous than implied by these studies, which examined major HDL subclasses obtained by ultracentrifugation or gradient electrophoresis. Isoelectric focusing, a high-resolution technique that achieves separations based on particle charge, reportedly resolved HDL_3 into as many as 12 separate peaks[23] and HDL_2 into at least 5,[24] suggesting that the common HDL_2/HDL_3 fractions are still heterogeneous. One of the newer gradient gel electrophoresis methods, resolving by particle size, obtains 10 bands in the HDL_2 class and 4 in the HDL_3 class.[25] These particles, when fractionated by density, charge, or size, contain varying combinations of apolipoproteins.[26, 27] The major distinction between particles involves those that contain apolipoprotein (apo) AI without apo AII and those that contain apo AI with apo AII. In addition, there are minor particles with only apo AII and combinations with other HDL proteins such as apo E. A particle within the HDL range containing only apo E was recently reported.[28]

These studies emphasize the considerable particle heterogeneity within the major HDL subclasses. Depending on the properties exploited in achieving separations, different subpopulations of HDL particles may be obtained, and particles of varying function may be included within the separated subclasses. Therefore, the apparent associations between CHD risk and the major HDL subclasses may vary by separation method. Nevertheless, quantitation of HDL subclasses does potentially improve assessment of CHD risk, although the choice of method and fraction is somewhat uncertain. The implications of HDL subclasses are perhaps in some respects analogous to those of total cholesterol (TC), whose value as a predictor of CHD risk can be improved by quantitating the individual lipoproteins, the risk-positive low-density lipoprotein cholesterol (LDL-C), and the risk-negative HDL-C.

Measurement of HDL subclasses has been a useful research tool in studies of lipoprotein metabolism, pathophysiology, and association with CHD risk. One must recognize that the variety of methods used in published clinical applications use the same nomenclature to refer to subclasses which in fact comprise different populations of particles. HDL is still generally separated into the two major subclasses, HDL_2 and HDL_3, with hydrated densities originally considered to be 1.063–1.125 Kg/L and 1.125–1.210 Kg/L, respectively.[12, 29] Inter-individual differences in HDL values are largely due to the larger and lighter HDL_2; the smaller, more dense HDL_3 fraction is relatively more consistent within and among individuals.[10]

The HDL_2 and HDL_3 fractions appear to be metabolically interrelated.[30] Patsch et al.[31] demonstrated that particles in the HDL_3 range are converted into HDL_2-like particles during lipolysis of very-low-density lipoprotein (VLDL) by hepatic lipoprotein lipase. Patsch and colleagues hypothesized that HDL_3 assimilates surface components, cholesterol, and phospholipids derived from lipolysis of triglyceride-rich VLDL particles, and that it is thereby converted to the larger and lipid-enriched HDL_2. The lipid enrichment of HDL_3 may also result from transfers from cell membranes as the particles transit the circulation. A parallel process in the conversion of HDL_3 to HDL_2 is the action of lecithin cholesterol acyltransferase (LCAT), which esterifies the free surface cholesterol, thereby moving it to the core. In the process of cholesterol removal, designated as reverse cholesterol transport, the esterified cholesterol can be transferred to low-density lipoprotein (LDL) for removal by the liver, or the HDL particles can be removed by specific receptor-mediated processes. These interconversions of HDL subclasses are considered reversible but are highly complex and have yet to be fully elucidated. (For additional information on this pathway, refer to Chapter 1.)

Factors that influence the plasma HDL-C concentration often have differential effects on the two major subclasses. For example, the increased HDL values in women are primarily due to increased HDL_2.[32] High-estrogen oral contraceptives increase HDL_2 values but have no discernible effect on HDL_3.[32] Exercise increases the HDL_2 and may also promote a slight decrease in the HDL_3.[33, 34] Thus, female gender, estrogen use, and exercise are all associated with increased HDL_2 and thereby with decreased risk of CHD. Alcohol intake is known to increase HDL values, and moderate intake is associated with a decreased risk of CHD.[35, 36] One might expect that this benefit is mediated by increasing the HDL_2 fraction, but in fact Haskell et al.[37] found that, among moderate drinkers, HDL_3 concentrations decreased with abstention from alcohol and increased upon resumption of drinking; there was no significant effect on the reputedly anti-atherogenic HDL_2 fraction. Haffner et al.[38] also reported that HDL_3, rather than HDL_2, was increased by alcohol consumption and decreased by smoking. These latter studies emphasize the complexities of the HDL subclass associations with metabolic factors and CHD risk.

OVERVIEW OF METHODS

Methods for separation of HDL subclasses are summarized in Table 12–1 and discussed in the following sections.

Ultracentrifugation Methods

The HDL subclasses were first identified by using analytical ultracentrifugation and observing their flotation rates through Schlieren patterns.[12] Since analytical ultracentrifugation is technically demanding and requires very expensive, specialized equipment, it is rarely used today, even in specialty laboratories. Although updated methods have been reported,[39] analytical ultracentrifugation is generally reserved for intensive research and is not considered practical for use in large-scale studies.

Table 12-1 ❖ Methods for Separation of HDL Subclasses

Property	*Method*
Density	Analytical ultracentrifugation
	Sequential density ultracentrifugation
	Density gradient ultracentrifugation:
	Swinging bucket rotors
	Vertical rotors
	Zonal rotors
Size	Gradient gel electrophoresis
	Chromatographic procedures
Charge and Size	Chemical precipitation:
	Heparin-Mn^{2+} and 15 kDa dextran sulfate
	Dextran sulfate (50 kDa) and Mg^{2+}
	Polyethylene glycol
	Polyethylene glycol and 15 kDa dextran sulfate
Apolipoprotein composition	Immunoaffinity fractionation
	Enzyme-linked immunosorbent assay
	Differential electroimmunoassay

Sequential Density Ultracentrifugation

Sequential density ultracentrifugation is more common and is considered by most to be the traditional standard for accuracy in HDL subclass quantitation.[12, 29] Total HDL can be isolated by flotation within the 1.063–1.21 Kg/L density interval after prior separation by flotation of VLDL and LDL at density 1.063 Kg/L. Specimens are adjusted to the indicated density by addition of a salt such as KBr and subjected to ultracentrifugation at forces upwards of 100,000 × g, during which lighter particles float to the top and heavier particles sink to the bottom. The fractions are recovered by careful pipetting, often in conjunction with tube slicing to separate the top and bottom fractions. The bottom fraction is adjusted to the next sequential density to cause flotation of the heavier particles. An advantage of sequential flotation methods is the ability to recover fractions on a preparative scale for physical or chemical analysis. The range of densities conferred by the heterogeneity in lipid to protein proportions makes density separation appropriate. Considerable versatility is afforded by the variety of ultracentrifuges and rotors available today, accommodating large volumes and many samples in conventional floor models and convenient table-top models suitable for micro-scale fractionation.

Sequential density ultracentrifugation does have its disadvantages, however, including the fact that only one fraction is isolated in each centrifugation step, thus requiring a sequential centrifugation step for each fraction. The labile lipoproteins may be altered by the harsh conditions, high salt, and high gravity, with loss of loosely bound apolipoproteins or deterioration during the course of a multi-step ultracentrifugal procedure.[40] In addition, the recovered HDL is likely to be contaminated with apo-B-containing lipoproteins in the low end of the density range. For example, the density range of lipoprotein(a) [Lp(a)] overlaps that of HDL. Sequential density ultracentrifugation is tedious and very time consuming, and the specialized equipment and proficiency required may not be available in many laboratories.

A more convenient variation of sequential density ultracentrifugation for quantitation of the HDL_2/HDL_3 subclasses involves a single fractionation of serum, usually performed at density 1.125 Kg/L.[41, 42] Cholesterol recovered in the bottom fraction is taken as a measure of HDL_3. Total HDL-C can be determined by a conventional precipitation method and the HDL_2-C is calculated by difference (total HDL-C - HDL_3-C.) An alternative procedure involves density adjustment and centrifugation of the HDL supernate obtained by a precipitation method.[43–46] A more recent evaluation[47] suggests that separation at a density of 1.100 Kg/L gives more specific fractionation in a single centrifugation.

As indicated previously, the major HDL subclasses, HDL_2 and HDL_3, were originally separated at a density of 1.125 Kg/L.[12, 29] In 1972, Kostner and Alaupovic[48] recommended that the optimal density for separation of the two major subclasses was 1.100 Kg/L. In a landmark study reported by Anderson et al.[49] in 1977, sequential density separation was used in combination with density gradient ultracentrifugation to subdivide the HDL_2 particles into two subclasses, HDL_{2b} (d = 1.063–1.100 Kg/L) and HDL_{2a} (d = 1.100–1.125 Kg/L). Using this method, the lightest subclass—HDL_{2b}—was clearly separated from HDL_{2a} and HDL_3, while the HDL_{2a} subclass showed appreciable overlap with HDL_3, suggesting that fractionation at a density of 1.125 Kg/L cross-contaminates the HDL_{2a} and HDL_3. These researchers decided to seek three major density subfractions because they had observed three major bands using an early version of gradient gel electrophoresis. This observation supports recent recommendations for a major fractionation at 1.100 Kg/L.[47, 48] There has been no general consensus on the choice of density, but there is considerable evidence that the lower density cutpoint is more specific.

Density Gradient Ultracentrifugation

Density gradient ultracentrifugation methods have been used to separate the major lipoprotein classes including HDL in a single spin,[50, 51] which for some purposes has an advantage over sequential methods. The major HDL subclasses, and even some of the minor subclasses, can also be separated under appropriate conditions. Density gradient fractionation methods have been described for fixed-angle, vertical, swinging-bucket, and zonal rotors. Separations may be made in continuous (smooth) gradients, or, more commonly, in discontinuous (stepwise) gradients. Bands can then be recovered based on rate of flotation (non-equilibrium) or after fractions reach their equilibrium densities. Fractions may be recovered by pipetting from the top of the tube or by puncturing the tube and collecting the effluent from the bottom in appropriate fractions. The isolated fractions are available for analysis and are usually quantified in terms of the cholesterol content and sometimes the amounts of other lipids and apolipoproteins or total protein. Since the density gradient methods give reasonably good resolution in a single centrifugation, they are less likely to alter the lipoproteins. Depending on the conditions of ultracentrifugation, the fractions may not be completely separated and may be contaminated with other plasma proteins. The number of specimens that can be processed at one time is determined by the capacity of the rotor, but is usually lower than with the sequential density techniques. Meticulous technique is required to achieve the reproducibility required for analytical purposes. Gradient ultracentrifugation is usually used in the research-oriented laboratories for small-scale studies.

Three density gradient procedures for isolating HDL subclasses are described here as examples of the technique. A representative approach for processing several specimens simultaneously employs a swinging-bucket rotor.[52-55] Specimens are adjusted to appropriate density and are often pre-stained with a lipophilic dye. They are then introduced into a stepwise gradient established with the tube in the vertical position. The tubes are placed into hinged buckets which allow the tubes to swing out to the horizontal during centrifugation, minimizing disruption of the gradient and affording good separation of the lipoproteins as they migrate along the length of the tube. Following centrifugation, typically for 20–30 h, the tubes return to the vertical and fractions are recovered for analysis.

A variation of this approach using a vertical rotor has been refined by Segrest et al.[56-58] into a complete analytical system designated as the vertical autoprofiler (VAP). Specimens are similarly introduced into a gradient, but tubes are placed into cavities in the vertical rotor which hold the tubes upright and at right angles to the gravitational force at speed. Rather than the tubes swinging out, the gradient reorients so that separation occurs more rapidly along the shorter side-to-side distance. As the rotor slows, the gradient shifts back to the length of the tube. Fractions may be recovered by puncturing the bottom of the tube. In the VAP procedure, the effluent drawn from the bottom of the tube is continuously analyzed for cholesterol using a continuous-flow chemistry analyzer. The effluent can be split, with a portion going to a fraction collector or used for continuous, simultaneous analysis of other constituents. The cholesterol profile of the effluent is deconvoluted by computerized analysis with quantitation of major lipoproteins and minor subclasses including HDL_2 and HDL_3.

The vertical rotor method has the advantage of giving a complete lipoprotein profile in a single spin, yet it does suffer the general disadvantages of sequential density ultracentrifugation. An additional problem is that some lipoprotein material may adhere to the sides of the tube and not be recovered.

Another gradient method that has value for large-scale preparative isolations but has also been used as a standard for quantitation involves zonal ultracentrifugation.[59–63] For separation of HDL subclasses, a sample is introduced into a spinning single-cavity rotor containing a three-step gradient. After centrifugation for 24 h, fractions are recovered from the slowly spinning rotor for analysis. This large-scale method has better resolving power than other density gradient methods, with three fractions in the HDL_2 range and five in the HDL_3.[60] The major fractions are designated as HDL_2, HDL_{3L}, and HDL_{3D}. Since this method accommodates only one specimen per run, it is generally reserved for detailed studies of compositional changes in metabolism and lipoprotein characterization. The highly specialized equipment and expertise required are available in only a few centers.

Gradient Gel Electrophoresis

In electrophoretic methods using a homogeneous gel, particles are separated on the basis of their charge and size; mobility is enhanced by charge and inhibited by size. When a gradient gel is used, the separation is based on size: particles migrate until their size restricts further penetration into the gradient.[26] Gradient gel electrophoresis (GGE) in 4–30% gels has been used to study HDL particle size profiles in normal human plasma.[64–66] HDL can be detected directly from plasma using lipid staining after electrophoresis, or a protein stain can be used with the ≤ 1.21 Kg/L density fraction isolated by ultracentrifugation. Blanche et al.[64] observed five HDL subclasses, designated $(HDL_{2a})_{gge}$, $(HDL_{2b})_{gge}$, $(HDL_{3a})_{gge}$, $(HDL_{3b})_{gge}$, and $(HDL_{3c})_{gge}$. The $(HDL_{2b})_{gge}$ particles correspond to the HDL_{2b} component obtained by gradient ultracentrifugation with a mean hydrated density of 1.090 Kg/L. The $(HDL_{3a})_{gge}$ and $(HDL_{3b})_{gge}$ particles correspond to HDL_{3L}, with a mean hydrated density of 1.140 Kg/L, and to HDL_{3D}, with a mean hydrated density of 1.160 Kg/L, identified by rate zonal ultracentrifugation.[60] The $(HDL_{2a})_{gge}$ and $(HDL_{3c})_{gge}$ particles are minor particles which are observed less frequently in normal subjects. More recently, Li et al.[25] have reported 14 peaks obtained with GGE in 4–30% gels: 7 peaks corresponding to the HDL_{2b} range, 3 peaks corresponding to HDL_{2a}, and 4 peaks corresponding to HDL_3. Another approach[67] uses a discontinuous gradient to resolve HDL_2 and HDL_3.

GGE is a relatively rapid and reproducible method for determining HDL subclasses which are in reasonably good agreement with those obtained by ultracentrifugation methods. However, because quantitation is based on staining either by lipophilic dyes or protein stains, results may not agree with ultracentrifugal quantitations based on cholesterol content. The particles may be only partially resolved and are available in very limited quantities for characterization. Nevertheless, GGE can provide highly useful information quite efficiently.

Chromatographic Methods

Various chromatographic methods also separate HDL subclasses based on particle size.[68, 69] A gel permeation column in high-performance liquid chromatography (HPLC) with quantitation of eluate fractions by enzymic cholesterol assay gave separations that corresponded reasonably well to those obtained by sequential density ultracentrifugation.[68] Resolution of two HDL_2 and three HDL_3 fractions, approximating those observed with GGE, was obtained with a Superose 6B column in fast protein liquid chromatography.[69]

Precipitation Methods

The most common methods for routine separation of the total HDL class involve chemical precipitation of VLDL and LDL with quantitation of HDL as cholesterol remaining in the supernate (see Chapter 6). The precipitation techniques are convenient and are amenable to processing large numbers of specimens without the need for expensive or specialized equipment; hence, it is not surprising that precipitation methods were adapted for quantitation of the major HDL subclasses. Selective precipitation exploits differences in size and charge properties of the lipoproteins. Since the HDL subclasses demonstrate relatively smaller differences in these properties than the major lipoproteins, their separation by precipitation is more tenuous. Nevertheless, with careful standardization and attention to technique, the precipitation methods have advantages, especially for large population studies.

One early method[70] involved the addition of heparin-Mn^{2+} for removal of VLDL/LDL, followed by addition to the HDL supernate of 15 kDa dextran sulfate to precipitate HDL_2. Cholesterol was assayed in the supernates and HDL_2 calculated by subtracting HDL_3-C from total HDL-C. The optimum dextran sulfate concentration was determined in comparisons with ultracentrifugation at a density of 1.125 Kg/L. In pooled sera, the amount of dextran sulfate required to achieve separation by precipitation equivalent to the ultracentrifugal fractionation varied over a twofold range. At the concentration of dextran sulfate considered to be optimal for patient specimens, precipitation results correlated reasonably well with ultracentrifugation. The authors concluded that the precipitation method is best suited for large-scale trials, and recommended optimizing concentrations for the conditions of a particular study.

Subsequent evaluations of the method have confirmed that this method does give convenient and reasonably accurate quantitation of the HDL_2/HDL_3 subclasses, but that it is not highly specific nor rugged.[71-75] Most researchers have endorsed the recommendation that the method be used for large-scale trials and not for routine evaluation of patients. One report cited interference from blood clotting proteins and the need for substantially different dextran sulfate concentrations for serum than for EDTA plasma.[76] A particular disadvantage of this method is that the manganese cation interferes with the common enzymic cholesterol assays; for this reason, the use of heparin and manganese for precipitation has diminished with increased use of the enzymic assays in both routine and research laboratories.

A similar precipitation procedure[77, 78] uses dextran sulfate (50 kDa) and Mg^{2+} to separate total HDL first and then HDL_3 in sequence. The first version precipitated HDL_2 from the HDL supernate, whereas a subsequent modification,[79] considered more convenient and equally accurate, precipitated VLDL, LDL, and HDL_2 together from a second plasma aliquot. Evaluators of the first version observed results in good agreement with zonal ultracentrifugation[62] and concluded that the convenient precipitation method is well suited for population studies. Since this method uses the dextran sulfate reagent, which is compatible with enzymic cholesterol assays and is increasingly common for routine HDL separations, the dextran sulfate technique is described in more detail later in this chapter.

HDL subclasses have also been separated in a dual precipitation procedure using polyethylene glycol (PEG) 6,000 to remove VLDL and LDL from serum, followed by addition of 15 kDa dextran sulfate to precipitate HDL_2 from the first supernate.[80] One evaluator observed agreement with ultracentrifugation,[81] although a subsequent evaluator urged caution in using the method for routine characterization of patients.[82] Another method used PEG for both steps, the first precipitating VLDL and LDL and the second precipitating VLDL, LDL, and HDL_2 from a second aliquot of serum.[83–85] A comparison between these two types of PEG methods revealed substantial disagreement.[86] Use of PEG for HDL separations is common in Europe but not in the United States. As a precipitant, PEG is much less effective than the polyanions (heparin and dextran sulfate), and the hundredfold higher concentrations required result in highly viscous solutions which are difficult to pipette reproducibly. The inexpensive, single reagent is an advantage, but there are batch-to-batch differences that must be considered in formulating a consistent reagent solution.

Particle Assays

HDL particles can be separated based on the particular protein constituents.[26-27] The major classes contain either apo AI alone or apo AI and apo AII together. These classes of particles have been fractionated by immuno-affinity techniques using antibodies specific for the two apolipoproteins.[26, 87] Antibodies to apo AII bind the latter particles, after which those particles with apo AI only can be bound by antibody to apo AI. The two types of particles have also been quantified by enzyme-linked immunosorbent assay (ELISA) in a sandwich format.[27] The antibody to apo AII coated on microtiter plates binds those particles that contain apo AII. A second enzyme-labeled antibody to apo AI quantitates the apo AI:AII species in terms of the apo AI content. Total apo AI is quantitated by a conventional ELISA and the apo-AI-only species is calculated by difference.

A differential electroimmunoassay is available commercially.[88] Specimens are electrophoresed through a gel containing a high concentration of antibody to apo AII and a normal concentration of antibody to apo AI. Particles with both apolipoproteins are precipitated in a small "rocket" near the origin, leaving those particles with apo AI only to migrate into the gel. The conventional rockets formed are then quantified compared to a standard. The apo AI:AII particles are calculated by difference.

Convenient Precipitation Method for Large-Scale Studies

This method is a variation of the dextran sulfate method for quantitation of total HDL-C, as described in Chapter 6.[77-78] HDL-C is quantitated by measuring cholesterol in the first supernate after precipitation of VLDL and LDL. HDL_3 is measured in a second supernate from a second aliquot of serum after precipitation of HDL_2, LDL, and VLDL. HDL_2 is calculated as the difference in cholesterol between the two supernates. Use of the method for characterization of CHD risk in patients in the routine laboratory is not recommended, but the procedure is considered suitable for research investigations.[62]

The dextran sulfate is of 50,000 kDa (Dextralip 50 from Genzyme Corp., Boston, MA). Reagent-grade magnesium chloride and sodium azide are available from several sources.

Solutions

Total HDL Reagent (DS1) Preparation: Weigh out 1.0 g dextran sulfate, 10.16 g $MgCl_2 \cdot 6H_2O$, and 50 mg NaN_3. Dissolve and mix well in deionized water. QS to 100 mL. It is not advisable to adjust pH, as fluctuations in ionic strength resulting from pH adjustment seem to contribute to more variability than the slight batch-to-batch differences in pH. The final working reagent contains 10 mg/mL dextran sulfate, 0.5 mol/L $MgCl_2$, and 0.05% NaN_3.

HDL_3 Reagent (DS2) Preparation: Weigh out 1.91 g dextran sulfate, 39.74 g $MgCl_2 \cdot 6H_2O$, and 50 mg NaN_3. Dissolve and mix well in deionized water. QS to 100 mL. The final working reagent contains 19.1 mg/mL dextran sulfate, 1.95 mol/L $MgCl_2$, and 0.05% NaN_3.

Two or more quality control pools are included in the precipitation steps and in the cholesterol analysis together with patient specimens. Additional pool(s) with low cholesterol in the HDL/HDL_3 range can be included in the analysis step only.

Precipitation Sequence

1. Allow specimens to equilibrate to room temperature and mix thoroughly before pipetting.
2. Pipette 0.5 mL specimen or control into appropriately labeled tubes. Pipette one aliquot for total HDL (DS1) and one for HDL_3 (DS2) determination. Volumes can be adjusted provided reagent volumes are adjusted accordingly.
3. Add 50 µL of the appropriate reagent (DS1 or DS2) to each tube and vortex for 5 sec.
4. Allow tubes to stand at room temperature for 5–30 min.
5. Centrifuge for 5 min. at 12,000 × g. Alternatively, centrifuge at 1500 × g for 30 min. or some intermediate combination of g and minutes.

6. Visually examine each supernate for turbidity. If supernate is clear, transfer the supernate—carefully, so as not to disturb the pellet—to a storage vial or sample cup.

   Note: Incomplete sedimentation of apo-B-containing lipoproteins is indicated by the presence of cloudy or turbid supernates in hypertriglyceridemic specimens. Cloudy supernates may be cleared by ultrafiltration with a 0.22 μm filter or by repeating the precipitation procedure with the original specimen diluted 1:1 with saline. Alternatively, the dilution can be made subsequently on turbid supernates. Simply mix equal volumes of the turbid supernate with an equal volume of saline solution containing precipitation reagent at the same concentration as the original specimen.

7. Perform the cholesterol analysis on clear supernates.

   Note: The cholesterol assay should be optimized for good performance in the low HDL/HDL_3 cholesterol concentration range. Analysis of the total HDL and HDL_3 supernates consecutively for each specimen may decrease analytical variation and improve precision in the calculation of HDL_2 (total HDL-C – HDL_3-C).

8. Correct cholesterol results in the supernates for reagent dilution by multiplying by the factor 1.1. If a turbid supernate was diluted with saline, further multiply the result by the dilution factor (e.g., × 2).

9. Calculate the HDL_2 cholesterol value as: Total HDL – HDL_3 = HDL_2

Since the separation of HDL subclasses is relatively sensitive to reaction conditions and technique, the method should be validated for individual laboratory conditions. Comparison may be made to ultracentrifugation or to an experienced laboratory. Standardization of conditions and technique are essential to achieving reproducible results.

SUMMARY

HDL is highly heterogeneous, consisting of particles with a range of size, density, and composition and with a variety of different functions. While the HDL class in its entirety is clearly anti-atherogenic and predictive of CHD risk, subclasses of particles may be more or less protective. Quantitation of specific subclasses may be useful in research applications and provide a better indication of CHD risk, just as measurement of the major lipoproteins, HDL and LDL, better assesses overall risk than TC alone. Drawbacks are that the available separation methods are technically difficult and tedious (e.g., ultracentrifugation); require specialized equipment and skills (e.g., GGE and column chromatography); or are highly sensitive to reaction conditions (e.g., precipitation). Therefore, the measurement of the subclasses is still somewhat tenuous and their predictive value is still uncertain. With further refinements in methods and validation in research investigations, measurement of certain HDL subspecies may eventually improve the prediction of CHD risk. ❖

Acknowledgments: The author wishes to thank Sherry Hamman, Shelley Tellefson, and Gayle Tjersland for their assistance in the preparation of this chapter, and Judy McNamara and Nader Rifai for their helpful comments on the manuscript.

REFERENCES

1. Miller GJ, Miller NE. Plasma high density lipoprotein concentration and development of ischemic heart disease. Lancet 1975;i:16–9.
2. Rhoads GG, Gulbrandsen CL, Kagan A. Serum lipoproteins and coronary heart disease in a population study of Hawaii Japanese men. N Engl J Med 1976; 294:293–8.
3. Gordon T, Castelli WP, Hjortland MC, et al. High density lipoprotein as a protective factor against coronary heart disease: The Framingham Study. Am J Med 1977;62;707–14.
4. Kannel WB, Castell WP, Gordon T. Cholesterol in the prediction of atherosclerotic disease: new perspectives based on The Framingham Study. Ann Intern Med 1979;90:85–91.
5. Miller NE, Thelle DS, Forde OH, et al. The Tromso Heart Study. High-density lipoprotein and coronary heart-disease: a prospective case-control study. Lancet 1977;1:965–8.
6. Abbott RD, Wilson PWF, Kannel WB, et al. High density lipoprotein cholesterol, total cholesterol screening, and myocardial infarction. The Framingham Study. Arteriosclerosis 1988;8/3:209–211.
7. Jacobs DR Jr., Mebane IL, Bangdiwala SI, et al. High density lipoprotein cholesterol as a predictor of cardiovascular disease mortality in men and women: the follow-up study of the Lipid Research Clinics prevalence study. Am J Epidemiology 1990;131/1:32–47.
8. NIH Consensus Development Conference. Triglyceride, high density lipoprotein, and coronary heart disease. Consensus statement. Bethesda, MD: National Institutes of Health, Feb. 26–28, 1992.
9. National Cholesterol Education Program. Report of the Expert Panel on Detection, Evaluation, and Treatment of High Blood Cholesterol in Adults. Arch Intern Med 1988;148:36–69.
10. Grundy SM, et al. Summary of the Second Report of the National Cholesterol Education Program (NCEP) Expert Panel on Detection, Evaluation and Treatment of High Blood Cholesterol in Adults (Adult Treatment Panel II). JAMA 1993; 269(23);3015–23.
11. National Cholesterol Education Program: Second report of the expert panel on detection, evaluation, and treatment of high blood cholesterol in adults. NIH Publication No. 93-3095. Bethesda, MD: National Institutes of Health, 1993.
12. DeLalla OF, Elliot HA, Gofman JW. Ultracentrifugal studies of high density serum lipoproteins in clinically healthy adults. Am J Physiol 1954;179:333–7.
13. Gofman JW, Young W, Tandy R. Ischemic heart disease, atherosclerosis, and longevity. Circulation 1966;34:679–97.
14. Miller NE, Hammett F, Saltissi S, et al. Relation of angiographically defined coronary artery disease to plasma lipoprotein subfractions and apolipoproteins. Br Med J 1981;282:1741-4.

15. Ballantyne FC, Clark RS, Simpson HS, et al. High density and low density lipoprotein subfractions in survivors of myocardial infarction and in control subjects. Metabolism 1982;31:433–9.
16. Nichols AV. Human serum lipoproteins and their interrelationships. Adv Biol Med Phys 1967;11:109–58.
17. Anderson DW, Nichols AV, Pan SS, Lindgren FT. High density lipoprotein distribution: resolution and determination of three major components in a normal population sample. Atherosclerosis 1978;29:161–79.
18. Glueck CJ, Fallat RW, Millett F et al. Familial hyperalphalipoproteinemia. Arch Intern Med 1975;135:1025–8.
19. Glueck CJ, Gartside P, Fallat RW, et al. Longevity syndromes: familial hypobeta and familial hyperalpha lipoproteinemia. J Lab Clin Med 1976;88:941–57.
20. Patsch W, Kuisk I, Glueck C, et al. Lipoproteins in familial hyperalphalipoproteinemia. Arteriosclerosis 1981;1:156–61.
21. Johanson J, Carlson LA, Landou C, et al. High density lipoproteins and coronary atherosclerosis. A strong inverse relation with the largest particles is confined to normotriglyceridemic patients. Arteriosclerosis and Thrombosis 1991;11(1):174–82.
22. Stampfer MJ, Sacks FM, Salvini S, et al. A prospective study of cholesterol, apolipoproteins, and the risk of myocardial infarction. The New England Journal of Medicine 1991;325(6):377–81.
23. Sundaram GS, Mackenzie SL, Sodhi HS. Preparative isoelectric focusing of human serum high-density lipoprotein (HDL3). Biochimica et Biophysica Acta 1974;337:196–203.
24. Mackenzie SL, Sundaram GS, Sodhi HS. Heterogeneity of human serum high-density lipoprotein (HDL2). Clin Chim Acta 1973;43:223–9.
25. Li Z, McNamara JR, Ordovas JM, et al. Analysis of high density lipoproteins by a modified gradient gel electrophoresis method. FASEB J 1993;7:A730.
26. Cheung MC, Albers JJ. Distribution of cholesterol and apolipoprotein A-I and A-II in human high density lipoprotein subfractions separated by CsCl equilibrium gradient centrifugation: evidence for HDL subpopulations with differing A-I/A-II molar ratios. J Lipid Res 1979;20:200–7.
27. Fruchart JC, Bard JM. Lipoprotein particle measurement: an alternative approach to classification of lipid disorders. Curr Opin Lipid 1991;2:362–6.
28. Huang Y, von Eckardstein A, Wi S, et al. A plasma lipoprotein containing only apolipoprotein E and with γ mobility on electrophoresis releases cholesterol from cells. Proc Natl Acad Sci USA 1994;91:1834–8.
29. Havel RJ, Eder HA, Bragdon JH. The distribution and chemical composition of ultracentrifugally separated lipoproteins in human serum. J Clin Invest 1955;34:1345.
30. Eisenberg S. Plasma lipoproteins. In: Albers JJ, Segrest JP, eds. Methods in enzymology, Vol. 129, Orlando, FL: Academic Press, 1986:347–66.
31. Patsch JR, Gotto AM, Olivercrona T, et al. Formation of high density lipoprotein 2-like particles during the lipolysis of very low density lipoprotein in vitro. Proc Nat Acad Sci 1978;75:4519–23.
32. Krauss RM. Regulation of high density lipoprotein levels. Med Clin North Am 1982;66:403–30.
33. Krauss RM, Wood PD, Giotas C, et al. Heparin-released plasma lipase activities and lipoprotein levels in distance runners [abstract]. Circulation 1979;59 & 60(Suppl 2):II–73.
34. Wood PD, Haskell WL, Blair SN, et al. Increased exercise level and plasma lipoprotein concentrations: a one-year, randomized, controlled study in sedentary middle-aged men. Metabolism 1983;32:31–9.

35. Klatsky AL, Friedman GD, Siegelaub AB. Alcohol consumption before myocardial infarction: results from the Kaiser-Permanente epidemiologic study of myocardial infarction. Ann Intern Med 1974;81:294–301.
36. Dyer AR, Stamler J, Paul O, et al. Alcohol consumption and 17-year mortality in the Chicago Western Electric Company study. Prev Med 1980;9:78–90.
37. Haskell WL, Camargo C, Williams PT, et al. The effect of cessation and resumption of moderate alcohol in take on serum high-density-lipoprotein subfractions. The New England Journal of Medicine 1984;310(13):805–10.
38. Haffner SM, Applebaum-Bowden D, Wahl PW, et al. Epidemiological correlates of high density lipoprotein subfractions, apolipoproteins A-I, A-II and D, and lecithin cholesterol acyltransferase: effects of smoking, alcohol and adiposity. Arteriosclerosis 1985;5:169–77.
39. Albers JJ, Warnick GR, Nichols AV. Laboratory measurement of HDL. In Miller NE, Miller GJ, eds. Clinical and metabolic aspects of high-density lipoproteins. New York: Elsevier Science Publishers B.V., 1984.
40. Kunitake ST, Kane JP. Factors affecting the integrity of high density lipoproteins in the ultracentrifuge. J Lipid Res 1982;23:936–40.
41. Kirstein P, Carlson K. Determination of the cholesterol content of high density lipoprotein subfractions HDL2 and HDL3, without contamination of Lp(a) in human plasma. Clin Chim Acta 1981;113:123–34.
42. Wallentin L, Fåhraes L. HDL_3 and HDL_2 determination by a combined ultracentrifugation and precipitation procedure. Clin Chim Acta 1981;116:199–208.
43. Eyre J, Hammett F, Miller NE. A micro-method for the rapid ultracentrifugal separation of human plasma high density lipoprotein subfractions, HDL_2 and HDL_3. Clin Chim Acta 1981;114:225–31.
44. März W, Groβ W. Ultracentrifugal determination of high-density lipoprotein subfractions HDL_2 and HDL_3 in a high capacity fixed angle rotor. Ärztl Lab 1988;34:265–70.
45. Rifai N, King ME. Further evidence for the heterogeneity of high density lipoprotein isolates: HDL_2:HDL_3 cholesterol ratios and the presence of apolipoprotein B. Annals of Clinical and Laboratory Science 1987;17(5):345–9.
46. Asayama K, Miyao A, Kato K. High-density lipoprotein (HDL), HDL_2, and HDL_3 cholesterol concentrations determined in serum of newborns, infants, children, adolescents, and adults by use of a micromethod for combined precipitation ultracentrifugation. Clin Chem 1990;36(1):29–131.
47. Demacker PN, van Sommeren-Zondag DF, Stalenhoef AF, et al. Ultracentrifugation in swinging-bucket and fixed-angle rotors evaluated for isolation and determination of high-density lipoprotein subfractions HDL2 and HDL3. Clin Chem 1983;29(4):656–63.
48. Kostner G, Alaupovic P. Studies of the composition and structure of plasma lipoproteins: separation and quantification of the lipoprotein families occurring in the high density lipoproteins of human plasma. Biochemistry 1972;11:3419–28.
49. Anderson DW, Nichols AV, Forte TM, et al. Particle distribution of human serum high density lipoproteins. Biochimica et Biophysica Acta 1977;493:55–68.
50. Redgrave TG, Roberts DCK, West CE. Separation of plasma lipoproteins by density gradient ultracentrifugation. Anal Biochem 1975;65:42–49.
51. Foreman JR, Karlin JB, Edelstein C, et al. Fractionation of human serum lipoproteins by single-spin ultracentrifugation: quantification of apolipoproteins B and A-I and lipid components. J Lipid Res 1977;18:759–67.
52. Chapman MJ, Godstein S, Lagrange D, et al. A density gradient ultracentrifugal procedure for the isolation of the major lipoprotein classes from human serum. J Lipid Res 1981;22:339–58.

53. Terpstra AHM, Woodward CJH, Sanchea-Munz FJ. Improved techniques for the separation of serum lipoproteins by density gradient ultracentrifugation: visualization by prestaining and rapid separation of serum lipoproteins from small volumes of serum. Anal Biochem 1981;111:149–57.
54. Nilsson J, Mannickarottu V, Edelstein C, et al. An improved detection system applied to the study of serum lipoproteins after single-step density gradient ultracentrifugation. Anal Biochem 1981;110:342–8.
55. Groot PHE, Scheek LM, Havekes L, et al. A one-step separation of human serum high density lipoproteins 2 and 3 by rate-zonal density gradient ultracentrifugation in a swinging bucket rotor. J Lipid Res 1982;23:1342–53.
56. Chung BH, Segrest JP, Cone JT, et al. High resolution plasma lipoprotein cholesterol profiles by a rapid, high volume semi-automated method. J Lipid Res 1981;22:1003–14.
57. Cone JT, Segrest JP, Chung BH, et al. Computerized rapid high resolution quantitative analysis of plasma lipoproteins based upon single vertical spin centrifugation. J Lipid Res 1982;23:923–35.
58. Cheung MC, Segrest JP, Albers JJ, et al. Characterization of high density lipoprotein subspecies: structural studies by single vertical spin ultracentrifugation and immunoaffinity chromatography. J Lipid Res 1987;28:913–29.
59. Patsch JR, Sailer S, Kostner G, et al. Separation of the main lipoprotein density classes from human plasma by rate-zonal ultracentrifugation. J Lipid Res 1974;15:356–66.
60. Patsch W, Schonfeld G, Gotto AM (Jr.), et al. Characterization of human high density lipoproteins by zonal ultracentrifugation. J Biol Chem 1980;255(7):3178–85.
61. Schmitz G, Assmann G. Isolation of human serum HDL1 by zonal ultracentrifugation. J Lipid Res 1982;23:903–10.
62. Patsch W, Brown SA, Morrisett JD, et al. A dual-precipitation method evaluated for measurement of cholesterol in high-density lipoprotein subfractions HDL_2 and HDL_3 in human plasma. Clin Chem 1989;35(2):265–70.
63. Patsch JR, Patsch W. Zonal ultracentrifugation. In: Albers JJ, Segrest JP, eds. Methods in enzymology, Vol. 129. Orlando, FL: Academic Press, 1986:3–26.
64. Blanche PJ, Gong EL, Forte TM, et al. Characterization of human high-density lipoproteins by gradient gel electrophoresis. Biochimica et Biophysica Acta 1981;665:408–19.
65. Nichols AV, Blanche PJ, Gong EL. Gradient gel electrophoresis of human plasma high density lipoproteins. In: L Lewis, Opplt JJ, eds. CRC handbook of electrophoresis, Vol. III. Boca Raton, FL: CRC Press, 1980:29–47.
66. Verdery RB, Benham DF, Baldwin HL, et al. Measurement of normative HDL subfraction cholesterol levels by Gaussian summation analysis of gradient gels. J Lipid Res 1989;30:1085–95.
67. Atger V, Malon D, Bertire MC, et al. Cholesterol distribution between high-density-lipoprotein subfractions HDL2 and HDL3 determined in serum by discontinuous gradient gel electrophoresis. Clin Chem 1991;37(7):1149–52.
68. Okazaki M, Itakura H, Shiraishi K, et al. Serum lipoprotein measurement: liquid chromatography and sequential floatation (ultracentrifugation) compared. Clin Chem 1983;29(5):768–73.
69. Clifton PM, MacKinnon AM, Barter PJ. Separation and characterization of high-density lipoprotein subpopulations by gel permeation chromatography. J Chromatography 1987;414:25–34.
70. Gidez LI, Miller GJ, Burstein M, et al. Separation and quantitation of subclasses of human plasma high density lipoproteins by a simple precipitation procedure. J Lipid Res 1982;23:1206–23.

71. Simpson HS, Ballantyne FC, Packard CJ. High-density lipoprotein subfractions as measured by differential polyanionic precipitation and rate zonal ultracentrifugation. Clin Chem 1982;28:2040–3.
72. Daerr WH, Windler EET, Rohwer HD, et al. Limitations of a new double-precipitation method for the determination of high density lipoprotein subfractions 2 and 3. Atherosclerosis 1983;49:211–13.
73. Martini S, Baggio G, Baroni L, et al. Evaluation of HDL_2 and HDL_3 cholesterol by a precipitation procedure in a normal population and in different hyperlipidemic phenotypes. Clin Chim Acta 1984;137:291–8.
74. Demacker PNM, Hak-Lemmers HLM, Hijmans AGM, et al. Evaluation of the dual-precipitation method for determination of cholesterol in high-density lipoprotein subfractions HDL_2 and HDL_3 in serum. Clin Chem 1986;32(5):819–25.
75. Cloey TA, Bachorik PS. Use of a dual-precipitation procedure for measuring high-density lipoprotein 3 (HDL_3) in normolipidemic serum. Clin Chem 1989;35(7): 1390–3.
76. Demacker PNM, Hak-Lemmers HLM, van Heijst PJ. Interference of blood clotting factors in the determination of HDL_2- and HDL_3-cholesterol by the dual precipitation method. Clin Chim Acta 1987;165:133–9.
77. Warnick GR, Benderson JM, Albers JJ. Quantitation of high-density-lipoprotein subclasses after separation by dextran sulfate and Mg^{2+} precipitation [abstract 128]. Clin Chem 1982;28(7):1574.
78. Bachorik PS, Albers JJ. Precipitation methods for quantification of lipoproteins. In: Albers JJ, Segrest JP, eds. Methods in enzymology, Vol. 129. Orlando, FL: Academic Press, 1986:78–100.
79. Nguyen T, Warnick GR. Improved methods for separation of total HDL and subclasses [abstract 097]. Clin Chem 1989;35(6):1086.
80. Lundberg B, Högström S., Pietiläinen P, et al. Separation of plasma high-density lipoprotein subclasses by a combined precipitation method using polyethylene glycol 6000 and dextran sulfate. Scan J Clin Lab Invest 1984;44:305–9.
81. Dias VC, Parsons HG, Boyd ND, et al. Dual-precipitation method evaluated for determination of high-density lipoprotein (HDL), HDL_2 and HDL_3 cholesterol concentrations. Clin Chem 1988;34(11):2322–7.
82. Demacker PNM. Differential determination of HDL-subfractions in clinical laboratories. Clin Chem 1989;35(4):701–2.
83. Kostner GM, Molinari E, Pichler P. Evaluation of a new HDL2/HDL3 quantitation method based on precipitation with polyethylene glycol. Clin Chim Acta 1983;148:139–47.
84. Brugger P, Kostner GM, Kullich WC, et al. Plasma concentrations of high-density lipoprotein (HDL)-2 and HDL-3 in myocardial infarction survivors and in control subjects. Clin Cardiol 1986;9:273–6.
85. Widhalm K, Pakosta R. Precipitation with polyethylene glycol and density-gradient ultracentrifugation compared for determining high-density lipoprotein subclasses HDL_2 and HDL_3. Clin Chem 1991;37(2):238–40.
86. Leino A, Viikari J, Koskinen P, et al. Problems with PEG-based precipitation methods in the determination of HDL_2- and HDL_3- cholesterol. Scan J Clin Lab Invest 1987;47:705–8.
87. James RW, Proudfoot A, Pometta D. Immunoaffinity fractionation of high-density lipoprotein subclasses 2 and 3 using anti-apolipoprotein A-I and A-II immunosorbent gels. Biochimica et Biophysica Acta 1989;1002:292–301.
88. Parra HJ, Mezdour H, Ghalim N, et al. Differential electroimmunoassay of human Lp A-I lipoprotein particles on ready-to-use plates. Clin Chem 1990;36(8):1431–5.

CHAPTER 13

Measurement and Clinical Significance of Low-Density Lipoprotein Subclasses

Melissa A. Austin, John E. Hokanson, and John D. Brunzell

INTRODUCTION

❖ The important relationship between increased plasma low-density lipoprotein (LDL) cholesterol and increased risk of coronary heart disease (CHD) has been definitively established. However, there is emerging evidence that subclasses of LDL, characterized by variations in the density, size, and chemical composition of LDL particles, may have important clinical significance as well. The heterogeneity of LDL particles has been recognized for many years,[1-3] and a number of techniques have been developed for detecting and characterizing the variation in LDL.[4] This chapter focuses on two techniques now widely used in research laboratories for measuring LDL size and density in large-scale studies: non-denaturing gradient gel electrophoresis (GGE) and density gradient ultracentrifugation (DGUC). Although the clinical implementation of these techniques is not routine, rapid progress is being made toward understanding the role of lipoprotein subclasses in risk of CHD.

GRADIENT GEL ELECTROPHORESIS OF LDL

Non-denaturing gradient gel electrophoresis using 2–16% polyacrylamide gradient gels is used to characterize LDL particles by size and shape. Two methods have been described for use with these gels, varying in electrophoretic conditions, stain, and data interpretation. This technique was first described by Krauss and Burke in 1972 and requires 5 µL of plasma or 5 µg of isolated LDL.[4, 5] The samples are electrophoresed at 125 V for 24 h and are stained either for lipid with Oil

Red O (Sigma Diagnostics, St. Louis, MO) or for protein with Coomassie Blue R-250.[5, 6] A quadratic calibration curve relating migration distance to particle diameter is constructed from high-molecular-weight standards (Pharmacia, Piscataway, NJ) and latex beads (Dow) run in another gel lane. Each lane is scanned at the appropriate wavelength to identify the LDL subclass peaks in the sample lanes and the standard peaks in the calibration lane. The diameter of the LDL particles in the sample lanes is then estimated by applying the calibration curve to the sample lanes. Although up to seven LDL peaks can be identified in this manner, most subjects have two or three LDL subclass peaks and shoulders on the densitometric scan (see Figure 13–1).

Using this technique, a dichotomous classification of LDL heterogeneity has been used to identify LDL subclass patterns, or phenotypes, denoted A and B.[7, 8] (See Figure 13–1.) LDL subclass phenotype A is characterized by a predominance of large LDL particles and skewing of the densitometric scan to the right, while LDL subclass phenotype B is characterized by a predominance of small LDL particles and skewing of the curve to the left.[7, 8] In general, the peak particle diameter corresponding to the major LDL subclass has an estimated diameter greater than 255Å in subjects with phenotype A, while those with phenotype B

Figure 13–1 ❖ Examples of LDL Subclass Phenotypes A and B as Determined by GGE of LDL from Individual Study Subjects

Phenotype A is characterized by a predominance of large, buoyant LDL, while phenotype B is characterized by a predominance of small, dense LDL. The estimated diameter (Å) of the major LDL subclass is also shown for each scan.

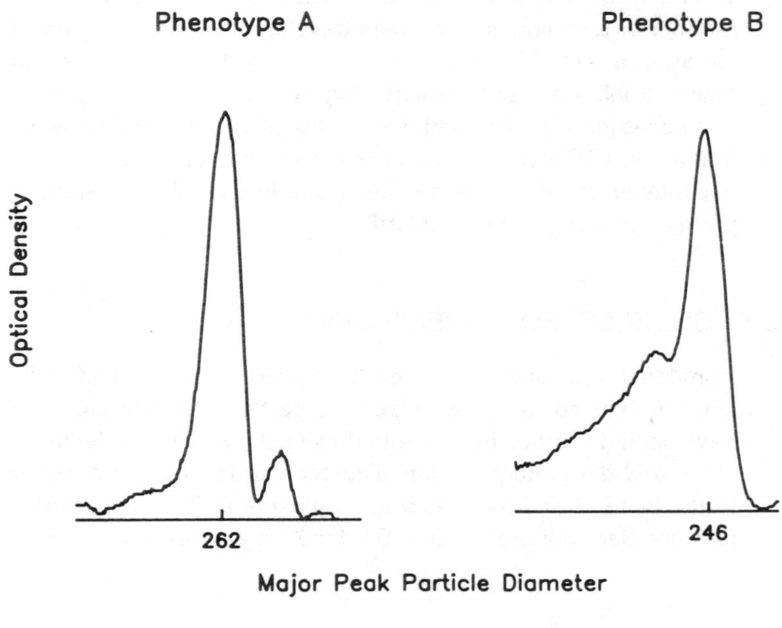

have peak particle diameter less than or equal to 255Å. In studies to date, 85–90% of study subjects clearly have either LDL subclass phenotype A or B, while the remainder have an intermediate phenotype.[8] In rare conditions, a study subject may have so little plasma LDL that the LDL subclass phenotype cannot be determined by GGE. A continuous variable can also be defined as LDL peak particle diameter (Å) based on these GGE scans, reflecting the estimated diameter or size of the major LDL subclass in individual subjects.[9]

These definitions of LDL subclass phenotypes have been used to demonstrate significant genetic influences on LDL heterogeneity.[10]

Two large family studies using the phenotype A and B definitions have used complex segregation analysis to conclude that LDL subclass phenotype B is influenced by a single major gene.[11, 12] The first study was based on primarily healthy families, while the second study was based on families with familial combined hyperlipidemia (FCHL). In both studies, the proposed gene controlling phenotype B was estimated to have a common allele frequency, a dominant or additive mode of inheritance, and reduced penetrance in young males and pre-menopausal women.[11, 12]

Two large studies of twins, one of women and the other of men, have confirmed the presence of genetic influences on LDL heterogeneity by reporting heritability estimates ranging from 0.39 to 0.55 for LDL peak particle diameter and LDL score, respectively.[13, 14] These results indicate that one-third to one-half of the variance is attributable to genetic influences, and were highly statistically significant in women but not in men due to the use of the "among components" estimate of heritability with little statistical power. However, these studies also demonstrate the importance of non-genetic environmental and behavioral influences on small, dense LDL.[10, 13]

Genetic linkage studies are currently in progress to determine the chromosomal location of the gene or genes that control small, dense LDL.[15, 16]

A second GGE procedure, also using 2–16% polyacrylamide gradient gels, has been developed to describe LDL heterogeneity by defining seven size subfractions.[17] Using this nomenclature, LDL_1 represents the largest subfraction of LDL particles and LDL_7 the smallest. In this approach, samples are subjected to electrophoresis at 460 V for 6 h and stained for lipid with Sudan Black B. A pooled plasma sample containing three of the seven major bands is also run on each gel to identify the sample LDL subfractions. Each lane is scanned with a laser densitometer and the resulting data are used to estimate the area under each peak as a percentage of the total band area.

When this analysis was performed using samples from adult men and women, the results demonstrated that smaller LDL subfractions were more common in men than in women, and were correlated with increasing age, plasma triglyceride (TG), and apolipoprotein (apo) B concentrations, as well as with decreased high-density lipoprotein (HDL) cholesterol and apo AI.[17, 18] A recent prospective analysis has confirmed these associations among 227 participants in the Framingham Offspring Study over a 3–4 y period using an overall "LDL score" calculated as the weighted average of the LDL subfraction numbers.[19] Changes in LDL score were most strongly correlated with changes in plasma TG ($r = -0.459$, $p < .0001$) and with HDL cholesterol ($r = 0.277$, $p < .0001$)

The methods and results described above were all based on nondenaturing polyacrylamide agarose gels obtained from Pharmacia. However, commercial production of these gels ceased early in 1991. As a result, Rainwater and colleagues have recently described a protocol to produce gradient gels for the size resolution of lipoproteins.[20] This protocol involves the use of a computer-controlled dual-pump system that results in a high degree of gradient reproducibility from lot to lot. After calibrating with size standards, peak particle diameters of protein-stained human lipoprotein estimated using these gels were highly correlated with the commercial gels. Thus, these new procedures can be used to produce gels that are suitable for analysis of lipoprotein subfraction phenotypes.

DENSITY GRADIENT ULTRACENTRIFUGATION

A number of density gradient ultracentrifugation procedures to characterize LDL subfractions have been developed.[21-25] One of these, single vertical-spin nonequilibrium DGUC, is designed to optimize the resolution of apo-B-containing lipoproteins using a modification of single vertical-spin DGUC.[22] A discontinuous salt gradient is produced by layering 1 mL of plasma adjusted to a density of 1.08 kg/L (total volume 5 mL) below 12 mL of a 1.006 kg/L NaCl solution in a Sorvall TV-865B tube (DuPont, Wilmington, DE). Samples are then centrifuged at 65,000 rpm for 90 min at 10° C and 38 fractions are collected from the bottom of the tube. Total cholesterol is measured in each fraction and the relative flotation

Figure 13–2 ❖ Relationship between LDL Peak Particle Diameter Estimated by GGE and LDL Buoyancy (Rf) by DGUC

Results from 23 middle-aged males with normal lipid values.[24] The correlation coefficient is 0.68 ($p < .01$)

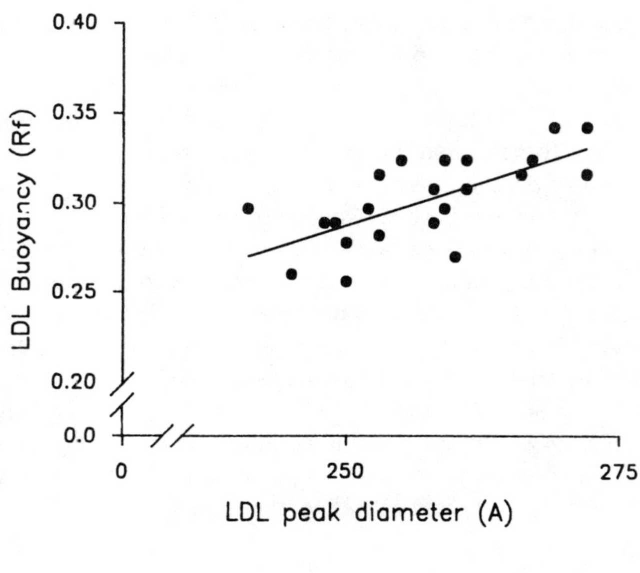

rate (Rf) is calculated as the fraction number containing the LDL cholesterol peak divided by the total number of fractions collected. The intra-assay coefficient of variation of the LDL Rf values has been estimated to be 0.2%.[26] Thus, DGUC provides a density profile of the spectrum of apo-B-containing lipoproteins in individual subjects across a range of 38 fractions.

As shown in Figure 13–2, the Rf of LDL by DGUC correlates with the peak particle diameter by GGE ($r = 0.68$), based on 23 middle-aged males with normal lipid values.[26] However, as shown in the left view of Figure 13–3, the mean cholesterol distribution by DGUC differs between subjects with LDL subclass phenotype B (open symbols) compared to those with phenotype A (solid symbols). A difference plot (right view of Figure 13–3) shows that very-low-density lipoproteins (VLDL), intermediate-density lipoproteins (IDL), and HDL all differ in the two

Figure 13–3 ❖ Mean Cholesterol Distribution by DGUC and Difference Plot Comparing Mean Values for Subjects with LDL Subclass Phenotype A and Phenotype B

Left view: Nonequilibrium DGUC of plasma from middle-aged males with normal lipid values.[26] Mean cholesterol values (mg/dL) are shown in each of 38 fractions in 9 males with LDL subclass phenotype A (solid symbols), and in 18 males with LDL subclass phenotype B (open symbols).

Right view: DGUC difference plot comparing mean values for subjects with LDL subclass phenotype B and those with phenotype A (B-A). Bars represent 95% confidence intervals of difference in means for each fraction. The plot shows that phenotype B subjects have more VLDL cholesterol, less buoyant LDL, more dense LDL, and less HDL cholesterol compared to phenotype A subjects.

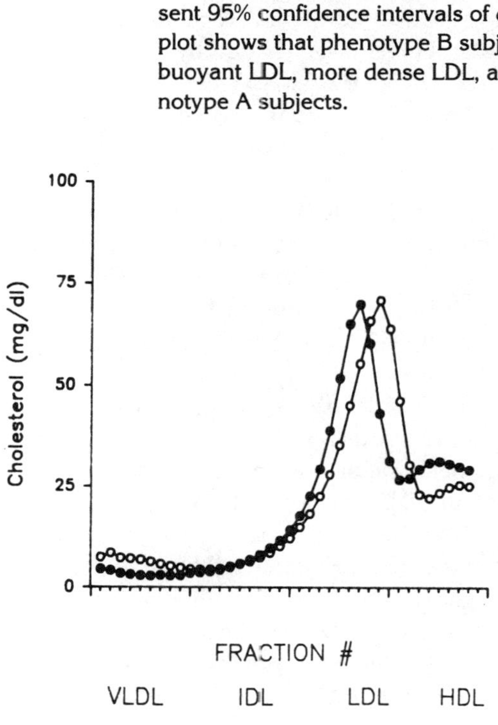

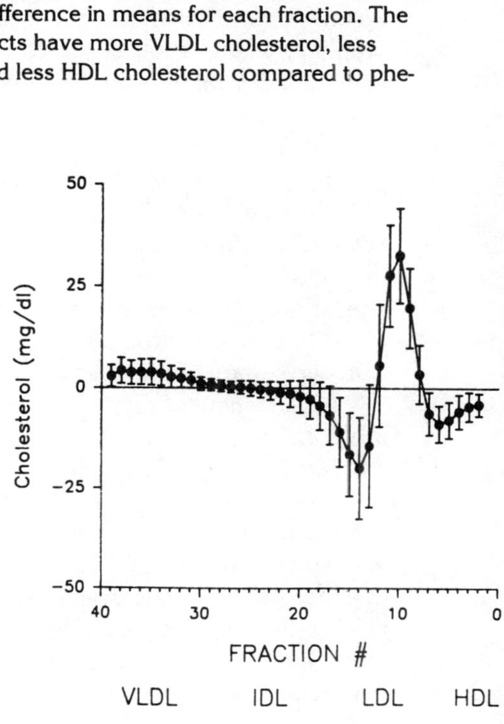

types of subjects, and that the LDL distribution in subjects with phenotype B is shifted toward the more dense particles.

This technique has recently been used to investigate the involvement of hepatic lipase (HL) in determining the physical characteristics of LDL.[26] In a sample of 23 normolipidemic males and a sample of 21 male patients with coronary artery disease (CAD), subjects with low HL activity had relatively buoyant LDL, while those with high HL activity had relatively dense LDL. Based on linear regression analysis, HL activity explained more than half of the variance in LDL buoyancy in each of the groups. These results support the hypothesis that HL may be involved in determining physicochemical characteristics of LDL particles.

Similar to the results based on GGE, plasma TG concentrations have been shown to be inversely associated with LDL peak buoyancy among normals and among subjects with familial hypercholesterolemia, familial hypertriglyceridemia, and FCHL.[27] Further, when 13 patients with well-characterized FCHL and with LDL subclass phenotype B were treated with gemfibrozil for 3 m, dense LDL persisted despite a 55% decrease in TG values.[28] However, changes in Rf were inversely correlated with changes in TG concentration, indicating an increase in buoyant LDL as well.

Another single-spin DGUC procedure using serum pre-stained with Coomassie Blue has been the basis of several clinical investigations.[23, 29, 30] In this approach, pre-stained serum in a density gradient is ultracentrifuged in a swinging-bucket rotor at 160,000 g for 19.5 h at 20° C. This results in a density gradient with up to three major LDL density subfractions, denoted LDL_1 (d = 1.030–1.033 kg/L), LDL_2 (d = 1.033–1.040 kg/L), and LDL_3 (d = 1.040–1.045 kg/L). After placing the tubes in a specially designed rack, they are photographed. Quantitation of the LDL subfractions is performed by scanning the resulting slides on a laser densitometer. Using the LKB GelScan program, LDL subfraction peaks are identified and the area under each curve is calculated, assuming Gaussian curves. Finally, the area under each LDL subfraction peak is expressed as percentage of the total area.

In a study of 131 healthy individuals, this DGUC technique identified three subfractions in most (88%) of the study subjects, while the remainder (12%) had two subfractions.[29] By defining a continuous LDL density variable using percentage of area and density midpoints for LDL_1, LDL_2, and LDL_3, LDL density was found to be associated with gender, serum TG, HDL cholesterol, and the ratio of LDL cholesterol to LDL apo B.[29]

A subsequent report of genetic studies defined a different continuous variable, "K," based on the same procedures but using relative peak heights rather than areas.[30] Specifically, percentage peak heights designated as %h1, %h2, and %h3 are calculated for LDL_1, LDL_2, and LDL_3, respectively. When an individual subfraction pattern was characterized by a predominance of buoyant LDL, K is defined such that $0 < K < 1$. When the pattern is characterized by a predominance of dense particles, K is defined such that $-1 < K < 0$. As a result, K ranges from −1 to +1, reflecting a spectrum of dense to buoyant LDL patterns. Among 159 family members, K was found to have an approximately normal distribution. However, the median value of K was lower for men than for women, decreased with age, and was lower for premenopausal women taking oral contraceptives than for those not taking oral contraceptives.[30]

Finally, this technique has been compared to GGE based on 41 healthy individuals.[31] Although the shapes of the densitometric scans differed using DGUC and GGE, the number of bands detected was the same for more than 90% of the study subjects. The density as determined by DGUC and the relative migration of the major LDL subfraction by GGE were also highly correlated ($r = 0.85$, $p < .0001$). In addition, this study demonstrated that both approaches give reproducible results when samples are stored at −80° C for up to 12 w.

CLINICAL SIGNIFICANCE

LDL Subclasses and Coronary Heart Disease

In addition to the studies noted above, the results from several case-control studies have established that small, dense LDL are associated with increased risk of CHD. One of the earliest such studies showed that the molecular weight of LDL was lower among 46 male CAD patients than among 47 male controls ($p < .01$) in the same age ranges.[32] In a study of 109 male and female myocardial infarction (MI) survivors age 65 or less, a gender-adjusted odds ratio of 3.0 ($p < .01$) was found for LDL subclass phenotype B (with a 95% confidence interval of 1.7 to 5.2), indicating a three-fold increased risk associated with small, dense LDL.[7] A similar result was found among male MI survivors under age 45, based on the distribution of LDL apo B over 15 density subfractions.[33] A small study using both GGE and DGUC showed that dense LDL were increased among male CAD patients compared to normal males, and similar results were seen using GGE.[25]

In the largest case-control study to date, the distribution of LDL scores as determined from GGE (described above) was compared between 275 men with documented CAD and 822 controls from the offspring cohort of the Framingham Heart Study.[34] Because the use of beta-blockers may be a confounding factor, cases were stratified as to whether subjects were or were not on beta-blockers. The mean LDL score was significantly higher (indicating smaller LDL) among both groups of cases compared to controls ($p < .001$), although the subjects who were on beta-blockers had smaller LDL than those who were not.

Two subsequent studies, one based on men and women with premature CAD[35] and one based on Finnish men,[36] further confirmed these results and also examined the chemical composition of LDL. The former study found that smaller, "cholesterol-poor" LDL particles were associated with CAD, independent of age, sex, smoking, diabetes, and LDL and HDL cholesterol concentrations.[35] The latter study reported a reduction in sialic acid content in dense LDL subfractions.[36]

Thus, although a variety of laboratory techniques were used in these studies to characterize LDL size and density, the results uniformly demonstrate that small, dense LDL particles are more common among coronary disease cases than controls. However, it is important to note that these studies were all based on Caucasian subjects, and no study to date has examined this association among other ethnic groups. Of equal importance is that these studies were all conducted cross-sectionally at one point in time. Thus, the temporal association of LDL subclass phenotype B with CHD risk remains to be established based on data from prospective cohort studies.

LDL Subclasses and Atherogenic Lipoproteins

Many studies, including several of the case-control studies noted above, have reported associations between small, dense LDL and other lipoprotein-related risk factors. Examples of three studies that have compared plasma concentrations of lipids and apolipoproteins among subjects with LDL subclass phenotypes A and B are presented in Table 13–1. Although the types of subjects examined in these studies were quite different, including members of primarily healthy families (n = 301),[8] members of families with FCHL (n = 173),[12] and a large sample of women twins (n = 682),[37] the magnitude of the lipid differences was remarkably similar. The largest difference is seen for plasma TG: subjects with LDL subclass phenotype B have mean concentrations that are approximately 80 mg/dL higher than those with phenotype A. In addition, LDL subclass phenotype B is simultaneously associated with significant decreases in HDL cholesterol, with differences of 7–9 mg/dL. These close relationships between LDL size and both plasma TG and HDL cholesterol have recently been confirmed cross-sectionally among Mexican-Americans,[38] and longitudinally among subjects from the Framingham Offspring Study.[19] Interestingly, LDL subclass phenotypes do not show as strong a relationship with LDL cholesterol concentrations.

LDL subclass phenotypes are also consistently associated with variations in apolipoprotein concentrations and with lipoprotein mass values. As shown in Table 13–1, phenotype B subjects have significantly increased mean apo B concentrations compared to subjects with phenotype A.[8, 12] In addition, phenotype B is associated with increased VLDL mass, IDL mass, and decreased HDL_2 mass[8] (data not shown in table). As a result of these associations, phenotype B has been termed an "atherogenic lipoprotein phenotype."[8] This constellation of lipid, lipoprotein, and apolipoprotein abnormalities may represent a marker for an underlying pathogenic process leading to increased risk of CHD.

Oxidative Susceptibility of LDL Subclasses

Although the mechanisms underlying the association of small LDL with risk of CAD are not yet established, several intriguing studies have reported that small LDL particles are more susceptible to oxidation than large LDL particles, at least *in vitro*. In the first such study, three density subfractions of LDL were examined in 11 healthy volunteers.[39] Based on the method described by Esterbauer,[40] the two most dense LDL subfractions were less well protected against oxidative modification than the most buoyant subfraction.[39] These results were confirmed in a subsequent study based on LDL from nine healthy males: oxidative susceptibility of LDL progressively increased with increasing density across five of six LDL subfractions.[41]

Similar findings have also been reported in patients with combined hyperlipidemia (increased cholesterol and TG) and with hypertriglyceridemia.[42, 43] In a

Table 13–1 ❖ Mean Concentrations of Lipids and Apolipoprotein Risk Factors by LDL Subclass Phenotype

	ADJUSTED MEANS (mg/dL)		
*Risk Factor**	*Phenotype A*	*Phenotype B*	*Difference in Means*
Triglyceride			
Healthy families	69	141	+70**
FCHL families	94	184	+90**
Women twins	83	161	+78**
HDL-Cholesterol			
Healthy families	46	37	−9**
FCHL families	48	41	−7**
Women twins	66	57	−9**
LDL-Cholesterol			
Healthy families	116	126	+10***
FCHL families	133	139	+6
Women twins	117	122	+5
Apolipoprotein B			
Healthy families	76	98	+12**
FCHL families	117	138	+21**

**Sources:*
Healthy families: Austin et al.[8] FCHL families: Austin et al.[12] Women twins: Selby et al.[37]
**$p < .01$
***$p < .05$
To convert cholesterol from mg/dL to mmol/L multiply by 0.0259.
To convert triglyceride from mg/dL to mmol/L multiply by 0.0113.

comparison of nine patients with combined hyperlipidemia who were under strict dietary control and five normal subjects, five LDL subfractions were isolated by DGUC.[42] In both groups, the most dense of the LDL subfractions (d = 1.050–1.063 kg/L) had significantly reduced oxidative resistance compared to the other four density subfractions. LDL isolated from 12 hypertriglyceridemic subjects has also been shown to have enhanced susceptibility to oxidation compared to LDL from age- and sex-matched controls.[43] Interestingly, in this study 2 m of clofibrate treatment resulted in both a more buoyant LDL profile and reduced susceptibility of LDL to oxidation.

Finally, the oxidative susceptibility of LDL from subjects specifically characterized to have LDL subclass phenotypes A and B has recently been investigated.[44] Although the major peak of LDL from phenotype B subjects was more susceptible to oxidative modification than the major peak from phenotype A subjects, this difference was attributable to the predominance of small, dense LDL in the phenotype B subjects.[44]

Taken together, the results of these studies suggest that small, dense LDL may contribute more to foam cell formation during atherogenesis than large, buoyant LDL, resulting in increased risk of CHD among individuals with small, dense LDL.

SUMMARY

A variety of methods are available for characterizing the heterogeneity of LDL particles. Two of the most common approaches, GGE and DGUC, have resulted in important advances in our understanding of the role of LDL subclasses in atherosclerosis. Using these techniques, several case-control studies have demonstrated that small, dense LDL are consistently associated with increased risk of CHD. These and other studies have shown that small LDL are characterized by an atherogenic lipoprotein profile, including increased plasma TG and apo B and decreased HDL cholesterol. Notably, both genetic and environmental influences on LDL heterogeneity have been demonstrated in studies of families and twins. The possibility that small LDL may contribute to CHD because of its increased oxidative susceptibility raises the possibility that small, dense LDL may be directly involved in atherosclerosis. Further studies are needed to determine if the risk associated with small, dense LDL can be reduced through targeted intervention strategies. ❖

Acknowledgments: This work was supported by NIH R-01 Grants HL46880 and HL49513 and NIH Program Project Grant HL-30086.

REFERENCES

1. Lindgren FT, Jensen LC, Wills RD, Freeman NK. Flotation rates, molecular weights and hydrated densities of the low density lipoproteins. Lipids 1969; 4:337–44.
2. Adams GH, Schumaker VN. Polydispersity of human low-density lipoproteins. Ann NY Acad Sci 1969;164:130–46.
3. Hammond MG, Fisher WR. The characterization of a discrete series of low density lipoproteins in the disease, hyper-pre-β-lipoproteinemia. J Biol Chem 1971; 246:5454–65.
4. Krauss RM, Blanche PJ. Detection and quantitation of LDL subfractions. Curr Opin Lipidol 1992;3:377–83.
5. Krauss RM, Burke DJ. Identification of multiple subclasses of plasma low density lipoproteins in normal humans. J Lipid Res 1982;23:97–104.
6. Nichols AV, Krauss RM, Musliner TA. Nondenaturing polyacrylamide gradient gel electrophoresis. In: Segrest JP, Albers JJ, eds. Methods in enzymology. Vol. 128, Plasma lipoproteins. Part A, Preparation, structure and molecular biology. Orlando, Florida: Academic Press, 1986:417–31.
7. Austin MA, Breslow JL, Hennekens CH, Buring JE, Willett WC, Krauss RM. Low-density lipoprotein subclass patterns and risk of myocardial infarction. JAMA 1988;260:1917–21.
8. Austin MA, King M-C, Vranizan KM, Krauss RM. The atherogenic lipoprotein phenotype (ALP): a proposed genetic marker for coronary heart disease risk. Circulation 1990;82:495–506.
9. Austin MA, Jarvik GP, Hokanson JE, Edwards KE. Complex segregation analysis of low-density lipoprotein peak particle diameter. Genet Epidemiol 1993;10:599–604.
10. Austin MA. Genetics of LDL subclass phenotypes. Curr Opin Lipidol 1993;4:125–31.

11. Austin MA, King M-C, Vranizan KMI, Newman B, Krauss RM. Inheritance of low density lipoprotein subclass patterns: results of complex segregation analysis. Am J Hum Genet 1988;43:838–46.
12. Austin MA, Brunzell JD, Fitch WL, Krauss RM. Inheritance of low density lipoprotein subclass patterns in familial combined hyperlipidemia. Arteriosclerosis 1990;10:520–30.
13. Austin MA, Newman B, Selby JV, Edwards K, Mayer EJ, Krauss RM. Genetics of low-density lipoprotein subclasses in women twins: concordance, heritability and commingling analysis. Arterioscler Thromb 1993;13:687–95.
14. Lamon-Fava S, Jimenez D, Christian JC, et al. The NHLBI Twin Study: heritability of apolipoprotein A-I, B, and low density lipoprotein subclasses and concordance for lipoprotein(a). Atherosclerosis 1991;91:97–106.
15. Nishina PM, Johnson JP, Naggert JK, Krauss RM. Linkage of atherogenic lipoprotein phenotype to the low density lipoprotein receptor locus on the short arm of chromosome 19. Proc Natl Acad Sci 1992;89:708–12.
16. Austin MA, Stephens K, Wijsman E., Walden C, Lortez C. Evidence against linkage of small, dense low-density lipoproteins (LDL) to the LDL receptor locus on chromosome 19 [Abstract]. Circulation 1993;88:1–268.
17. McNamara JR, Campos H, Ordovas JM, Peterson J, Wilson PWF, Schaefer EJ. Effect of gender, age, and lipid status on low density lipoprotein subfraction distribution. Results of the Framingham Offspring Study. Arteriosclerosis 1987; 7:483–90
18. Campos H, Blijlevens E, McNamara JR, et al. LDL particle size distribution: results from the Framingham Offspring Study. Arterioscler Thromb 1992;12:1410–19.
19. McNamara JR, Jenner JL, Li Z, Wilson PWF, Schaefer EJ. Change in LDL particle size is associated with change in plasma triglyceride concentration. Arterioscler Thromb 1992;12:1284–90.
20. Rainwater DL, Andres DW, Ford AI, Lowe WF, Blanche PJ, Krauss RM. Production of polyacrylamide gradient gels for the electrophoretic resolution of lipoproteins. J Lipid Res 1992;33:1876–81.
21. Lee DM, Downs D. A quick and large-scale density gradient subfractionation method for low density lipoproteins. J Lipid Res 1982;23:14–27.
22. Chung BH, Segrest JP, Marjory JR, et al. Single vertical spin density gradient ultracentrifugation. In Segrest JP, Albers JJ, eds. Methods in enzymology. Vol. 128, Plasma lipoproteins. Part A, Preparation, structure and molecular biology. Orlando, Fla, Academic Press, 1986:181–209.
23. Swinkels DW, Hak-Lemmers HLM, Demacker PNM. Single spin density gradient ultracentrifugation method for the detection and isolation of light and heavy low density lipoprotein subfractions. J Lipid Res 1987;28:1233–9.
24. Chapman MJ, Laplaud PM, Luc G, et al. Further resolution of the low density lipoprotein spectrum in normal human plasma: physicochemical characteristics of discrete subspecies separated by density gradient ultracentrifugation. J Lipid Res 1988;29:442–58.
25. Griffin BA, Caslake MJ, Yip B, Tait GW, Packard CJ, Shepherd J. Rapid isolation of low density lipoprotein (LDL) subfractions from plasma by density gradient ultracentrifugation. Atherosclerosis 1990;83:59–67.
26. Zambon A, Austin MA, Brown GB, Hokanson JE, Brunzell JD. Effect of hepatic lipase on LDL in normal men and those with coronary artery disease. Arterioscler Thromb 1993;13:147–53.
27. Failor RA, Hokanson JE, Brunzell JD. Low density lipoproteins in familial hyperlipidemias. In: Fidge NH, Necitel PH, eds. Atherosclerosis VII. Amsterdam: Elsevier Sciences, 1986:156–63.

28. Hokanson JE, Austin MA, Zambon A, Brunzell JD. Plasma triglyceride and LDL heterogeneity in familial combined hyperlipidemia. Arterioscler Thromb 1993; 13:427–34.
29. Swinkels DWL, Demacker PNM, Hendriks JCM, van't Laar A. Low density lipoprotein subfractions and relationship to other risk factors for coronary artery disease in healthy individuals. Arteriosclerosis 1989;9:604–13.
30. de Graaf J, Swinkels DW, de Haan AFJ, Demacker PNM, Stalenhoef AFH. Both inherited susceptibility and environmental exposure determine low density lipoprotein subfraction pattern distribution in healthy, Dutch families. Am J Hum Genet 1992:51:1295–1310.
31. Dormans TPJ, Swinkels DW, de Graaf J, Hendriks JCM, Stalenhoef AFH, Demacker PNM. Single-spin density gradient ultracentrifugation vs. gradient gel electrophoresis: two methods for detecting low-density lipoprotein heterogeneity compared. Clin Chem 1991;37:853–8.
32. Crouse JR, Parks JS, Schey HM, Kahl FR. Studies of low density lipoprotein molecular weight in human beings with coronary artery disease. J Lipid Res 1985; 26:566–74.
33. Tornvall P, Karpe F, Carlson LA, Hamsten A. Relationship of low density lipoprotein subfractions to angiographically defined coronary artery disease in young survivors of myocardial infarction. Atherosclerosis 1991;90:67–80.
34. Campos H, Genest JJ Jr., Blijlevens E, et al. Low density lipoprotein particle size and coronary artery disease. Arterioscler Thromb 1992;12:187–95.
35. Coresh J, Kwiterovich PO Jr., Smith HH, Bachorik P. Association of plasma triglyceride and LDL particle diameter, density and chemical composition with premature coronary artery disease in men and women. J Lipid Res 1993;34: 1687–97.
36. Jaakkola O, Solakivi T, Tertov VV, Orekhov AN, Miettinen TA, Nikkari T. Characteristics of low-density lipoprotein subfractions from patients with coronary artery disease. Coronary Artery Disease 1993;4:379–85.
37. Selby JV, Austin MA, Newman B, Mayer EJ, Krauss RM. LDL subclass phenotypes and the insulin resistance syndrome in women. Circulation 1993;88:381–7.
38. Haffner SM, Mykkanen L, Valdez RA, Paidi M, Stern MP, Howard BV. LDL size and subclass pattern in a biethnic population. Arterioscler Thromb 1993;13: 1623–30.
39. de Graaf J, Hak-Lemmers HLM, Hectors MPC, Demacker PNM, Hendriks JCM, Stalenhoef AFH. Enhanced susceptibility to in vitro oxidation of the dense low density lipoprotein subfraction in healthy subjects. Arterioscler Thromb 1991;11: 298–306.
40. Esterbauer H, Striegl G, Puhl H, Rotheneder M. Continuous monitoring of in vitro oxidation of human low density lipoprotein. Free Rad Res Commun 1989;6:67–75.
41. Tribble DL, Holl LG, Wood PD, Krauss RM. Variations in oxidative susceptibility among six low density lipoprotein subfractions of differing density and particle size. Atherosclerosis 1992;93:189–99.
42. Dejager S, Bruckert E, Chapman J. Dense low density lipoprotein subspecies with diminished oxidative resistance predominate in combined hyperlipidemia. J Lipid Res 1993;34:295–308.
43. de Graaf J, Hendriks JCM, Demacker PNM, Stalenhoef AFH. Identification of multiple dense LDL subfractions with enhanced susceptibility to in vitro oxidation among hypertriglyceridemic subjects: normalization after clofibrate treatment. Arterioscler Thromb 1993;13:712–19.
44. Chait A, Brazg RL, Tribble DL, Krauss RM. Susceptibility of small, dense, low-density lipoproteins to oxidative modification in subjects with the atherogenic lipoprotein phenotype, pattern B. Am J Med 1993;94:350–6.

CHAPTER 14

Lipoprotein(a): Structure, Measurement, and Clinical Significance

Santica M. Marcovina, Daniel M. Levine, and Giuseppe Lippi

INTRODUCTION

❖ In 1963, Kåre Berg observed that rabbits immunized with human low-density lipoproteins (LDL) produced an antiserum which reacted with an antigenic component that appeared to be present within the LDL fraction of some but not all of the individuals tested.[1] Berg called this newly discovered antigen lipoprotein(a) [Lp(a)], and he found that the presence of this antigen was genetically determined by an autosomal dominant mode of inheritance.[2] Subsequent studies provided evidence that Lp(a) is a specific family of lipoprotein particles whose protein moiety is comprised of at least two major proteins, a single copy of apo B-100 linked to a single copy of a protein of variable mass which has not been found in any other lipoprotein. This protein, which is responsible for the peculiar characteristics of Lp(a), is called apolipoprotein(a) [apo(a)].

Over the past several decades, numerous studies have established that increased Lp(a) concentrations are associated with increased risk of cardiovascular and cerebral vascular disease. Although the mechanism of Lp(a) atherogenicity or thrombogenicity has yet to be clarified, numerous structural, genetic, and metabolic studies have provided some intriguing possibilities to explain its association with atherosclerosis.

STRUCTURE OF LIPOPROTEIN(a)

Lp(a) is formed by the assembly of a lipoprotein which is structurally very similar to LDL in protein and lipid composition with a carbohydrate-rich, highly hydrophilic protein named apo(a). One molecule of apo B-100 in Lp(a) is covalently linked to apo(a) by a disulfide bridge, although several other non-covalent interactions may play a role in the association of the two proteins. Lp(a) has peculiar characteristics and, unlike the other human plasma lipoproteins, it cannot be classified on the basis of its physical-chemical properties as defined by the classical lipoprotein separation methods such as density gradient ultracentrifugation and electrophoretic separation. In fact, in electrophoresis Lp(a) has a pre-beta mobility similar to very-low-density lipoproteins (VLDL),[3] and in ultracentrifugation its density spans both the LDL and high-density lipoprotein (HDL) range, with most of the Lp(a) found within the 1.050–1.100 kg/L density range.[4] However, Lp(a) can be detected in small amounts in all density intervals in the range of 1.000–1.210 kg/L, [5,6] and the concentration of Lp(a) appears to significantly increase in the triglyceride-rich $<$ 1.006 kg/L density fraction after a fatty meal.[7,8] In studies performed on Lp(a) from different individuals, Fless et al.[5] not only confirmed the broad inter-individual heterogeneity of Lp(a) particles, but also found that Lp(a) exhibits intra-individual density heterogeneity due to differences in the protein to lipid ratio in Lp(a) particles.

Considering that apo(a)-free Lp(a) obtained by reduction of the disulfide bond between apo(a) and apo B-100 has physical-chemical properties similar if not identical to LDL, it is evident that the peculiar characteristics and the size and density heterogeneity of Lp(a) are almost entirely accounted for by the presence of its distinct protein component, apo(a). Apo(a) is a carbohydrate-rich protein which exhibits both intra- and inter-individual variations in size. The mean carbohydrate content has been estimated to be 28% by weight with a molar ratio of 3:7:5:4:7 for mannose, galactose, galactosamine, glucosamine, and sialic acid, respectively.[9] Although differences in apo(a) size are primarily related to differences in the length of the polypeptide chain,[10,11] a contribution of carbohydrate heterogeneity to the size heterogeneity cannot be excluded. The analysis of the apo(a) cDNA isolated from human liver libraries led to the discovery that apo(a) has a high degree of homology with plasminogen.[12] Plasminogen contains a protease domain and one copy each of five domains called kringles, designated as kringle 1 through kringle 5. A kringle is a highly conserved, tri-loop polypeptide structure stabilized by three disulfide bridges also found in other proteins involved in hemostasis and fibrinolysis.[13,14] The tri-loop structure is similar in shape to the danish pastry called a kringle, from which its name derives.

The basic structure of apo(a) is graphically represented in Figure 14–1. Apo(a) contains a kringle domain and a carboxyl-terminal protease domain with 85% amino acid homology with the plasminogen protease domain. The kringle domain is composed of 11 distinct kringle types, ten of which are similar but not identical to each other and to plasminogen kringle 4, and one which shares over 85% amino acid homology with plasminogen kringle 5. The ten kringles are identified as kringle 4 (K4) type 1 through type 10, and their homology with plasminogen K4 ranges between 78% and 88%.[15,16] Apo(a) K4 type 1 and types 3 through 10 appear to have only one copy present per apo(a) particle, while K4 type 2 is present in multiple repeats varying in number from as few as 3 to as many as 40.[17,18] The number of K4 type 2 repeats is therefore responsible for

Figure 14–1 ❖ *Schematic Representation of the Structure of Apo(a)*

This small-size apo(a) is formed by a protease domain, a kringle 5 domain (K5), and ten basic types of kringle 4 (K4), identified as type 1 through type 10 (T1-T10). K4 type 2 is present in four repeats. In the insert is an enlargement of K4 type 2 with the amino acid sequence, the three internal disulfide bonds, and putative glycosylation sites.

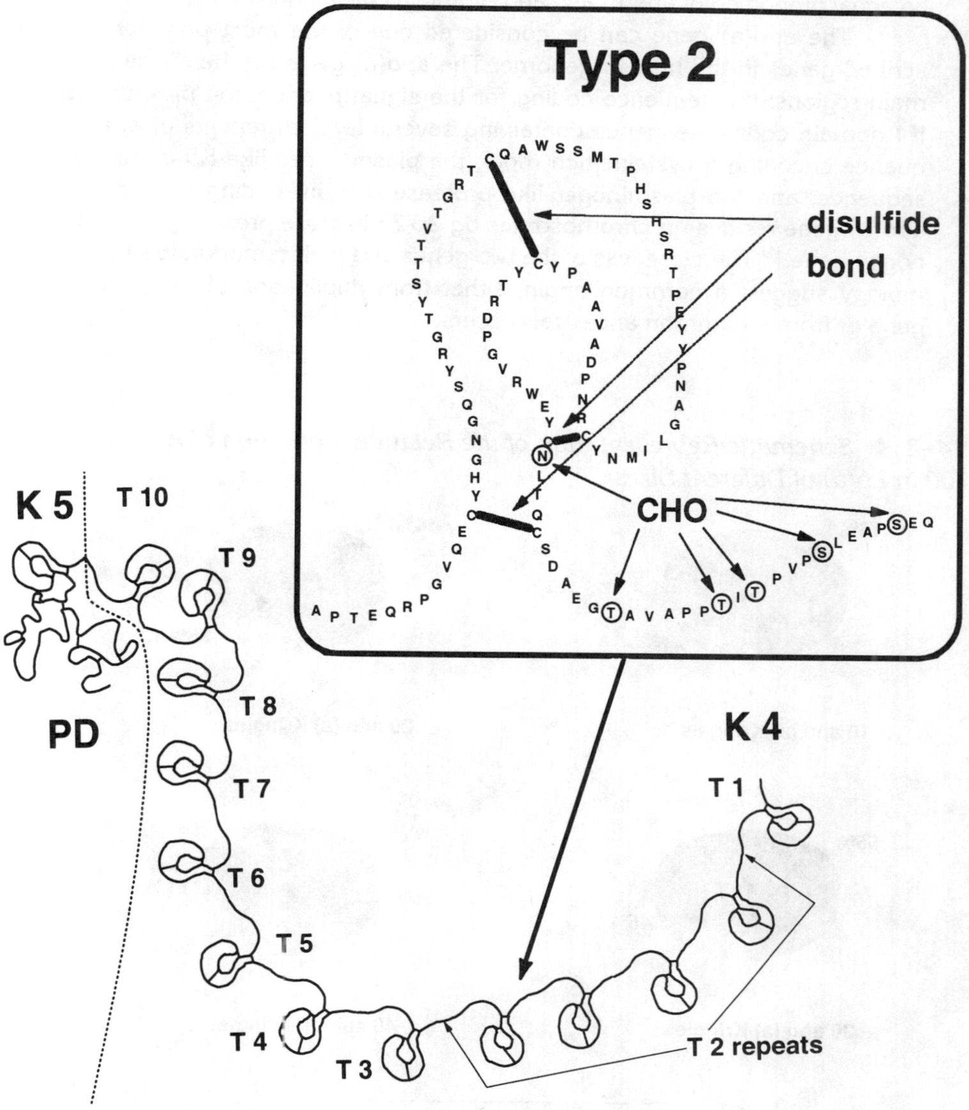

the size heterogeneity of apo(a) and ultimately of Lp(a) and for the different apo(a) isoforms.

Based on the cloning sequence,[12] the molecular mass of apo(a) protein varies from 187 kD for an apo(a) which contains 12 K4 domains, to 662 kD for an apo(a) which contains 50 K4 domains. Therefore, the weight ratio of apo(a) to apo B varies significantly in Lp(a) particles. For example, Lp(a) particles with 20 apo(a) K4 domains have about 64% of their protein mass as apo B and 36% as apo(a). In contrast, large apo(a) particles containing 40 apo(a) K4 domains contain approximately an equal proportion of apo(a) and apo B protein (see Figure 14–2).

The apo(a) gene can be considered one of the most polymorphic transcribed genes in the human genome. The apo(a) gene can be divided into four main regions: the sequence coding for the signal peptide, the plasminogen-like K4 domain coding sequence containing several tandem repeats of a 5.5 Kb sequence encoding a cysteine-rich motif, the plasminogen-like K5 domain coding sequence, and the plasminogen-like protease domain coding sequence.[12] The apo(a) gene resides on chromosomes 6q 26-27 in close proximity to the plasminogen gene.[19] The closeness of the two genes and their remarkable sequence homology suggest a common origin, either from duplication of the plasminogen gene or from a common ancestral precursor.

Figure 14–2 ❖ Schematic Representation of the Relative Proportion of Apo(a) and Apo B-100 in Lp(a) of Different Sizes

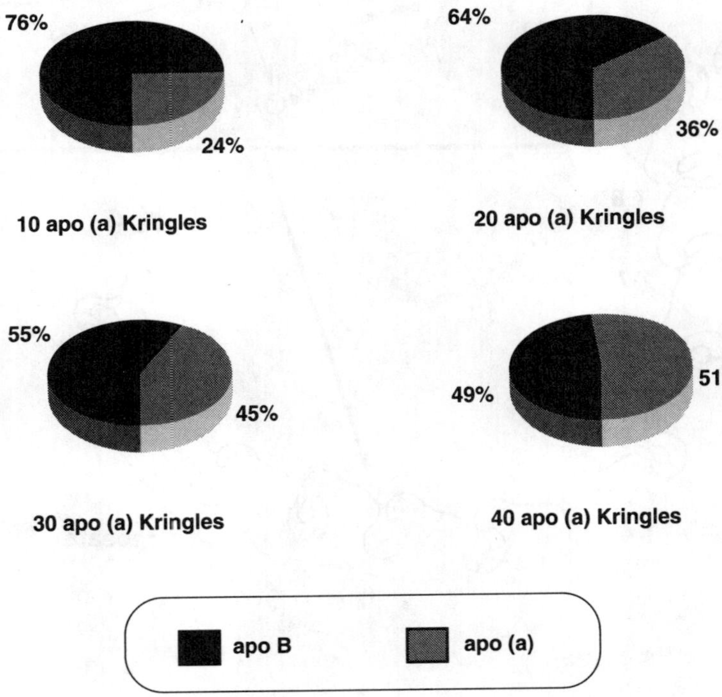

GENETICS OF LIPOPROTEIN(a)

Early family studies using diffusion in agarose gel indicated the presence of Lp(a) segregated in a Mendelian autosomal dominant fashion determined by two alleles, Lp(a^+) and Lp(a^0) or null allele.[2] At that time, Lp(a) was found in only 30–40% of the subjects studied in different populations.[2] Numerous studies in which quantitative immunochemical methods have been used to measure Lp(a) have firmly established that Lp(a) concentrations are strongly influenced by genetic factors.[20–22] Discovery of genetic size polymorphism of apo(a) by Utermann and colleagues provided new insights into the genetic control of Lp(a). Apo(a) isoforms were determined by the mobility of apo(a) in sodium dodecylsulphate polyacrylamide gel electrophoresis (SDS-PAGE) under reducing conditions. Six different apo(a) isoforms were identified: F, B, S1, S2, S3, and S4.[23] The apo(a) isoforms are specified by alleles at a single apo(a) locus. In addition, a null allele has been postulated to account for the segregation of apo(a) isoforms in families and in individuals with no detectable apo(a).[24] These studies indicated that apo(a) isoforms are specified at a single locus by multiple autosomal alleles co-dominantly expressed. Additionally, it was shown that the size of the apo(a) isoforms were, in general, inversely correlated with the Lp(a) plasma concentrations. Gaubatz, et al.[25] resolved 11 different apo(a) isoforms using PAGE combined with a sensitive detection system. Kamboh et al.,[26] using an improved separation technique, were able to distinguish as many as 23 apo(a) isoforms by SDS agarose gel electrophoresis followed by immunoblotting. Marcovina et al.[27] developed a modification of the procedure of Kamboh et al. using agarose gel electrophoresis followed by immunoblotting with ^{125}I-labelled monoclonal antibody and autoradiography. This approach allowed the identification of 34 apo(a) isoforms in a biracial population and is entirely consistent with the observations of Lackner, et al.,[17] in which a total of 34 apo(a) alleles could be distinguished. By using pulsed-field gel electrophoresis and genomic blotting,[28] 94% of Caucasians are heterozygous at the apo(a) locus.[29] Examination of Lp(a) concentrations and apo(a) genotypes in 48 nuclear Caucasian families indicates that the apo(a) gene accounts for greater than 90% of the variation in plasma Lp(a) concentrations.[29] The number of K4 repeats in the apo(a) gene accounted for about 70% of the Lp(a) variation, and the other factors at the apo(a) locus accounted for about 20% of the inter-individual variation in plasma Lp(a) concentrations.

MEASUREMENT OF LIPOPROTEIN(a)

The high size heterogeneity of apo(a), the distinct protein component of Lp(a), its association with apo B-100 to form a single macromolecular complex, and the amino acid sequence homology with plasminogen constitute major challenges to the development of suitable immunoassays for the quantitative determination of Lp(a) in human plasma. To obtain purified apo(a), reducing agents are used to dissociate apo(a) from apo B. However, it has been demonstrated that the use of reducing agents results in a marked decrease of apo(a) immunoreactivity, probably due to the cleavage of the intrachain disulfide bonds of the kringle domains.[30] The fact that the structure of isolated apo(a) is immunochemically different from apo(a) in Lp(a) has been further demonstrated by the observation that monoclonal antibodies raised against purified apo(a) have limited reactivity with native Lp(a).[31] Therefore, antibodies against apo(a) are generated by immunizing animals with intact Lp(a). If the generated antibodies are polyclonal, they need to be absorbed against apo B. Additionally, their immunoreactivity should also be tested against plasminogen in the assay format in which the antibodies are intended to be used, and, if cross-reactivity is detected, the antibodies should be absorbed against plasminogen. Similarly, when monoclonal antibodies against apo(a) are produced, they should be selected to react to epitopes present only in apo(a), so as to prevent cross-reactivity with plasminogen.

A major problem in the generation of antibodies for measuring apo(a) derives from the size heterogeneity of apo(a), which is due to the variable number of K4 type 2 repeats present in the various apo(a) polymorphs.[17, 18] In fact, repeated antigenic determinants are present in variable numbers in different apo(a) particles, and therefore the immunoreactivity of the antibodies directed to these repeated epitopes will vary depending on the size of apo(a). As a consequence, it is expected that immunoassays using polyclonal antibodies or monoclonal antibodies directed to epitopes present in K4 type 2 will tend to underestimate apo(a) concentration in subjects with apo(a) of a smaller size than the apo(a) present in the assay calibrator, and to overestimate the apo(a) concentration of the larger apo(a) particles.

The problem of variable immunoreactivity of the antibodies can be solved by selecting monoclonal antibodies directed to apo(a) antigenic determinants not expressed in K4 type 2. However, the use of monoclonal antibodies will limit the choice of the assay format mainly to enzyme-linked immunosorbent assay (ELISA). Oxidative modification of Lp(a) may produce significant changes in Lp(a) conformation,[32] and the same may occur with carbohydrate removal. The immunoreactivity of some of the monoclonal antibodies produced in our laboratory was differentially affected by carbohydrate removal and Lp(a) oxidation. The removal of sialic acid, for example, resulted in total loss of immunoreactivity for one monoclonal antibody, while the immunoreactivity of other antibodies was significantly higher (unpublished observations). Therefore, to avoid technical artifacts and lack of comparability of values, the monoclonal antibodies to be used in Lp(a) immunoassays should be selected after a careful characterization of their epitope specificity and immunochemical properties.

Specific recommendations for selecting and characterizing monoclonal antibodies for measuring apo AI and B are contained in a document of the International Federation of Clinical Chemistry.[33] Most of the recommendations made in this document are also applicable to monoclonal antibodies to apo(a).

A second approach to solve the effect of the size heterogeneity of apo(a) on the antibody immunoreactivity has been proposed by several investigators.[34–38] This approach involves the use of a sandwich-type ELISA in which the Lp(a) particles in plasma are captured by monoclonal or polyclonal antibodies against apo(a), and then Lp(a) concentration is measured by enzyme-conjugated monoclonal or polyclonal antibodies against the apo B component of Lp(a). It should be noted that even though the immunoreactivity of apo B in Lp(a) has been found to be only modestly lower than that in LDL,[39] studies performed with monoclonal antibodies clearly indicate that the epitope expression of apo B in Lp(a) is different from that in LDL.[40, 41] Therefore, the anti-apo B antibodies to be used in Lp(a) assays, particularly monoclonal antibodies generated against LDL–apo B, should be carefully characterized.

Conceptually, it is true that Lp(a) assays using anti-apo B antibodies or anti-apo(a) monoclonal antibodies directed to epitopes present in apo(a) domains other than K4 type 2 are not affected by variation in Lp(a) size. However, the results of Lp(a) concentration in different samples will depend on the isoforms in the assay calibrator. In the immunoassays, the concentration of an analyte is obtained by comparing the immunoreactivity of the samples with that of the calibrator, which contains a known concentration of the analyte. Therefore, the composition of the assay calibrator should be very similar to that of the samples being analyzed. Considering the high heterogeneity of apo(a) size, this requirement is impossible to obtain; therefore, the composition of the calibrator in the immunoassays is an arbitrary choice that will influence the values obtained in the samples.

Following the approach of Albers et al.,[42] historically the values assigned to Lp(a) assay calibrators, and consequently Lp(a) concentrations in the measured samples, have been expressed in terms of total lipoprotein mass. Obtaining an accurate determination of Lp(a) particle mass is difficult in that it requires the summation of the independent determination of Lp(a) protein, lipid, and carbohydrate components. For these reasons, the expression of Lp(a) in terms of total lipoprotein mass has been discouraged and its expression in terms of Lp(a) protein has been suggested.[43]

Because Lp(a) is a heterogeneous complex of two proteins, the total protein mass is determined in the primary standard, usually by the Lowry method. However, considering the size heterogeneity of apo(a), the ratio of apo(a) to apo B in Lp(a) particles is highly variable, as illustrated in Figure 14–2. Therefore, it is evident that, independent of whether the values are expressed in terms of lipoprotein or protein mass, the ratio of apo(a) to apo B in the calibrator will affect the results for the samples. This concept is illustrated by the following example.

Lp(a) analyses are performed by the same ELISA, with an anti-apo-B antibody enzyme-conjugate used for detection, calibration with two different Lp(a) preparations, and total protein mass determined by the Lowry method. The first preparation is an Lp(a) in which apo B is 70% of the total protein mass, and the second preparation is an Lp(a) in which apo B is 50% of the total protein mass. Considering the higher proportion of apo B in the first preparation, it is evident that the anti-apo-B antibodies will have a higher reactivity than with the second preparation. As a consequence, the same samples will have lower Lp(a) values in the assay calibrated with the first standard than those obtained by the assay calibrated with the second standard. Therefore, it is misleading to state that Lp(a) concentration may be accurately quantified regardless of the apo(a) isoforms with values expressed in terms of Lp(a) mass.[37] The only potentially accurate measurements, independent of apo(a) size polymorphism, are those obtained by methods using monoclonal antibodies specific to epitopes that are not present in

K4 type 2, or antibodies to apo B using calibrators with value assigned in terms of nanomoles per liter of apo(a) or nanomoles per liter of apo B, depending on the antibodies used. A particle concentration fluorescence immunoassay for Lp(a) with values expressed in nanomoles of apo B per liter has recently been reported.[38] Considering that in 20 years of measurement of Lp(a) and in the large number of clinical and epidemiological studies Lp(a) has been expressed in terms of Lp(a) mass or Lp(a) protein, it will be difficult to relate to previous findings Lp(a) values expressed in terms of number of particles per unit volume.

Despite all of the above-mentioned problems, numerous immunochemical methods for the measurement of Lp(a) in a wide variety of formats, some of which are commercially available, have been reported.[43, 44] Albers et al. were the first to report the development of a radial immunodiffusion (RID) and subsequently of a radioimmunoassay (RIA) for the measurement of Lp(a).[42, 45] Other assay formats include electroimmunodiffusion,[46, 47] ELISA,[34–37, 48–50] latex immunoassay,[51] immunonephelometric assay,[52] immunoturbidimetric assay,[53] and fluorescence assay.[38]

Albers and Marcovina recently compared Lp(a) values obtained by two direct binding ELISA methods developed in their laboratory and seven commercially available methods, all calibrated with the same fresh-frozen serum with an intermediate size apo(a) isoform.[54] Of the seven commercially available methods, four were ELISA and three were turbidimetric assays. Reasonable agreement between the Lp(a) values obtained by these methods was found, except that the samples having apo(a) of a smaller size than the apo(a) in the calibrator were significantly higher by the apo B detection methods. However, despite the significant differences in antibody source and in assay format, a common calibrator was able to decrease the differences in Lp(a) values obtained by the different methods. The results of two international surveys of Lp(a) measurements recently reported indicated that the lack of a common primary standard and poor assay precision were the main factors responsible for the high inter-laboratory variation observed in the surveys.[55] A potential approach to bypass the problems of the immunochemical determination of Lp(a) could be to quantify Lp(a) by its cholesterol content. To this end, two methods were recently reported. In one method, cholesterol is enzymatically measured by a continuous flow analyzer in lipoprotein classes separated by single spin vertical rotor ultracentrifugation.[56] The second reported method is based on the use of lectin affinity to separate Lp(a) from other lipoproteins with measurement of Lp(a) cholesterol by enzymatic assay.[57] Both of these methods have the potential for use in screening subjects with high Lp(a) concentrations. However, further studies are required to compare the clinical significance of Lp(a) cholesterol with that of Lp(a) mass or Lp(a) particle number.

At the present time, there is no standardization for measurements of Lp(a). Albers et al.[43] summarized the problems that must be addressed in order to standardize Lp(a) measurement:

1. careful selection and characterization of the antibodies and optimization of the assay format

2. selection of the apo(a) polymorph in the primary standard, reference material, and assay calibrators
3. development of a common, validated method for isolation of Lp(a) to be used as primary standard
4. development of suitable reference material
5. expression of apo(a) values
6. development of a reference method.

Until a valid standardization program is implemented, clinical chemistry laboratories should standardize their assays to selected lipid research laboratories with long-standing reputations for measuring Lp(a). Manufacturers of Lp(a) test kits should include standardization as one of their primary design goals.

POPULATION STUDIES

Lp(a) concentrations are almost entirely genetically determined, and variation at the apo(a) locus contributes to this heritability.[29, 58] Numerous studies[59–66] also reported that Lp(a) values vary greatly between ethnic groups and that the distribution is highly skewed toward the low concentrations. In contrast, in the Black population the distribution is almost Gaussian, and the median Lp(a) value in Blacks is nearly three times as high as in the Whites.[66] In both Black and White infants, Lp(a) concentrations are lower at birth and rise to adult concentrations by age 2 y.[67] A more recent study performed in a consecutive series of 1032 babies reported longitudinal changes in Lp(a) concentrations during the first year of life.[68] Lp(a) concentrations measured during the first postnatal week predicted the concentrations at 8.5 months, at which time Lp(a) concentrations were of the same order of magnitude as those of the parents.

After the initial rise in early childhood, Lp(a) concentrations were reported to be stable throughout adulthood, with no significant gender-related differences.[42, 60] However, results from four major population-based studies do not entirely confirm these findings. In the CARDIA study,[66] Lp(a) concentrations were determined in a group of 4165 young adults comprising approximately the same number of Blacks and Whites and men and women. In this study, no gender-related difference in Lp(a) concentrations was found among Whites. Among Blacks, a small but statistically significant ($p < .05$) difference was found, with concentrations of Lp(a) higher in women than in men. In the PROCAM study,[69] Lp(a) concentrations were measured in 1865 White males (age 20–64 y) and 819 females (age 20–59 y). The gender-related mean values were not significantly different. However, in men, Lp(a) concentrations remained constant with age, while Lp(a) concentrations in postmenopausal women were 2.1 mg/dL higher. The same observation was made in the ARIC study,[65] in which Lp(a) concentrations were determined in a large cohort of Black and White subjects. Premenopausal women from both races had lower Lp(a) concentrations than postmenopausal women. Additionally, it was found that both Black and White postmenopausal women taking hormone replacement therapy had lower Lp(a) concentrations than women not on hormone therapy.[70] Among 1284 White men and 1394 White women participating in the Framingham Offspring study,[71] no gender-related differences in Lp(a) values were found. Although Lp(a) concentrations were 8% higher in postmenopausal women than in premenopausal women, this difference was not statistically significant. Furthermore, there was no significant difference in Lp(a) concentration between postmenopausal women on estrogen therapy and those not taking estrogen. The findings of this study contradict previous reports [65, 69, 70] and are rather surprising in view of the fact that a decrease in Lp(a) concentrations in women on hormone replacement therapy has been reported in several additional studies.[72–75]

Data from all the population-based studies indicate that Lp(a) concentrations are race-related, with Blacks having the most striking difference compared to other races, both in terms of Lp(a) values and distribution. Additionally, it is likely that differences in Lp(a) concentration may be observed among ethnic groups within the same race. Therefore, population-based reference values, as well as the concentration at which Lp(a) is considered to be a risk factor, should be established for each ethnic group.

THROMBOTIC AND ATHEROGENIC ACTIVITIES OF LIPOPROTEIN(a)

Components of Lp(a) have similarities to both LDL and plasminogen, suggesting that Lp(a) may represent a bridge between the fields of atherosclerosis and thrombosis.[76] Recent studies have demonstrated that Lp(a) inhibits fibrinolysis by competing with plasminogen for binding to fibrin[77–80] and to cell membranes.[78, 81, 82] The presence of both apo(a) and apo B has been documented in atheromatous plaques in arteries and vein grafts,[83–88] with Lp(a) co-localizing with fibrin.[89] Lp(a) has been shown to regulate the expression of plasminogen activator inhibitor-1 (PAI-1) in endothelial cells, which in turn inhibits the generation of plasmin by complexing and inactivating tissue-type plasminogen activator (t-PA).[90] In contrast, one study reported that Lp(a) did not inhibit fibrinolysis,[91] while another reported enhancement of clot lysis.[92]

Harpel et al.[93, 94] have shown that sulfhydryl compounds such as homocysteine, known to be associated with atherosclerosis and thromboembolic disease, and glutathione increase the affinity of Lp(a) for fibrin. They further propose that following vascular injury, partial plasmin degradation of the thrombus formed at the vessel wall produces a substrate that binds Lp(a). Increased concentrations of homocysteine and glutathione leaking from red cells and platelets in the thrombus increase the affinity of Lp(a) for fibrin. During the repair process, the Lp(a)-fibrin complex is incorporated into the vessel intima, contributing to the formation of an atheromatous plaque. Lp(a) is also thought to promote the growth of plaques by inhibiting transforming growth factor beta (TGF-β), thereby promoting proliferation of smooth muscle cells.[95] It has also been suggested that formation of an insoluble complex of Lp(a) with calcium in atherosclerotic lesions may also contribute to plaque growth.[96] The incorporation of Lp(a)-derived cholesterol into plaque has also been documented.[97] Thus, Lp(a) is thought to contribute to the continuous growth of the lesion as well as the acute thrombotic event that occurs after a plaque ruptures by inhibiting the activation of plasminogen by t-PA. It is still possible, however, that localization of apo(a) in plaque may represent a secondary phenomenon, and that Lp(a) is not causally related to lesion development. Scanu has suggested that the action of Lp(a) at the cell surface may be related to focal events rather than plasma concentrations.[98]

Results from these and other studies have prompted investigators to examine the relationship of Lp(a) concentration in plasma and the success of thrombolytic therapy. Five studies have shown that baseline Lp(a) concentrations do not affect the outcome of thrombolytic therapy.[99–103] In addition, blood concentrations of Lp(a) do not correlate with α-2-antiplasmin-plasmin inhibitor complex[104] or D-dimer values,[105] both measures of fibrinolysis *in vivo*. Reports of Lp(a) concentrations during thrombolytic therapy are contradictory. Qiu et al. reported no effect of streptokinase or single-chain form tPA therapy on plasma Lp(a) concentrations in unstable angina,[101] while Hegele et al., reported acute reduction of Lp(a) concentrations by double-chain form tPA.[106] Comparison of these studies is difficult because one study sampled Lp(a) concentrations 6 h after therapy while the other sampled after 12 h, and different forms of t-PA were used. Di Lorenzo et al. reported acute lowering of Lp(a) concentrations by heparin in patients undergoing percutaneous transluminal coronary angioplasty (PTCA),[107] but Hegele et al. observed that heparin alone had no effect.[106] All of these studies were done with small numbers of patients and must be repeated with sample sizes adequate to represent the different isoforms of apo(a), some of which may have different effects on fibrinolysis *in vivo*; differences in fibrinolytic activity for different apo(a) isoforms was recently demonstrated *in vitro*.[108]

CLINICAL SIGNIFICANCE OF LIPOPROTEIN(a)

Lp(a) Concentrations and Coronary Heart Disease

In the early 1970s, several studies reported the association between coronary heart disease (CHD) and a plasma lipoprotein that was detected by electrophoresis and named pre-beta-1 lipoprotein because of its mobility relative to LDL in gels.[109–111] Subsequent studies demonstrated that pre-beta-1 was Lp(a) and confirmed its association with CHD.[112–114] In 1977, Albers et al. were the first to report the association between high concentrations of Lp(a), determined by an immunochemical method, and CHD.[45] The distribution of Lp(a) was measured in 90 myocardial infarction (MI) survivors and in their healthy spouses, and was found to be shifted to higher concentrations in MI survivors than in the spouse controls. In another study of MI survivors performed in 1981, Kostner et al. found that the relative risk for MI in subjects with Lp(a) concentrations > 30 mg/dL, expressed in terms of total Lp(a) mass, were 1.75-fold higher than in subjects with Lp(a) concentrations > 30 mg/dL.[115]

More recently, Dahlen et al. studied 307 patients who underwent coronary angiography and reported that the coronary lesion score significantly correlated with Lp(a) values. Multivariate analysis demonstrated that high Lp(a) concentrations were associated significantly and independently with CHD.[116] In a study aimed at evaluating Lp(a) in the context of other risk factors, Armstrong et al. found that Lp(a) was an even stronger predictor of CHD if other risk factors were also present.[117]

Interest in Lp(a) increased with the discovery that apo(a) has a high homology with plasminogen,[12] and numerous clinical studies subsequently reported the association of Lp(a) with CHD,[118-122] MI,[123-125] stenosis following coronary artery bypass graft surgery (CABG),[126, 127] restenosis after PTCA,[128-130] progression of CHD in familial hypercholesterolemia (FH),[131-134] and stroke.[135-142] One study reported no significant difference in Lp(a) concentrations between CABG patients and controls, although a trend toward higher Lp(a) concentrations with increasing severity of disease was reported.[143]

Two preliminary studies examined the utility of acute lowering of Lp(a) and LDL concentrations in patients by LDL-apheresis before and after PTCA for preventing restenosis.[144, 145] In one study, in patients with Lp(a) concentration of > 30 mg/dL, restenosis rates were lower in those patients in whom a reduction of 50% or more of the initial Lp(a) concentration was achieved.[144] In the second study, restenosis was observed in only one out of ten patients treated once every 2 w for 3–4 m; three patients pheresed for 2 y did not restenose.[145] Similarly, Walzl et al. reported that plasmapheretic reduction of lipids, including Lp(a), in patients shortly after a stroke contributed to improved neurological recovery.[146] These are the first studies to show an improvement in clinical outcome as a result of lowering Lp(a) concentrations.

Only a few studies have looked at the relationship of Lp(a) phenotype to CHD. Kark et al. reported a significantly higher prevalence of low-molecular-weight apo(a) isoforms in female, but not in male, MI survivors compared to controls.[124] Similarly, two studies reported a higher prevalence of low-molecular-weight isoforms in FH patients with CHD compared to those without disease.[131, 132] Guo et al. reported the interesting finding that in FH patients heterozygous for apo(a) isoform, the different isoforms resided on distinct lipoprotein particles.[132] Further studies with larger sample sizes are needed to better define the contribution of Lp(a) concentrations and apo(a) isoforms to risk of disease in men and women and in different ethnic groups.

Among numerous case-control studies, virtually all have shown a strong association between Lp(a) and CHD. In contrast, relatively few prospective studies of the significance of increased Lp(a) concentrations have been conducted, and the results are quite discordant. In fact, while three of these prospective studies, all conducted in men, concluded that Lp(a) concentration is an independent risk factor for subsequent MI or CHD,[122, 147, 148] two major published studies reached opposite conclusions. In a nested case-control study of the Helsinki Heart Study participants, neither the distribution of Lp(a) concentrations nor the mean or median Lp(a) concentrations were significantly different between subjects with and without coronary events.[149] In the largest prospective study conducted thus far, based on a 5-y follow-up of 14,916 participants in the Physician's Health Study, Ridker et al. found no evidence of association between Lp(a) and risk of future MI.[150] Two provocative editorials have critically evaluated the possible explanations for these discordant findings and underlined the limitations of the present studies.[151, 152]

Historically, the large majority of clinical studies relating Lp(a) to risk of disease have been performed in White populations. More recently, a number of investigators have begun to evaluate the clinical significance of Lp(a) in other ethnic groups.[58, 61, 153–157] Sandholzer et al.[61] reported the results of a multicenter prospective case-control study that looked at apo(a) phenotypes and Lp(a) concentrations in six ethnic populations. These authors reported a consistent association of low-molecular-weight polymorphs with CHD across populations. However, plasma samples from all of the populations were not stored in the same way; the German samples were improperly stored at –20° C for 2 y and therefore were not analyzed for Lp(a) concentration, and the Israeli samples were stored at –20° C for 6 m. In addition, in one of the six populations, samples were from nonfasting individuals. This may be an important consideration if particles enriched with triglyceride containing Lp(a) react differently with antibodies used in Lp(a) assays. Abe and Noma found that the distribution of Lp(a) concentrations, apo(a) phenotype, and allele frequencies in healthy Japanese were not significantly different from White European populations, but were significantly different from other Asian populations, including Chinese, Indians, and Malaysians.[155] Further findings were that Lp(a) concentrations were significantly higher in patients with CHD compared to healthy controls, that there was a higher frequency of double-banded phenotypes in CHD patients, that the frequency of apo(a) alleles was not different from those in healthy individuals, and that Lp(a) concentrations in CHD patients were higher than those in healthy people who had the same phenotype.[156]

The most interesting and intriguing results are those obtained in the few studies in which the clinical significance of Lp(a) was evaluated in Black and White subjects. In The Bogalusa Heart Study, Lp(a) concentrations in White and Black children and incidence of MI in their parents were used as a measure of future risk for the children. Lp(a) concentrations in White children with parental MI were significantly higher than in White children without parental MI. In addition, the prevalence of parental MI in Whites was significantly higher in subjects with Lp(a) concentrations > 25 mg/dL. No relationship of Lp(a) concentration to incidence of parental MI was observed in Blacks. Guyton et al.[59] and Sorrentino et al.[157] studied the relationship between Lp(a) concentrations and CHD in Blacks and Whites and reported identical conclusions. White and Black patients with CHD had higher Lp(a) concentrations than Blacks and Whites without CHD. Although the Lp(a) concentrations in Blacks averaged twice those of Whites, the incidence of CHD was not different in the two racial groups, and Blacks did not appear to experience increased atherosclerotic progression and mortality. Based on these findings, Guyton et al. concluded that the atherogenicity of Lp(a) in Blacks is decreased or is counterbalanced by other factors.[59]

It is clear from these studies that the clinical significance of Lp(a) concentration and isoform in each ethnic group must be determined separately. Additionally, basic studies are required to investigate the reasons for the apparent difference in atherogenicity of Lp(a) in Blacks.

Lp(a) Concentrations in Patients with Diabetes

The literature is contradictory in describing the potential role of Lp(a) as a contributing risk factor to CHD in diabetes mellitus. Numerous groups have reported that Lp(a) concentrations are increased in patients with insulin-dependent diabetes (IDDM),[158-162] and that Lp(a) concentration was[158, 160, 163] or was not[164-166] related to the degree of glycemic control. However, other studies reported that Lp(a) concentrations were not increased in IDDM patients compared to healthy control subjects,[164, 167-169] and one study found no significant difference in apo(a) phenotypes as well.[170] There seems to be general agreement, with two exceptions, [160, 171] that Lp(a) concentrations in non-insulin dependent (NIDDM) diabetes are not increased.[163, 164, 172-174] Increased concentrations of Lp(a) have also been reported in patients suffering from diabetic retinopathy.[175-177]

Three studies recently reported that other risk factors surpassed the significance of Lp(a) in diabetic patients.[178-180] Ritter et al. studied the relationship between Lp(a) concentration and late diabetic complications, including diabetic polyneuropathy, autonomic neuropathy, nephropathy, peripheral occlusive disease, diabetic gangrene, CHD, and retinopathy. No significant increases in Lp(a) correlated with any of these complications except for retinopathy, and this correlation was not significant when the duration of diabetes was factored into logistic regression analysis.[178] Similarly, in a study of 45 identical twin pairs discordant for IDDM, Dubrey et al. reported no relationship between genetic susceptibility to IDDM and apo(a) phenotype or Lp(a) concentration. Lp(a) concentrations in diabetics were similar to those found in the control population.[179] In the third study, Császár et al. reported no significant differences between Lp(a) concentration or phenotype frequencies between diabetic and non-diabetic subjects within two ethnic populations in Austria and Hungary.[180]

These studies collectively suggest that Lp(a) may not be a risk factor related to diabetes directly. Rather, Lp(a) concentrations may be increased in some individuals independent of the diabetic condition, and may pose an additional risk for atherosclerotic disease. Further, the incidence of increased concentrations of Lp(a) in the diabetic population may be the same as in the normal population.

Lp(a) Concentrations in Patients with Renal Disease

A number of studies have reported increased Lp(a) concentrations in patients with IDDM and proteinuria,[181-184] while others found no such association.[160, 165, 185] In three studies of subjects with NIDDM, one study found that microalbuminuria was associated with increased Lp(a) concentrations,[186] while the remaining two studies reported the opposite finding.[187, 188] Several authors reported increased Lp(a) concentrations in non-diabetic patients with various renal diseases;[189, 190] Stenvinkel reported significant reduction of Lp(a) concentrations in patients with nephrotic syndrome after remission.[191] Recently, Sato et al. demonstrated co-localization of apo(a) and apo B-100 in the mesangial area and capillary loops in glomeruli from patients with glomerular diseases, and proposed that

Lp(a) might advance glomerular disease by interfering with activation of TGF-β, thereby interfering with repair of injured cells or tissue.[192]

Most of these studies did not control for ethnicity; among the few that did, sample sizes within ethnic groups were too small to be meaningful. Most of these studies shared a similar finding that Lp(a) concentrations were higher in both diabetic and non-diabetic patients with proteinuria compared to those without proteinuria. These studies suggest that metabolic conditions in patients with renal disease and proteinuria may cause plasma Lp(a) concentrations to rise. In addition, increased Lp(a) concentrations in renal disease may contribute to the increased risk of CHD associated with renal disease. Larger studies are needed to define the true role, if any, of Lp(a) in renal disease.

Lp(a) Concentrations in Patients Receiving Renal Replacement Therapy

A number of recent studies have found significantly higher concentrations of Lp(a) in patients receiving hemodialysis (HD), continuous ambulatory peritoneal dialysis (CAPD), or transplant therapy compared to healthy control subjects.[193-201]

Irish et al. reported significantly higher concentrations of Lp(a) in patients with chronic renal disease and in those undergoing CAPD compared to healthy controls, but found no difference between control subjects and patients undergoing HD or transplant therapy.[202] Black and Wilcken measured Lp(a) before and after kidney transplantation in 20 patients with end-stage renal failure and found a significant decrease in Lp(a) after transplantation[203] to a concentration comparable to values reported by others for renal transplant patients.[193, 202] The decrease in Lp(a) was significantly correlated with an increase in creatinine clearance. It has been suggested that cyclosporine may raise Lp(a) concentrations in transplant patients,[204, 205] but this conclusion has been contested.[206-208]

Cressman et al. reported that Lp(a) concentration and the presence of a previous CHD-related clinical event were the only independent risk factors for predicting CHD in HD patients in a multiple logistic regression model.[194] In a follow-up study, Lp(a) concentration was an independent predictor of fatal events attributable to CHD.[209]

Dieplinger et al. reported increased Lp(a) values in patients with end-stage renal disease compared to controls, but found no difference in frequency of apo(a) isoforms between the groups. Concentrations of Lp(a) were increased two- to threefold in patients with high-molecular-weight apo(a) isoforms, while in patients with small isoforms, Lp(a) concentrations were comparable to those found in controls.[210] These results suggest that non-genetic factors related to renal insufficiency are responsible for increased Lp(a) concentrations in this population. Haffner et al. proposed that increased Lp(a) concentrations in chronic renal failure must occur early or contribute to its development, since Lp(a) concentration did not correlate with creatinine concentrations, was similarly increased in different ethnic groups, and was not affected by diet, HD, or CAPD treatment.[211] The release of tumor necrosis factor alpha, interleukin 1, and interleukin 6,[212] along with increased concentrations of C-reactive protein,[213] have been reported in hemodialysis patients. The release of cytokines has been attributed to chronic exposure to dialysis membranes.[214] Interleukin 6 is a potent activator of the synthesis of acute phase proteins by hepatocytes.[215, 216] Lp(a) has been shown to behave like an acute phase protein after MI, stroke, or CABG surgery,[82, 137, 217–219] and may also play a role in inflammation.[220, 221] These findings may explain why Lp(a) concentrations are increased in patients undergoing chronic hemodialysis. Similar cytokine production has been reported in CAPD patients.[222] These studies provide evidence that increased Lp(a) concentrations may contribute to the atherosclerotic disease associated with renal disease.

Requirements for Clinical Studies of Lp(a) and Atherosclerotic Disease

The lack of standardization of both Lp(a) measurement and apo(a) phenotype determination makes it difficult to compare clinical studies. Further, until the development of Lp(a) assays which are not sensitive to apo(a) size, no conclusion can be reached regarding the independent effects of Lp(a) concentration and apo(a) phenotype on the risk for atherosclerotic disease. The following guidelines are proposed for designing future studies of Lp(a) and atherosclerotic disease:

1. Sample size, estimated by appropriate power calculations, must be large enough to adequately represent the numerous apo(a) isoforms[27] and the most common of the possible polymorphic combinations. This will prevent possible selection bias due to phenotype.
2. Isoform studies should use the technique with the highest resolution[26, 27] and, when a standardized approach for apo(a) phenotyping is developed, it should be utilized exclusively.
3. Antibodies used for the measurement of Lp(a) concentrations in clinical studies must be characterized with regard to the epitope recognized, and a clear demonstration of the effect of isoform size must be demonstrated in order to be able to interpret Lp(a) concentrations.
4. The ethnic composition of all study populations must be defined.
5. To avoid possible confounding effects of the acute phase response on Lp(a) concentrations, Lp(a) should not be measured during an active inflammatory state.
6. Medians should be reported rather than means.
7. Nonparametric statistics must be used for analysis of Lp(a) data in ethnic groups with non-Gaussian distributions.

SCREENING FOR LIPOPROTEIN(a) CONCENTRATIONS

Based on the contradictory results of the epidemiologic studies and the problems in the immunochemical measurement of Lp(a), screening of the general population for Lp(a) concentration is not recommended. However, we do recommend screening patients with CHD or family history of CHD or stroke, siblings and offspring of individuals with increased concentrations of Lp(a), patients with FH, and patients with renal dysfunction accompanied by proteinuria. However, considering the numerous methodological problems, Lp(a) measurement on this high-risk group of subjects should be performed only by specialized lipid research laboratories. These laboratories should have population-based reference values determined by their method for different ethnic groups. To enable the physician to estimate the patient's risk, Lp(a) values should be reported with the indication of the corresponding percentile of the general population. Based on the outcome of the clinical studies, it appears reasonable to consider patients with Lp(a) values above the 80th percentile at increased risk for CHD.

Therapeutic regimens for the groups of patients with a predisposition for developing atherosclerotic disease with increased Lp(a) concentrations should be designed to minimize the lipid- and non-lipid-related risk factors that can be modified, since standard lipid-lowering therapy does not affect Lp(a) concentrations. Diet, hydroxymethylglutaryl-CoA (HMG-CoA) reductase inhibitors, bile acid sequestrants, and probucol have been shown not to lower Lp(a) concentrations.[223] Niacin and neomycin, separately and in combination, and bezafibrate have been moderately effective in lowering Lp(a) concentrations,[224-226] while niacin in high doses of 3–6 g daily has been shown to significantly lower Lp(a).[227] These dosages however, are difficult for patients to tolerate.

The most effective way to decrease Lp(a) concentrations by 50% or more is by LDL- or Lp(a)-apheresis procedures. However, these techniques are expensive and are reserved for the extreme cases of hetero- and homozygous FH.[228-230] Because Lp(a) concentrations are not easily decreased, evidence that decreasing Lp(a) concentration is beneficial has not been established. Although it is a logical therapeutic goal to decrease Lp(a) in patients with multiple risk factors, specific goals or targets have not been established due to the lack of reference ranges for different ethnic groups. The studies discussed earlier[228-230] provided preliminary evidence that decreasing Lp(a) concentrations utilizing apheresis procedures before and after PTCA and during recovery after stroke may improve outcome.

SUMMARY

Lp(a), a cholesterol-rich plasma lipoprotein which is genetically controlled, is thought to be an independent risk factor for premature atherosclerosis. Although Lp(a) is thought to represent a bridge between the fields of atherosclerosis and thrombosis, studies of the effect of Lp(a) on fibrinolysis *in vitro* have been somewhat inconsistent, and no effect on fibrinolysis has been demonstrated *in vivo*. Standardization of Lp(a) measurement and apo(a) phenotype determination must be developed in order to provide clinical investigators with the tools they need to elucidate the true relationship of Lp(a) to atherosclerotic disease.

Currently all measurements of Lp(a) concentration reported in clinical studies in the literature are confounded by isoform size, making it difficult to come to a sound conclusion regarding the role of Lp(a) concentration in disease. Measurement of Lp(a) concentrations with methods not sensitive to isoform size and determination of apo(a) phenotypes in different ethnic groups are required before specific recommendations can be made regarding screening of the general population for Lp(a). At the present time, screening for Lp(a) concentration is recommended for patients with, or with a family history of, CHD, stroke, MI, or FH and renal dysfunction accompanied by proteinuria. Management of patients with high Lp(a) concentrations should be directed at minimizing all other lipid- and nonlipid risk factors for atherosclerotic disease. ❖

Acknowledgments: The authors wish to express their gratitude to Dr. John J. Albers for reviewing the manuscript and providing helpful suggestions and Dr. Peter C. Harpel for reviewing the section on fibrinolysis. Special thanks to Hal Kennedy for computer generation of the figures, and to Dorthy Westmoreland for typing the manuscript. Dr. Marcovina's original work cited in this review was supported by Program Project Grant HL 30086 from the National Heart, Lung, and Blood Institute, U.S. Public Service.

REFERENCES

1. Berg K. A new serum type system in man: the Lp system. Acta Pathol Microbiol Scand 1963;59:362–82.
2. Berg K. The Lp system [Review]. Ser Haemotol 1968;1:111–36.
3. Albers JJ, Cabana VG, Warnick GR, Hazzard WR. Lp(a) lipoprotein: relationship to sinking pre-β lipoprotein, hyperlipoproteinemia, and apolipoprotein B. Metabolism 1975;24:1047–54.
4. Gaubatz JW, Heideman C, Gotto AM Jr, Morrisett JD, Dahlen GH. Human plasma lipoprotein(a): structural properties. J Biol Chem 1983;258:4582–9.
5. Fless GM, Rolih CA, Scanu AM. Heterogeneity of human plasma lipoprotein(a): isolation and characterization of the lipoprotein subspecies and their apoproteins. J Biol Chem 1984;259:11470–8.
6. Fless GM. Heterogeneity of particles containing the apo B-apo(a) complex. In: Scanu AM, ed. Lipoprotein(a). San Diego: Academic Press, 1990:41–51.
7. Pfaffinger D, Schuelke J, Kim C, Fless GM, Scanu AM. Relationship between apo(a) isoforms and Lp(a) density in subjects with different apo(a) phenotype: a study before and after a fatty meal. J Lipid Res 1991;32:679–83.
8. Cohn JS. Lam CWK, Sullivan DR, Hensley WJ. Plasma lipoprotein distribution of apolipoprotein(a) in the fed and fasted states. Atherosclerosis 1991;90:59–66.
9. Fless GM, ZumMallen ME, Scanu AM. Physicochemical properties of apolipoprotein(a) and lipoprotein(a-) derived from the dissociation of human plasma lipoprotein(a). J Biol Chem 1986;261:8712–18.
10. Hixson JE, Britten ML, Manis GS, Rainwater DL. Apolipoprotein(a) (Apo(a)) glycoprotein isoforms result from size differences in apo(a) mRNA in baboons. J Biol Chem 1989;264:6013–16.
11. Koschinsky ML, Beisiegel U, Henne-Bruns D, Eaton DL, Lawn RM. Apolipoprotein(a) size heterogeneity is related to variable number of repeat sequences in its mRNA. Biochemistry 1990;29:640–4.
12. McLean JW, Tomlinson JE, Kuang W-J, et al. cDNA sequence of human apolipoprotein(a) is homologous to plasminogen. Nature 1987;330:132–7.
13. Magnusson S, Sottrup-Jensen L, Petersen TE, Dudek-Wojciechowska G, Claeys H. Homologous "kringle" structures common to plasminogen and prothrombin:

Substrate specificity of enzymes activating prothrombin and plasminogen. In: Ribbons DW, Brew K. eds. Proteolysis and physiological regulation. New York: Academic Press, 1976:203–38.

14. Castellino FJ, Beals JM. The genetic relationships between the kringle domains of human plasminogen, prothrombin, tissue plasminogen activator, urokinase, and coagulation factor XII. J Mol Evol 1987;26:358–69.
15. Guevara J Jr, Knapp RD, Honda S, Northup SR, Morrisett JD. A structural assessment of the apo(a) protein of human lipoprotein(a). Proteins 1992;12:188–99.
16. Guevara J Jr, Jan AY, Knapp R, Tulinsky A, Morrisett JD. Comparison of ligand-binding sites of modeled apo(a) kringle-like sequences in human lipoprotein(a). Arterioscler Thromb 1993;13:758–70.
17. Lackner C, Cohen JC, Hobbs HH. Molecular definition of the extreme size polymorphism in apolipoprotein(a). Hum Mol Genet 1993;2:933–40.
18. van der Hoek YY, Wittekoek ME, Beisiegel U, Kastelein JJ, Koschinsky ML. The apolipoprotein(a) kringle IV repeats which differ from the major repeat kringle are present in variably-sized isoforms. Hum Mol Genet 1993;2:361–66.
19. Frank SL, Klisak I, Sparkes RS, et al. The apolipoprotein(a) gene resides on human chromosome 6q26-27, in close proximity to the homologous gene for plasminogen. Hum Genet 1988;79:352–6.
20. Albers JJ, Wahl P, Hazzard WR. Quantitative genetic studies of the human plasma Lp(a) lipoprotein. Biochem Genet 1974;11:475–86.
21. Iselius L, Dahlen G, DeFaire U, Lundman T. Complex segregation analysis of the Lp(a)/pre-beta 1-lipoprotein trait. Clin Genet 1981;20:147–51.
22. Morton NE, Berg K, Dahlen G, Ferrel RE, Rhoads GG. Genetics of the Lp lipoprotein in Japanese-Americans. Genet Epidemiol 1985;2:113–21.
23. Utermann G, Menzel HJ, Kraft HG, Duba HC, Kemmler HG, Seitz C. Lp(a) glycoprotein phenotypes: inheritance and relation to Lp(a)-lipoprotein concentrations in plasma. J Clin Invest 1987;80:458–65.
24. Utermann G, Duba C, Manzel HJ. Genetics of the quantitative Lp(a) lipoprotein trait. II. Inheritance of Lp(a) glycoprotein phenotypes. Hum Genet 1988;78:47–50.
25. Gaubatz JW, Ghanem KI, Guevara J Jr, et al. Polymorphic forms of human apolipoprotein(a): inheritance and relationship of their molecular weights to plasma concentrations of lipoprotein(a). J Lipid Res 1990;31:603–13.
26. Kamboh MI, Ferrell RE, Kottke BA. Expressed hypervariable polymorphism of apolipoprotein(a). Am J Hum Genet 1991;49:1063–74.
27. Marcovina SM, Zhang ZH, Gaur VP, Albers JJ. Identification of 34 apolipoprotein(a) isoforms: differential expression of apolipoprotein(a) alleles between American Blacks and Whites. Biochem Biophys Res Commun 1993;191:1192–6.
28. Lackner C, Boerwinkle E, Leffert CC, Rahmig T, Hobbs HH. Molecular basis of apolipoprotein(a) isoform size heterogeneity as revealed by pulsed-field gel electrophoresis. J Clin Invest 1991;87:2153–61.
29. Boerwinkle E, Leffert CC, Lin J, Lackner C, Chiesa G, Hobbs HH. Apolipoprotein(a) gene accounts for greater than 90% of the variation in plasma lipoprotein(a) concentrations. J Clin Invest 1992;90:52–60.
30. Scanu AM, Pfaffinger D, Fless GM, Makino K, Eisenbart J, Hinman J. Attenuation of immunologic reactivity of lipoprotein(a) by thiols and cysteine-containing compounds: structural implications. Arterioscler Thromb 1992;12:424–9.
31. Guo H-C, Armstrong VW, Luc G, et al. Characterization of five mouse monoclonal antibodies to apolipoprotein(a) from human Lp(a): evidence for weak plasminogen reactivity. J Lipid Res 1989;30:23–37.
32. Naruszewicze M, Giroux L-M, Davignon J. Oxidative modification of Lp(a) causes changes in the structure and biological properties of apo(a). Chem Phys Lipids 1994;67/68:167–74.

33. Marcovina SM, Curtiss LK, Milne R, Albers JJ. Selection and characterization of monoclonal antibodies for measuring plasma concentrations of apolipoproteins A-I and B. Scientific Division, Committee on Apolipoproteins, Working Group on antibody reagents, IFCC document. JIFCC 1990;2:138–44.
34. Rainwater DL, Manis GS. Immunochemical characterization and quantitation of lipoprotein(a) in baboons. Development of an assay depending on two antigenically distinct proteins. Atherosclerosis 1988;73:23–31.
35. Fless GM, Snyder ML, Scanu AM. Enzyme-linked immunoassay for Lp(a). J Lipid Res 1989;30:651–62.
36. Vu-Dac N, Mezdour H, Parra HJ, Luc G, Luyeye I, Fruchart JC. A selective bi-site immunoenzymatic procedure for human Lp(a) lipoprotein quantification using monoclonal antibodies against apo(a) and apo B. J Lipid Res 1989;30:1437–43.
37. Taddei-Peters WC, Butman BT, Jones GR, Venetta TM, Macomber PF, Ransom H. Quantification of lipoprotein(a) particles containing various apolipoprotein(a) isoforms by a monoclonal anti-apo(a) capture antibody and a polyclonal anti-apolipoprotein B detection antibody sandwich enzyme immunoassay. Clin Chem 1993;39:1382–9.
38. Kottke BA, Bren ND. A particle concentration fluorescence immunoassay for Lp(a). Chem Phys Lipids 1994;67/68:249–56.
39. Fless GM, Pfaffinger DJ, Eisenbart JD, Scanu AM. Solubility, immunochemical, and lipoprotein binding properties of apo B-100-apo(a), the protein moiety of lipoprotein(a). J Lipid Res 1990;31:909–18.
40. Zawadzki Z, Tercé F, Seman LJ, et al. The linkage with apolipoprotein(a) in lipoprotein(a) modifies the immunochemical and functional properties of apolipoprotein B. Biochemistry 1988;27:8474–81.
41. Gries A, Fievet C, Marcovina S, et al. Interaction of LDL, Lp(a) and reduced Lp(a) with monoclonal antibodies against apo B. J Lipid Res 1988;29:1–8.
42. Albers JJ, Hazzard WR. Immunochemical quantification of human plasma Lp(a) lipoprotein. Lipids 1974;9:15–26.
43. Albers JJ, Marcovina SM, Lodge MS. The unique lipoprotein(a): properties and immunochemical measurement. Clin Chem 1990;36:2019–26.
44. Labeur C, Shepherd J, Rosseneu M. Immunological assays of apolipoproteins in plasma: methods and instrumentation. Clin Chem 1990;36:591–7.
45. Albers JJ, Adolphson JL, Hazzard WR. Radioimmunoassay of human plasma Lp(a) lipoprotein. J Lipid Res 1977;18:331–8.
46. Molinari E, Pichler P, Krempler F, Kostner G. A rapid screening method for pathological lipoprotein Lp(a) concentrations by counterimmunoelectrophoresis. Clin Chim Acta 1983;128:373–8.
47. Gaubatz JW, Cushing GL, Morrissett JD. Quantitation, isolation, and characterization of human lipoprotein(a). Methods Enzymol 1986;129:167–86.
48. Labeur C, Michiels G, Bury J, Usher DC, Rosseneu M. Lipoprotein(a) quantified by an enzyme-linked immunosorbent assay with monoclonal antibodies. Clin Chem 1989;35:1380–4.
49. Abe A, Maeda S, Makino K, et al. Enzyme-linked immunosorbent assay of lipoprotein(a) in serum and cord blood. Clin Chim Acta 1988;177:31–40.
50. Wong WLT, Eaton DL, Berloui A, Fendly B, Hass PE. A monoclonal antibody-based enzyme-linked immunosorbent assay of lipoprotein(a). Clin Chem 1990; 36:192–7.
51. Vu-Dac N, Chekkor A, Parra H, Duthilleul P, Fruchard JC. Latex immunoassay of human serum Lp(a+) lipoprotein. J Lipid Res 1985;26:267–9.
52. Cazzolato G, Prakasch G, Green S, Kostner GM. The determination of lipoprotein Lp(a) by rate and endpoint nephelometry. Clin Chim Acta 1983;135:203–8.

53. Levine DM, Sloan BJ, Donner JE, Lorenz JD, Henzerling R. Automated measurement of Lp(a) by immunoturbidimetric analysis. Int J Clin Lab Res 1992; 22:173–8.
54. Albers JJ, Marcovina SM. Standardization of Lp(a) measurement. Chem Phys Lipids 1994;67/68:257–63.
55. Labeur C, Rosseneu M, Henderson O. International Lp(a) standardization. Chem Phys Lipids 1994;67/68:265–70.
56. Kulkarni KR, Garber DW, Marcovina SM, Segrest JP. Quantification of cholesterol in all lipoprotein classes by the VAP-II method. J Lipid Res 1994;35:159–68.
57. Seman LJ, Jenner JL, McNamara JR, Schaefer EJ. Quantification of lipoprotein(a) in plasma by assaying cholesterol in lectin-bound plasma fraction. Clin Chem 1994;40:400–3.
58. Austin MA, Sandholzer C, Selby JV, Newman B, Krauss RM, Utermann G. Lipoprotein(a) in women twins: Heritability and relationship to apolipoprotein(a) phenotypes. Am J Hum Genet 1992;51:829–40.
59. Guyton JR, Dahlen GH, Patsch W, Kautz JA, Gotto AM, Jr. Relationship of plasma lipoprotein Lp(a) levels to race and to apolipoprotein B. Arteriosclerosis 1985;5:265–72.
60. Helmhold M, Bigge J, Muche R, et al. Contribution of the apo(a) phenotype to plasma Lp(a) concentration shows considerable ethnic variation. J Lipid Res 1991;32:1919–28.
61. Sandholzer C, Saha N, Kark JD, et al. Apo(a) isoforms predict risk for coronary heart disease: a study in six populations. Arterioscler Thromb 1992;12:1214–26.
62. Cobbaert C, Kesteloot H. Serum lipoprotein(a) levels in racially different populations. Am J Epidemiol 1992;136:441–9.
63. Parra HJ, Luyeye I, Bouramoue C, Demarquilly C, Fruchart JC. Black-White differences in serum Lp(a) lipoprotein levels. Clin Chim Acta 1987;168:27–31.
64. Srinivasan SR, Dahlen GH, Jarpa RA, Webber LS, Berenson GS. Racial (Black-White) difference in serum lipoprotein(a) distribution and its relation to parental myocardial infarction in children. Bogalusa Heart Study. Circulation 1991; 84:160–7.
65. Brown SA, Hutchinson R, Morrisett J, et al. Plasma lipid, lipoprotein cholesterol, and apoprotein distributions in selected US communities: The Atherosclerosis Risk in Communities (ARIC) Study. Arterioscler Thromb 1993;13:1139–58.
66. Marcovina SM, Albers JJ, Jacobs DR, Jr, et al. Lipoprotein(a) concentrations and apolipoprotein(a) phenotypes in Caucasians and African Americans: The CARDIA study. Arterioscler Thromb 1993;13:1037–45.
67. Rifai N, Heiss G, Doetsch K. Lipoprotein(a) at birth, in Blacks and Whites. Atherosclerosis 1992;92:123–9.
68. Wilcken DEL, Wang XL, Dudman NPB. The relationship between infant and parent Lp(a) concentrations. Chem Phys Lipids 1994;67/68:299–304.
69. Sandkamp M, Assmann G. Lipoprotein(a) in PROCAM participants and young myocardial infarction survivors. In: Scanu AM, ed. Lipoprotein(a). San Diego: Academic Press, 1990:205–9.
70. Nabulsi AA, Folsom AR, White A, et al. Association of hormone-replacement therapy with various cardiovascular risk factors in postmenopausal women. N Engl J Med 1993;328:1069–75.
71. Jenner JL, Ordovas JM, Lamon-Fava S, et al. Effects of age, sex, and menopausal status on plasma lipoprotein(a) levels: The Framingham Offspring Study. Circulation 1993;87:1135–41.
72. Henriksson P, Angelin B, Berglund L. Hormonal regulation of serum Lp(a) levels: opposite effects after estrogen treatment and orchidectomy in males with prostatic carcinoma. J Clin Invest 1992;89:1166–71.

73. Soma MR, Osnago-Gadda I, Paoletti R, et al. The lowering of lipoprotein(a) induced by estrogen plus progesterone replacement therapy in postmenopausal women. Arch Intern Med 1993;153:1462–8.
74. van der Mooren MJ, Demacker PN, Thomas CM, Rolland R. Beneficial effects on serum lipoproteins by 17 β-oestradiol-dydrogesterone therapy in postmenopausal women: a prospective study. Eur J Obstet Gynecol Reprod Biol 1992; 47:153–60.
75. Seed M, Crook D. Postmenopausal hormone replacement therapy, coronary heart disease and plasma lipoproteins. Curr Opin Lipidol 1994;5:48–58.
76. Scanu AM. Lipoprotein(a): A potential bridge between the fields of atherosclerosis and thrombosis. Arch Pathol Lab Med 1988;112:1045–7.
77. Harpel PC, Gordon BR, Parker TS. Plasmin catalyzes binding of lipoprotein(a) to immobilized fibrinogen and fibrin. Proc Natl Acad Sci USA 1989;86:3847–51.
78. Gonzalez-Gronow M, Edelberg JM, Pizzo SV. Further characterization of the cellular plasminogen binding site: evidence that plasminogen 2 and lipoprotein(a) compete for the same site. Biochemistry 1989;28:2374–7.
79. Loscalzo J, Weinfeld M, Fless GM, Scanu AM. Lipoprotein(a), fibrin binding, and plasminogen activation. Arteriosclerosis 1990;10:240–5.
80. Rouy D, Grailhe P, Nigon F, Chapman J, Angles-Cano E. Lipoprotein(a) impairs generation of plasmin by fibrin-bound tissue-type plasminogen activator: in vitro studies in a plasma milieu. Arterioscler Thromb 1991;11:629–38.
81. Miles LA, Fless GM, Levin EG, Scanu AM, Plow E. A potential basis for the thrombotic risks associated with lipoprotein(a). Nature 1989;339:301–3.
82. Hajjar KA, Gavish D, Breslow JL, Nachman RL. Lipoprotein(a) modulation of endothelial cell surface fibrinolysis and its potential role in atherosclerosis. Nature 1989;339:303–5.
83. Walton KW, Hitchens J, Magnani HN, Kahn M. A study of methods of identification and estimation of Lp(a) lipoprotein and of its significance in health, hyperlipidaemia and atherosclerosis. Atherosclerosis 1974;20:323–46.
84. Cushing, GL, Gaubatz JW, Nava ML, et al. Quantitation and localization of apolipoproteins(a) and B in coronary artery bypass vein grafts at re-operation. Arteriosclerosis 1989;9:593–603.
85. Rath M, Niendorf A, Reblin T, Dietel M, Krebber HJ, Beisiegel U. Detection and quantification of lipoprotein(a) in the arterial wall of 107 coronary artery bypass patients. Arteriosclerosis 1989;9:579–92.
86. Niendorf A, Rath M, Wolf K, et al. Morphological detection and quantification of lipoprotein(a) deposition in atheromatous lesions of human aorta and coronary arteries. Virchows Archiv Pathol Anat 1990;417:105–11.
87. Beisiegel U, Niendorf A, Wolf K, Reblin T, Rath M. Lipoprotein(a) in the arterial wall. Eur Heart J 1990;11(supplement E):174–83.
88. Hoff HF, O'Neil J, Yashiro A. Partial characterization of lipoproteins containing apo(a) in human atherosclerotic lesions. J Lipid Res 1993;34:789–98.
89. Wolf K, Rath M. Niendorf A, Beisiegel U, Dietel M. Morphological co-localization of apoprotein(a) and fibrin(ogen) in human coronary atheromas. Circulation 1989; 80:II–522.
90. Etingin OR, Hajjar DP, Hajjar KA, Harpel PC, Nachman RL. Lipoprotein(a) regulates plasminogen activator inhibitor-1 expression in endothelial cells: a potential mechanism in thrombogenesis. J Biol Chem 1991;266:2459–65.
91. Halvorsen S, Skjonsberg OH, Berg K, Ruyter R, Godal HC. Does Lp(a) lipoprotein inhibit the fibrinolytic system? Thromb Res 1992;68:223–32.
92. Mao SJ, Tucci MA. Lipoprotein(a) enhances plasma clot lysis in vitro. FEBS Lett 1990;267:131–4.

93. Harpel PC, Chang VT, Borth W. Homocysteine and other sulfhydryl compounds enhance the binding of lipoprotein(a) to fibrin: a potential biochemical link between thrombosis, atherogenesis and sulfhydryl compound metabolism. Proc Natl Acad Sci USA 1992;89:10193–7.
94. Harpel PC, Borth W. Fibrin, lipoprotein(a), plasmin interactions: a model linking thrombosis and atherogenesis. Ann NY Acad Sci 1992;667:233–8.
95. Kojima S, Harpel PC, Rifkin DB. Lipoprotein(a) inhibits the generation of transforming growth factor beta: an endogenous inhibitor of smooth muscle cell migration. J Cell Biol 1991;113:1439–45.
96. Yashiro A, O'Neil J, Hoff HF. Insoluble complex formation of lipoprotein(a) with low density lipoprotein in the presence of calcium ions. J Biol Chem 1993; 268:4709–15.
97. Smith EB, Cochran S. Factors influencing the accumulation in fibrous plaques of lipid derived from low density lipoprotein. II. Preferential immobilization of lipoprotein(a) (Lp(a)). Atherosclerosis 1990;84:173–81.
98. Scanu AM. Structural basis for the presumptive atherothrombogenic action of lipoprotein(a). Facts and speculations. Biochem Pharmacol 1993;46:1675–80.
99. Armstrong VW, Neubauer C, Schutz E, Tebbe U. Lack of association between raised serum Lp(a) concentration and unsuccessful thrombolysis after acute myocardial infarction. Lancet 1990;336:1077.
100. von Hodenberg E, Kreuzer J, Hautmann KJM, Nordt T, Kubler W, Bode C. Effects of lipoprotein(a) on success rate of thrombolytic therapy in acute myocardial infarction. Am J Cardiol 1991;67:1349–53.
101. Qiu S, Theroux P, Genest J Jr, Solymoss BC, Robitaille D, Marcil M. Lipoprotein(a) blood concentrations in unstable angina pectoris, acute myocardial infarction and after thrombolytic therapy. Am J Cardiol 1991;67:1175–9.
102. von Hodenberg, Pestel E, Kreuzer J, Freitag M, Bode C. Effects of lipoprotein(a) on thrombolysis. Chem Phys Lipids 1994;67/68:381–5.
103. Tranchesi B, Santos-Filho RS, Vinagre C, et al. Lipoprotein(a) concentrations do not influence the outcome of rt-PA therapy in acute myocardial infarction. Ann Hemato 1991;62;141–2.
104. Oshima S, Uchida K, Yasu T, Uno K, Nonogi H, Haze K. Transient increase of plasma lipoprotein(a) in patients with unstable angina pectoris: does lipoprotein(a) alter fibrinolysis? Arterioscler Thromb 1991;11:1772–7.
105. Donders SH, Lustermans FA, Van Wersch JW. Coagulation factors and lipid composition of the blood in treated and untreated hypertensive patients. Scand J Clin Lab Invest 1993;53:179–86.
106. Hegele RA, Freeman MR, Langer A, Connelly PW, Armstrong PW. Acute reduction of lipoprotein(a) by tissue-type plasminogen activator. Circulation 1992;85: 2034–8.
107. Di Lorenzo M, Salvini P, Levi-Della-Vida M, Maddaloni E. Acute reduction of lipoprotein(a) by tissue-type plasminogen activator. Circulation 1993;87:1052–3.
108. Hervio L, Chapman MJ, Thillet J, Loyau S, Anglés-Cano E. Does apolipoprotein(a) heterogeneity influence lipoprotein(a) effects on fibrinolysis? Blood 1993;82: 392–7.
109. Dahlen G, Ericson C, Furberg C, Lundkvist L, Svardsudd K. Angina of effort and an extra pre-beta lipoprotein fraction. Acta Med Scan 192;531(Suppl):6.
110. Papadopoulos NM, Bedynek JL. Serum lipoprotein patterns in patients with coronary atherosclerosis. Clin Chim Acta 1973;44:153.
111. Insull W, Najmi M, Vloedman DA. Plasma pre-beta lipoprotein subfractions in diagnosis of coronary artery disease. Circulation 1972;45 (suppl II):II–170.
112. Berg K, Dahlen G, Frick MH. Lp(a) lipoprotein and pre-beta1-lipoprotein in patients with coronary heart disease. Clin Genet 1974;6:230–5.

113. Dahlen G, Berg K, Gillnas T, Ericson C. Lp(a) lipoprotein/pre-beta1-lipoprotein in Swedish middle-aged males and in patients with coronary heart disease. Clin Genet 1975;7:334–41.
114. Dahlen G, Frick MH, Berg K, Valle M, Wiljasalo M. Further studies of Lp(a) lipoprotein/pre-beta1-lipoprotein in patients with coronary heart disease. Clin Genet 1975;8:183–9.
115. Kostner GM, Avogaro P, Cazzolato G, Marth E, Bittolo-Bon G, Qunici GB. Lipoprotein Lp(a) and the risk for myocardial infarction. Atherosclerosis 1981;38:51.
116. Dahlen GH, Guyton JR, Mohammad A, Farmer JA, Kautz JA, Gotto AM. Association of levels of lipoprotein Lp(a), plasma lipids, and other lipoproteins with coronary artery disease documented by angiography. Circulation 1986;74:758–65.
117. Armstrong VW, Cremer P, Eberle E, et al. The association between serum Lp(a) concentrations and angiographically assessed coronary atherosclerosis. Atherosclerosis 1986;62:249–57.
118. Hearn JA, DeMaio SJ Jr, Roubin GS, Hammarstrom M, Sgoutas D. Predictive value of Lipoprotein(a) and other serum lipoproteins in the angiographic diagnosis of coronary artery disease. Am J Cardiol 1990;66:1176–80.
119. Genest J Jr, Jenner JL, McNamara JR, et al. Prevalence of lipoprotein(a) [Lp(a)] excess in coronary artery disease. Am J Cardiol 1991;67:1039–45.
120. Genest JJ Jr, Martin-Munley SS, McNamara JR, et al. Familial lipoprotein disorders in patients with premature coronary artery disease. Circulation. 1992;85: 2025–33.
121. Genest JJ Jr, McNamara JR, Ordovas JM, et al. Lipoprotein cholesterol, apolipoprotein A-I and B and Lipoprotein(a) abnormalities in men with premature coronary artery disease. J Am Coll Cardiol 1992;19:792–802.
122. Rosengren A, Wilhelmsen L, Eriksson E, Risberg B, Wedel H. Lipoprotein (a) and coronary heart disease: a prospective case-control study in a general population sample of middle aged men. BMJ 1990;301:1248–51.
123. Sandkamp M, Funke H, Schulte H, Kohler E, Assmann G. Lipoprotein(a) is an independent risk factor for myocardial infarction at a young age. Clin Chem 1990;36:20–23.
124. Kark JD, Sandholzer C, Friedlander Y, Utermann G. Plasma Lp(a), apolipoprotein(a) isoforms and acute myocardial infarction in men and women: a case-control study in the Jerusalem population. Atherosclerosis 1993;98:139–51.
125. Graziani MS, Zanolla L, Righetti G, et al. Lipoprotein(a) concentrations are increased in patients with myocardial infarction and angiographically normal coronary arteries. Eur J Clin Chem Clin Biochem 1993;31:135–7.
126. Hoff HF, Beck GJ, Skibinski CI, et al. Serum Lp(a) level as a predictor of vein graft stenosis after coronary artery bypass surgery in patients. Circulation 1988;77: 1238–44.
127. Solymoss BC, Marcil M, Wesolowska E, Lesperance J, Pelletier LC, Campeau L. Risk factors of venous aortocoronary bypass graft disease noted at late symptom-directed angiographic study. Can J Cardiol 1993;9:80–84.
128. Hearn JA, Donohue BC, Baalbaki H, et al. Usefulness of serum lipoprotein(a) as a predictor of restenosis after percutaneous transluminal coronary angioplasty. Am J Cardiol 1992;69:736–9.
129. Tenda K, Saikawa T, Maeda T, et al. The relationship between serum lipoprotein(a) and restenosis after initial elective percutaneous transluminal coronary angioplasty. Jpn Circ J 1993;57:789–95.
130. Shimizu Y, Nishikawa H, Motoyasu M, et al. Influence of Lp(a) on restenosis after coronary angioplasty. Kokyu To Junkan 1991;39:687–90.

131. Seed M, Hopplicher F, Reaveley D, et al. Relation of serum lipoprotein(a) concentration and apolipoprotein(a) phenotype to coronary heart disease in patients with familial hypercholesterolemia. N Engl J Med 1990;322:1494–9.
132. Guo HC, Chapman MJ, Bruckert E, Farriaux JP, De-Gennes JL. Lipoprotein Lp(a) in homozygous familial hypercholesterolemia: density profile, particle heterogeneity and apolipoprotein(a) phenotype. Atherosclerosis 1991;86:69–83.
133. Hegele RA, Connelly PW, Cullen-Dean G, Rose V. Elevated plasma lipoprotein(a) associated with abnormal stress thallium scans in children with familial hypercholesterolemia. Am J Cardiol 1993;72:402–6.
134. Sorensen KE, Celermajer DS, Georgakopoulos D, Hatcher G, Betteridge DJ, Deanfield JE. Impairment of endothelium-dependent dilation is an early event in children with familial hypercholesterolemia and is related to the lipoprotein(a) level. J Clin Invest 1994;93:50–55.
135. Koltringer P, Jurgens G. A dominant role of lipoprotein(a) in the investigation and evaluation of parameters indicating development of cervical atherosclerosis. Atherosclerosis 1985;58:187–98.
136. Jurgens G, Koltringer P. Lipoprotein(a) in ischemic cerebrovascular disease: a new approach to the assessment of risk for stroke. Neurology 1987;37:513–15.
137. Woo J, Lam CWK, Kay R, Wong HY, Teoh R, Nicholls MG. Acute and long-term changes in serum lipids after acute stroke. Stroke 1990;21:1407–11.
138. Woo J, Lau E, Lam CWK, et al. Hypertension, lipoprotein(a), and apolipoprotein A-I as risk factors for stroke in the Chinese. Stroke 1991;22:203–8.
139. Pedro-Botet J, Senti M, Nogues X, et al. Lipoprotein and apolipoprotein profile in men with ischemic stroke: role of lipoprotein(a), triglyceride-rich lipoproteins, and apolipoprotein E polymorphism. Stroke 1992;23:1556–62.
140. Shintani S, Kikuchi S, Hamaguchi H, Shiigai T. High serum lipoprotein(a) concentrations are an independent risk factor for cerebral infarction. Stroke 1993;24:965–9.
141. Jovicic A, Ivanisevic V, Ivanovic I. Lipoprotein(a) in patients with carotid atherosclerosis and ischemic cerebrovascular disorders. Atherosclerosis 1993; 98:59–65.
142. Nagayama M, Shinohara Y, Nagayama T. Lipoprotein(a) and ischemic cerebrovascular disease in young adults. Stroke 1994;25:74–78.
143. Averna MR, Barbagallo CM, Ocello S, et al. Lp(a) concentrations in patients undergoing aorto-coronary bypass surgery. Eur Heart J 1992;13:1405–9.
144. Yamaguchi H, Lee YJ, Daida H, et al. Effectiveness of LDL-apheresis in preventing restenosis after percutaneous transluminal coronary angioplasty (PTCA): LDL-apheresis angioplasty restenosis trial (L-ART). Chem Phys Lipids 1994; 67/68:399–403.
145. Kanemitsu S, Tekekoshi N, Murakami E. Effects of LDL apheresis on restenosis after angioplasty. Chem Phys Lipids 1994;67/68:339–43.
146. Walzl M. Lechner H, Walzl B, Schied G. Improved neurological recovery of cerebral infarctions after plasmapheretic reduction of lipids and fibrinogen. Stroke 1993;24:1447–51.
147. Sigurdsson G, Baldursdottir A, Sigvaldason H, Agnarsson U, Thorgeirsson G, Sigfusson N. Predictive value of apolipoproteins in a prospective survey of coronary artery disease in men. Am J Cardiol 1992;69:1251–4.
148. Schaefer EJ, Lamon-Fava S, Jenner JL, et al. Lipoprotein(a) levels and risk of coronary heart disease in men. JAMA 1994;271:999–1003.
149. Jauhiainen M, Koskinen P, Ehnholm C, et al. Lipoprotein (a) and coronary heart disease risk: a nested case-control study of the Helsinki Heart Study participants. Atherosclerosis 1991;89:59–67.
150. Ridker PM, Hennekens CH, Stampfer MJ. A prospective study of lipoprotein(a) and the risk of myocardial infarction. JAMA 1993;270:2195–9.

151. Barnathan ES. Has lipoprotein "little" (a) shrunk? JAMA 1993;270:2224–5.
152. Gurewich V, Mittleman M. Lipoprotein(a) in coronary heart disease: is it a risk factor after all? JAMA 1994;271:1025–6.
153. Kim JQ, Song JH, Lee MM, et al. Evaluation of Lp(a) as a risk factor of coronary artery disease in the Korean population. Ann Clin Biochem 1992;29:226–8.
154. Sandholzer CH, Boerwinkle E, Saha N, Tong MC, Utermann G. Apolipoprotein(a) phenotypes, Lp(a) concentration and plasma lipid leve s in relation to coronary heart disease in a Chinese population: evidence for the role of the apo(a) gene in coronary heart disease. J Clin Invest 1992;89:1040–46.
155. Abe A, Noma A. Studies on apolipoprotein(a) phenotypes. Part 1. Phenotype frequencies in a healthy Japanese population. Atherosclerosis 1992;96:1–8.
156. Abe A, Noma A, Lee YJ, Yamaguchi H. Studies on apolipoprotein(a) phenotypes. Part 2. Phenotype frequencies and Lp(a) concentrations in different phenotypes in patients with angiographically defined coronary artery diseases. Atherosclerosis 1992;96:9–15.
157. Sorrentino MJ, Vielhauer C, Eisenbart JD, Fless GM, Scanu AM, Feldman T. Plasma lipoprotein (a) protein concentration and coronary artery disease in black patients compared with white patients. Am J Med 1992;93:658–62.
158. Bruckert E, Davidoff P, Grimaldi A, et al. Increased serum levels of lipoprotein(a) in diabetes mellitus and their reduction with glycemic control. JAMA 1990;263: 35–36.
159. Haffner SM, Tuttle KR, Rainwater DL. Decrease of lipoprotein(a) with improved glycemic control in IDDM subjects. Diabetes Care. 1991;14:302–7.
160. Guillausseau PJ, Peynet J, Chanson P, et al. Lipoprotein(a) in diabetic patients with and without chronic renal failure. Diabetes Care 1992;15:976–9.
161. Couper JJ, Bates DJ, Cocciolone R, et al. Association of lipoprotein(a) with puberty in IDDM. Diabetes Care, 1993;16:869–73.
162. Salzer B, Stavljenic A, Jurgens G, Dumic M, Radica, A. Polymorphism of apolipoprotein E, lipoprotein(a), and other lipoproteins in children with type I diabetes. Clin Chem 1993;39:1427–32.
163. Nagashima K, Yutani S, Miyake H, Onigata K, Yagi H, Kuroume T. Lipoprotein(a) levels in Japanese children with IDDM. Diabetes Care 1993;16:846.
164. Ritter MM, Richter WO, Lyko K, Schwandt P. Lp(a) serum concentrations and metabolic control. Diabetes Care 1992;15:1441–2.
165. Maser RE, Usher D, Becker DJ, Drash AL, Kuller LH, Orchard TJ. Lipoprotein(a) concentration shows little relationship to IDDM complications in the Pittsburgh Epidemiology of Diabetes Complications Study cohort. Diabetes Care 1993; 16:755–8.
166. Heller FR, Jamart J, Honore P, et al. Serum lipoprotein(a) in patients with diabetes mellitus. Diabetes Care 1993;16:819–23.
167. Levitsky LL, Scanu AM, Gould SH. Lipoprotein(a) levels in black and white children and adolescents with IDDM. Diabetes Care 1991;14:283–7.
168. Austin A, Warty V, Janosky V, Arslanian S. The relationship of physical fitness to lipid and lipoprotein(a) levels in adolescents with IDDM. Diabetes Care 1993; 16:421–5.
169. Winocour PH, Durrington PN, Bhatnagar D, et al. A cross-sectional evaluation of cardiovascular risk factors in coronary heart disease associated with type 1 (insulin-dependent) diabetes mellitus. Diabetes Res Clin Pract 1992;18:173–84.
170. Klausen IC, Schmidt EB, Lervang HH, Gerdes LU, Ditzel J, Faergeman O. Normal lipoprotein(a) concentrations and apolipoprotein(a) isoforms in patients with insulin-dependent diabetes mellitus. Eur J Clin Invest 1992;22:538–41.

171. Ramirez LC, Arauz-Pacheco C, Lackner C, Albright G, Adams BV, Raskin P. Lipoprotein(a) levels in diabetes mellitus: relationship to metabolic control. Ann Intern Med 1992;117:42–7.
172. Joven J, Vilella E. Serum levels of lipoprotein(a) in patients with well-controlled non-insulin-dependent diabetes mellitus. JAMA 1991;265:1113–14.
173. Wolffenbuttel BH, Leurs PB, Sels JP, Rondas-Colbers GJ, Menheere PP, Nieuwenhuijzen Kruseman AC. Improved blood glucose control by insulin therapy in type 2 diabetic patients has no effect on lipoprotein(a) concentrations. Diabet Med 1993;10:427–30.
174. Velho G, Erlich D, Turpin E, et al. Lipoprotein(a) in diabetic patients and normoglycemic relatives in familial NIDDM. Diabetes Care 1993;16:742–7.
175. Maioli M, Tonolo G, Pacifico A, et al. Raised serum apolipoprotein (a) in active diabetic retinopathy. Diabetologia 1993;36:88–90.
176. Muller HM, Diekstall FF, Schmidt E, Marz W, Canzler H, Demeler U. Lipoprotein (a): a risk factor for retinal vascular occlusion. Ger J Ophthalmol 1992;1:338–41.
177. Tomikawa S, Mezawa M, Yoshida Y, Saiga T, Shimizu Y. Lipoprotein (a) and sclerotic changes in retinal arterioles. Nippon Ganka Gakkai Zasshi 1993;97:967–74.
178. Ritter MM, Loscar M, Richter WO, Schwandt P. Lipoprotein(a) in diabetes mellitus. Clin Chim Acta 1993;214:45–54.
179. Dubrey SW, Reaveley DA, Leslie DG, O'Donnell M, O'Connor BM, Seed M. Effect of insulin-dependent diabetes mellitus on lipids and lipoproteins: a study of identical twins. Clin Sci 1993;84:537–42.
180. Császár A, Dieplinger H, Sandholzer C, et al. Plasma lipoprotein (a) concentration and phenotypes in diabetes mellitus. Diabetologia 1993;36:47–51.
181. Takegoshi T, Haba T, Hirai J, et al. Alterations of lipoprotein(a) in patients with diabetic nephropathy. Atherosclerosis 1990;83:99–100.
182. Jenkins AJ, Steele JS, Janus ED, Best JD. Increased plasma apolipoprotein(a) concentrations in IDDM patients with microalbuminuria. Diabetes 1991;40:787–90.
183. Winocour PH, Bhatnagar D, Ishola M, Arrol S, Durrington PN. Lipoprotein(a) and microvascular disease in type I (insulin-dependent) diabetes. Diabet Med 1991;8:922–7.
184. Kapelrud H, Bangstad HJ, Dahl-Jorgensen K, Berg K, Hanssen KF. Serum Lp(a) lipoprotein concentrations in insulin dependent diabetic patients with microalbuminuria. BMJ 1991;303:675–8.
185. Gall MA, Rossing P, Hommel E, et al. Apolipoprotein(a) in insulin-dependent diabetic patients with and without diabetic nephropathy. Scand J Clin Lab Invest 1992;52:513–21.
186. Jenkins AJ, Steele JS, Janus ED, Santamaria JD, Best JD. Plasma apolipoprotein (a) is increased in type 2 (non-insulin-dependent) diabetic patients with microalbuminuria. Diabetologia 1992;35:1055–9.
187. Haffner SM, Morales PA, Gruber MK, Hazuda HP, Stern MP. Cardiovascular risk factors in non-insulin-dependent diabetic subjects with microalbuminuria. Arterioscler Thromb 1993;13:205–10.
188. Irish AB, Simons LA, Simons J. Lipoprotein(a) concentration in diabetes: relationship to proteinuria and diabetes control. Aust NZ J Med 1992;22:329–33.
189. Takegoshi T, Kotoh C, Haba T, et al. A study of the clinical significance of lipoprotein(a) in nephrotic syndrome. Jpn J Med 1991;30:21–5.
190. Thomas ME, Freestone A, Varghese Z, Persaud JW, Moorhead JF. Lipoprotein(a) in patients with proteinuria. Nephrol Dial Transplant 1992;7:597–601.
191. Stenvinkel P, Berglund L, Heimburger O, Pettersson E, Alvestrand A. Lipoprotein(a) in nephrotic syndrome. Kidney Int 1993;44:1116–23.
192. Sato H, Suzuki S, Ueno M, et al. Localization of apolipoprotein(a) and B-100 in various renal diseases. Kidney Int 1993;43:430–5.

193. Heimann P, Josephson MA, Felner SK, Thistlethwaite JR Jr, Stuart FP, Dasgupta A. Elevated lipoprotein(a) levels in renal transplantation and hemodialysis patients. Am J Nephrol 1991;11:470–4.
194. Cressman MD, Heyka RJ, Paganini EP, O'Neil J, Skibinski CI, Hoff HF. Lipoprotein(a) is an independent risk factor for cardiovascular disease in hemodialysis patients. Circulation 1992;86:475–82.
195. Barbagallo CM, Averna MR, Scafidi V, Galione A, Notarbartolo A. Increased lipoprotein(a) levels in subjects with chronic renal failure on hemodialysis. Nephron 1992;62:471–2.
196. Webb AT, Reaveley DA, O'Donnell M, O'Connor B, Seed M, Brown EA. Lipoprotein (a) in patients on maintenance haemodialysis and continuous ambulatory peritoneal dialysis. Nephrol Dial Transplant 1993;8:609–13.
197. Thillet J, Faucher C, Issad B, Allouache M, Chapman J, Jacobs C. Lipoprotein(a) in patients treated by continuous ambulatory peritoneal dialysis. Am J Kidney Dis 1993;22:226–32.
198. Takegoshi T, Kitoh C, Shimada T, Kawai K, Yamazaki Y, Mabuchi H. Alterations of Lp(a) lipoprotein in patients with chronic renal failure treated by continuous ambulatory peritoneal dialysis. Nippon Jinzo Gakkai Shi 1993;35:757–63.
199. Anwar N, Bhatnagar D, Short CD, et al. Serum lipoprotein (a) concentrations in patients undergoing continuous ambulatory peritoneal dialysis. Nephrol Dial Transplant 1993;8:71–74.
200. Webb AT, Plant M, Reaveley DA, et al. Lipid and lipoprotein(a) concentrations in renal transplant patients. Nephrol Dial Transplant 1992;7:636–41.
201. Murphy BG, McNamee P, Duly E, Henry W, Archbold P, Trinick T. Increased serum apolipoprotein(a) in patients with chronic renal failure treated with continuous ambulatory peritoneal dialysis. Atherosclerosis 1992;93:53–57.
202. Irish AB, Simons LA, Savdie E, Hayes JM, Simons J. Lipoprotein(a) concentrations in chronic renal disease states, dialysis and transplantation. Aust NZ J Med 1992;22:243–8.
203. Black IW, Wilcken DE. Decreases in apolipoprotein(a) after renal transplantation: implications for lipoprotein(a) metabolism. Clin Chem 1992;38:353–7.
204. Webb AT, Reaveley DA, O'Donnell M, O'Connor B, Seed M, Brown EA. Does cyclosporin increase lipoprotein(a) concentrations in renal transplant recipients? Lancet 1993;341:268–70.
205. Hilbrands LB, Demacker PN, Hoitsma AJ. Cyclosporin and serum lipids in renal transplant recipients. [Discussion] Lancet 1993;341:765–6.
206. Kronenberg F, Konig P, Lhotta K, et al. Cyclosporin and serum lipids in renal transplant recipients. [Discussion] Lancet 1993;341:765.
207. Irish A. Cyclosporin and serum lipids in renal transplant recipients. [Discussion] Lancet 1993;341:766.
208. Segarra A, Chacon P, Vilardell M, Piera LL. Cyclosporin and serum lipids in renal transplant recipients. [Discussion] Lancet 1993;341:766.
209. Cressman MD, Abood D, O'Neil J, Hoff HF. Lp(a) and premature mortality during chronic hemodialysis treatment. Chem Phys Lipids 1994;67/68:419–27.
210. Dieplinger H, Lackner C, Kronenberg F, et al. Elevated plasma concentrations of lipoprotein(a) in patients with end-stage renal disease are not related to the size polymorphism of apolipoprotein(a). J Clin Invest 1993;91:397–401.
211. Haffner SM, Gruber KK, Aldrete G Jr, Morales PA, Stern MP, Tuttle KR. Increased lipoprotein(a) concentrations in chronic renal failure. J Am Soc Nephrol 1992; 3:1156–62.
212. Cavaillon JM, Poignet JL, Fitting C, Delons S. Serum interleukin-6 in long-term hemodialyzed patients. Nephron 1992;60:307–13.

213. Docci D, Bilancioni R, Buscaroli A, et al. Elevated serum levels of C-reactive protein in hemodialysis patients. Nephron 1990;56:364–7.
214. Mege JL, Olmer M, Purgus R, et al. Haemodialysis membranes modulate chronically the production of TNF alpha, IL1beta and IL6. Nephrol Dial Transplant 1991; 6:868–75
215. Le JM, Vilcek J. Interleukin-6: a multifunctional cytokine regulating immune reactions and the acute phase protein response. Lab Invest 1989;61:588–602.
216. Kishimoto T. The biology of interleukin-6. Blood 1989;74:1–10.
217. Sonoda M, Sakamoto K, Miyauchi T, et al. Changes in serum lipoprotein(a) and C4b-binding protein levels after acute myocardial infarctions. Jpn Circ J 1992; 56:1214–20.
218. Slunga L, Johnson O, Dahlen GH, Eriksson S. Lipoprotein(a) and acute-phase proteins in acute myocardial infarction. Scand J. Clin Lab Invest 1992;52:95–101.
219. Cobbaert C, Sergeant P, Meyns B, Szecsi J, Kesteloot H. Time course of serum Lp(a) in men after coronary artery bypass grafting. Acta Cardiol 1992;47:529–42.
220. Rantapaa-Dahlqvist S, Wallberg-Jonsson S, Dahlen G. Lipoprotein(a), lipids, and lipoproteins in patients with rheumatoid arthritis. Ann Rheum Dis 1991;50:366–8.
221. Maeda S, Abe A, Seishima M, Makino K, Noma A, Kawade M. Transient changes of serum lipoprotein(a) as an acute phase protein. Atherosclerosis 1989;78:145–50.
222. Steinhauer HB, Brugger U, Atmanspacher R, Lubrich-Birkner I, Schollmeyer P. Effect of CAPD dialysate on the release of eicosanoids and cytokines from human peritoneal macrophages. Adv Perit Dial 1992;8:47–52.
223. Brewer HB. Effectiveness of diet and drugs in the treatment of patients with elevated Lp(a) levels. In: Scanu AM, ed. Lipoprotein(a). Academic Press: New York 1990;211–20.
224. Gurakar A, Hoeg JM, Kostner GM, Papadopoulos NM, Brewer HB Jr. Levels of lipoprotein Lp(a) decline with neomycin and niacin treatment. Atherosclerosis 1985;57:293–301.
225. Kostner GM. The affection of lipoprotein-a by lipid lowering drugs. In: Widholm K, Naito HK, eds. Recent aspects of diagnosis and treatment of lipoprotein disorders: impact on prevention of atherosclerotic diseases. New York: Alan R. Liss 1988:255–63.
226. Pelegri A, Romero R, Senti M, Nogues X, Pedro-Botet J, Rubies-Prat J. Effect of bezafibrate on lipoprotein(a) and triglyceride-rich lipoproteins, including intermediate-density lipoproteins, in patients with chronic renal failure receiving haemodialysis. Nephrol Dial Transplant 1992;7:623–6.
227. Carlson LA, Hamsten A, Asplund A. Pronounced lowering of serum levels of lipoprotein Lp(a) in hyperlipidemic subjects treated with nicotinic acid. J Intern Med 1989;226:271–6.
228. Armstrong VW, Schleef J, Thiery J, et al. Effect of HELP-LDL-apheresis on serum concentrations of human lipoprotein(a): kinetic analysis of the post-treatment return to baseline levels. Eur J Clin Invest 1989;19:235–40.
229. Pokrovsky SN, Adamova IY, Afanasieva OY, Benevolenskaya GF. Immunosorbent for selective removal of lipoprotein(a) from human plasma: in vitro study. Artif Organs 1991;15:136–46.
230. Gordon BG, Kelsey SF, Bilheimer DW, et al. Treatment of refractory familial hypercholesterolemia by low-density lipoprotein apheresis using an automated dextran sulfate cellulose adsorption system: The Liposorber Study Group. Am J Cardiol 1992;70:1010–16.

CHAPTER 15

Immunological Assays of Apolipoproteins AII, AIV, CI, CII, and CIII in Plasma: Methods and Applications

Christine Labeur and Maryvonne Rosseneu

INTRODUCTION

❖ In atherosclerosis research, quantification of both the lipid and protein content of the lipoproteins has gained importance in the past ten years.[1-3] All of the major apolipoproteins have been isolated, purified, and sequenced, and most of their physiological functions have been elucidated.

Given the metabolic role of apolipoproteins as lipid carriers, as ligands for receptors, and as enzyme activators, the measurement of the apolipoprotein concentrations in plasma should usefully complement that of the lipids with which they are associated. Because several of these apolipoproteins are present on the same lipoprotein particles, they are best measured in whole plasma by immunological techniques requiring the use of monoclonal antibodies or polyclonal antisera.[4-6] These methods have already been applied to the quantification of most of the apolipoproteins in plasma, including apolipoprotein (apo) AI, AII, AIV, B, CI, CII, CIII, D, E, and lipoprotein(a) [Lp(a)].[7-9] In general terms, the plasma concentration of the smaller apolipoproteins—apo CI, CII, CIII, AII, AIV, and E—are within the range of 5–50 mg/dL. This concentration range offers few difficulties for most of the current immunological techniques.

The various methodological aspects of immunological assays for apolipoproteins AII, AIV, CI, CII, and CIII and their current applications in research and clinical chemistry laboratories are described in this chapter.

GENERAL ASPECTS OF APOLIPOPROTEIN QUANTIFICATION

Immunological assays of apolipoproteins require some specific considerations.[4-6] Because of the lipid-binding properties of the apolipoproteins and their presence on the surface of lipoprotein particles, some of the antigenic sites are susceptible to partial or complete masking by lipids.[4, 5, 10] Therefore, accurate quantification of apolipoproteins requires pretreatment of the plasma or lipoprotein samples to completely expose the antigenic sites by delipidation or, on a more practical basis, by exposure to enzymes, denaturants, or detergents.[10] The presence of a given apolipoprotein on several lipoprotein fractions raises corollary problems. This effect is less pronounced for the smaller apolipoproteins, such as apo CII and CIII, which are probably equally exposed at the surface of all particles.[11, 12]

Antisera for Apolipoprotein Assays

Immunological assays of apolipoproteins seem quite sensitive to the nature of the antisera used in the measurement; these vary with the kind of antigen used in the immunization procedure, the species of host animal, and the dose of antigen injected. Significant differences have been reported between antisera. Monoclonal antibodies have been investigated as a possible approach to circumventing this problem. Apparently, however, the exposure of epitopes may vary with the lipoprotein particles in different individuals. Optimal quantification of apolipoproteins will therefore require a mixture of several monoclonal antibodies.

Standards for Apolipoprotein Assays

Selection and quantification of a suitable primary standard has also been a major problem in the development and standardization of apolipoprotein assays. Purified apolipoproteins have been chosen as primary standards for the apolipoproteins considered here. These purified apolipoprotein fractions were obtained from plasma lipoproteins separated by ultracentrifugal flotation.[7-10] For apo AII isolation, the high-density lipoproteins (HDL) were isolated from normolipemic plasma, while for purification of the C apolipoproteins, the very-low-density lipoprotein (VLDL) fraction was recovered from hypertriglyceridemic plasma. All lipoproteins were subsequently delipidated with ether/ethanol,[9] and the apo HDL and apo VLDL proteins were obtained. Apo AII was further isolated from other apo HDL proteins by ion-exchange chromatography on a DEAE Sepharose column.[10] Apo VLDL was fractionated by gel filtration on a Sephacryl S200 column,[11, 12] yielding three major fractions. Apo CIII was isolated from the apo C-containing fraction by ion-exchange chromatography on a DEAE Sepharose column,[12] while apo CII was recovered from the same fraction by chromatofocusing.[11]

A different purification strategy had to be applied for apo AIV, as this apolipoprotein is only partially associated with lipoproteins in plasma.[13] Apo AIV was purified from lymph chylomicrons obtained from a patient with lymph carcinoma,[13] or was recovered from plasma after adsorption on a triglyceride (TG) emulsion.[14] Further purification was performed by preparative gel electrophoresis in sodium dodecyl sulphate, and the protein was recovered by electroelution.[13] The mass of the primary standard was determined either by protein quantitation according to the method of Lowry[15] with bovine serum albumin as a standard, or preferably by quantitative amino acid analysis or by phenylalanine quantitation by high-performance liquid chromatography (HPLC).[16]

TECHNIQUES FOR THE QUANTIFICATION OF APO AII, AIV, CI, CII, AND CIII

Techniques for the quantification of apo AII, AIV, CI, CII and CIII include radioimmunoassay (RIA); enzyme immunoassays, particularly enzyme-linked immunosorbent assay (ELISA); electroimmunoassay (EIA); radial immunodiffusion (RID); and immunoprecipitin assays, including nephelometric and turbidimetric procedures. The characteristics of these assays are summarized in Table 15-1 and their advantages and disadvantages are given in Table 15-2.

Radioimmunoassay (RIA)

This technique, which can detect as little as a few nanograms of protein, has been applied to the assay of apo AII,[17-20] AIV,[21, 22] CI,[23, 24] CII,[23, 25, 26] and CIII.[23, 25] RIAs are of either the liquid or the solid-phase type and use either monoclonal antibodies or polyclonal antisera. However, RIAs require the iodination of pure apolipoprotein, a procedure that might decrease the reactivity of the standard compared to that of the lipoproteins, owing to self-association and aggregation of the isolated apolipoprotein. Another drawback of this technique is the need for high dilutions, which might lead to substantial experimental errors. The equipment needed for the RIA assay includes a gamma-counter for radioactivity measurement and an automated dilutor and pipetting station for precise handling of the high dilutions and small volumes. Calculation of the concentrations of the unknown samples is preferably carried out by a curve-fitting procedure after log-logit transformation of the measured values.

These assays can be easily automated, but are not convenient for clinical laboratories, owing to the safety requirements inherent to the handling and disposal of radioactive samples.

Enzyme Immunoassay

Several types of enzyme immunoassays, either direct or competitive, have been reported for the apolipoproteins. The sensitivity of these assays is similar to that of RIAs, but no radioactive tracers are required. These assays have enabled the quantification of apo AII,[27-29] AIV,[13] CI,[30-32] CII,[11, 30, 31] and CIII.[11, 30, 31] Most assays begin with the preparation of affinity-purified antibodies, which are subsequently labeled with either peroxidase or alkaline phosphatase, using either periodate or other classical coupling procedures.[9]

Table 15–1 ❖ Characteristics of Immunological Assays for Apolipoproteins*

	RIA	*ELISA*	*EIA*	*RID*	*Nephel-ometry*	*Turbid-imetry*
			APO AII			
Sensitivity (ng)	0.5	1	20	5	30	20
Intra-assay CV (%)	7	8	5	5	5	5
Inter-assay CV (%)	9	7	8	—	7	6
Normal values (mg/dL) mean ± SD	25 ± 4	35 ± 5	76 ± 2	29 ± 2	37 ± 9	38 ± 3
			APO AIV			
Sensitivity (ng)	5.0	0.2	20	—	—	—
Intra-assay CV (%)	5	4	5	—	—	—
Inter-assay CV (%)	7	8	9	—	—	—
Normal values (mg/dL) mean ± SD	17 ± 3	14 ± 5	14 ± 4	—	—	—
			APO CI			
Sensitivity (ng)	1	5	50	—	—	—
Intra-assay CV (%)	7	3	5	—	—	—
Inter-assay CV (%)	9	5	8	—	—	—
Normal values (mg/dL) mean ± SD	5 ± 3	6 ± 2	6 ± 1	—	—	4 ± 1
			APO CII			
Sensitivity (ng)	1	0.3	45	—	—	10
Intra-assay CV (%)	6	3	6.5	—	—	3.3
Inter-assay CV (%)	10	8	8.2	—	—	6.2
Normal values (mg/dL) mean ± SD	4 ± 1	33 ± 1	4 ± 2	—	—	4 ± 1
			APO CIII			
Sensitivity (ng)	0.3	10	20	—	15	—
Intra-assay CV (%	9	3.6	6	—	2.2	2.8
Inter-assay CV (%)	13	4.3	8	—	6.3	5.2
Normal values (mg/dL) mean ± SD	15 ± 6	16 ± 3	12 ± 4	—	12 ± 3	11 ± 3

*RIA = Radio Immunoassay
ELISA = Enzyme-linked Immunosorbent Assay
EIA = Electroimmunoassay
RID = Radial Immunodiffusion
CV = Coefficient of Variance
SD = Standard Deviation
— = No data available

Table 15-2 ❖ Advantages and Disadvantages of Immunological Assays for Apolipoproteins

*Technique**	*Advantages*	*Disadvantages*
RIA	High sensitivity High specificity	High dilution Uses isotope
ELISA	High sensitivity High specificity No isotope	High dilution
RID	Easy Low dilution No isotope	Incubation time 48 h Not easy to automate
EIA	Low dilution No isotope	No monoclonal antibody 10 samples per plate Not easy to automate
Nephelometry	Low dilution No isotope Easy to automate	Special equipment No monoclonal antibody Specific polyclonal antibodies required
Turbidimetry	Low dilution No isotope Easy to automate	No monoclonal antibody Specific polyclonal antibodies required

*RIA = Radioimmunoassay
ELISA =Enzyme-linked Immunosorbent Assay
RID = Radial Immunodiffusion
EIA - Electroimmunoassay

Recently, an enzyme-linked immunosorbent assay (ELISA) procedure involving monoclonal antibodies was reported for apo AII.[29] The nature of the monoclonal antibody seems crucial for the recognition of all apo AII present in plasma, so that a combination of several antibodies seems preferable to only one.

Two types of ELISA assays are currently used: the "sandwich" type and the competitive type. In the sandwich-type ELISAs, a polystyrene plate is coated with affinity-purified antibodies prepared against the apolipoprotein to be measured. After blocking all residual binding sites on the plate with either albumin or casein,[9] the plasma is incubated with the immobilized antibody for 2 h at 37° C. After extensive washing, the antibody-antigen complex on the plate is incubated with a second detecting antibody which is enzyme labeled. The amount of bound conjugated antibody is a direct measure of the antigen concentration. This assay is very sensitive, with a detection limit of 0.5–1 ng antigen and has the major advantage of easy and stable coating of the plates with affinity-purified antibodies.[9, 10]

In the competitive-type of assay, the plates are coated with the antigen and competition for the conjugated antibody takes place between the plasma antigen and the antigen immobilized on the plate. This assay is about 10–20 times less sensitive than the sandwich-type ELISA. Its major drawback for apolipoprotein measurement is the requirement for purified apolipoproteins for plate coating. As these apolipoproteins normally occur as lipid-bound components of lipoproteins, their antigenic properties when immobilized on the plate might differ from those in their native state.

ELISA assays, especially those of the sandwich type, are well suited to the quantification of apolipoproteins; they require a minimal amount of antiserum compared to nephelometry, radial immunodiffusion (RID), or electroimmunoassay (EIA), and have the great advantage of not being affected by sample turbidity.[10] They are also easy to automate, and their accuracy and reproducibility compares well to that of other assays, provided the dilutions are made accurately. These assays should become more widely used in the future, especially for quantification of the smaller apolipoproteins such as apo CII, and CIII, whose plasma concentrations are around 10 mg/dL or less. ELISA processes can be partially or fully automated, including sample dilution, dispensing the samples into the plates, and plate washing. The plates can be read after the colorimetric reaction, either manually or by an automated reader which is connected to a personal computer or hard-wired to its own computing device. Fully automated systems perform all operations automatically after programming for the application of the samples, the incubation times, wash cycles, volumes of reagents, etc.

For computing results, several types of calculations are available, depending upon the type of assay. In the sandwich-type direct assay, the curve can be fitted to a four-parameter equation[9] and the concentrations of the samples calculated from the fitted curve. For the competitive-type assay (as in RIAs), a log-logit transformation is used to linearize the curve before the concentration of test samples is calculated.

Electroimmunoasssay (EIA)

This technique has also been used to quantify most apolipoproteins, including apo AII,[33, 34] AIV,[35-37] CI,[38] CII,[34, 38-40] and CIII.[34, 40, 41] The coefficient of variation (CV) for this assay is significantly higher than that obtained using other techniques. This method is difficult to automate and is therefore more limited in its application.

Radial Immmunodiffusion (RID)

This method has been used to assay apo AII,[42-45] CI,[45] CII,[44, 45] and CIII.[44, 45] Although RID is simple in principle, its applicability depends upon the ability to obtain sharp precipitin rings of sufficient diameter. However, silver staining of the immunoprecipitates can enhance the sensitivity tenfold, bringing the threshold level of detection to 10 ng.[45] The migration of VLDL is slow in comparison to that of low-density lipoprotein (LDL), and 72–96 h diffusion is required to reach an equilibrium. Although the technique is easy to perform, its accuracy is questionable; moreover, it is not readily applied to large population studies that generate a large number of samples.

Immunoprecipitin Assays

This type of assay includes both nephelometric and turbidimetric procedures, which have been applied to the measurement of most apolipoproteins. Nephelometric assays have been described for apo AII[46, 47] and for apo CIII.[48] Both apolipoproteins were assayed by end-point nephelometry,[46–48] where the concentration is proportional to the turbidity after the formation of the antigen antibody complex is complete.

In the turbidimetric assay, the absorption of the antigen antibody complex is measured at 340 nm in a spectrophotometer. In the nephelometric assay, the intensity of the scattered light is measured at an angle of 30–90° to that of the incident beam. These techniques are fast, accurate, and reproducible.[10, 16] However, they have drawbacks in that they are extremely sensitive to the turbidity of the sample and require the use of enzyme or detergent when TG concentrations exceed 200 mg/dL. Measuring apo CII, CIII, and AIV in hyperlipemic samples by means of turbidimetry can be inaccurate. Turbidimetry has been used to quantify apo AII,[49, 50, 52] CII,[50–52] and CIII.[50, 52] This technique has been adapted to most of the available clinical analyzers and is especially suited to the measurement of large sample numbers with good reproducibility.

CLINICAL APPLICATIONS

Apolipoproteins and Cardiovascular Disease

Apolipoprotein assays have found three major fields of application: in the assessment of risk for cardiovascular disease in myocardial infarction survivors, in patients with evidence of coronary heart disease (CHD) as diagnosed by coronary angiography, and in offspring of patients with CHD. Apo AI and B were the major parameters measured in most case-control studies, showing that the concentrations of these apolipoproteins differ significantly from the values measured in a control group.[8, 53] These studies were further extended to the measurement of apo AII, CII, and CIII.[10, 16]

In a case-control study[54] comparing 70 male survivors of myocardial infarction to an equal number of healthy controls matched for age and body-mass index, apo B was shown to be significantly increased and apo AI significantly decreased in the cases compared to controls. A multivariate analysis indicated that the apo AI/B ratio, HDL-cholesterol (HDL-C), and the apo AI/AII ratio contributed independently to the discrimination of cases from controls, with an overall classification of 82%. Similar results were obtained by other investigators (see Table 15–3). In most of these studies, the apo AI/B ratio was the best discriminator between cases and controls, while apo AII was a less powerful parameter.

Apolipoproteins were further measured in a group of patients with coronary lesions by coronary angiography (see Table 15–3). In most studies, a statistically significant negative correlation between the apo AI/B ratio and the severity of CHD was reported.[55–59] Other investigators found that apo AI plasma concentrations relate better to the presence or absence of CHD than to the severity of the disease, while apo AII was a poor discriminator.[56, 59]

Table 15-3 ❖ Lipids and Apolipoprotein Measurements in Patients with Coronary Heart Disease

Author	*Variables*	*Principal Findings*
De Backer et al.[54]	apo AI, AII, B TC, TG, HDL-C	apo AI, AI/B, HDL-C lower apo B higher in cases
Leitersdorf et al.[53]	apo AI, AII, B, E HDL-C, TC, TG	apo AI lower apo B, E higher in cases
Riesen et al.[55]	apo AI, AII, B HDL-C, LDL-C, TC, TG	apo AI, AII lower apo B, LDL-C, TC higher in cases
Miller et al.[56]	apo AI, AII, E LDL_2-C, HDL_2-C, TC, TG	HDL_2-C lower in severe cases
Noma et al.[57]	apo AI, AII, B LDL-C, HDL-C, TC, TG	LDL/HDL-C higher in cases apo B/AI, B/AII higher in cases
Kottke et al.[59]	apo AI, AII, B TC, HDL-C, TG	apo AI, AII, B best discriminators
Sedlis et al.[58]	apo AI, AII, B, E HDL-C, TC, TG	apo AI/B best discriminator apo E higher in cases with total occlusion
Labeur et al.[60]	apo AI, AII, B HDL-C, TC, TG	apo AI/B best discriminator

In a study of 2020 patients aged 31–70 y who underwent coronary angiography, the patients were classified as having one-, two-, or three-vessel disease according to the extent of the lesions.[60] When compared to age-matched controls free of angiographically detected lesions, a significant decrease in the apo AI/B ratio was observed in the cases with two- and three-vessel disease. Further, there was a tendency toward a decrease of the apo AI/B ratio with an increasing severity of the lesions. Among the other apolipoproteins and lipids, apo B was the best discriminative parameter, while apo AI, AII, total cholesterol (TC), and TG did not differ significantly between the four groups of patients.[60]

More recent studies which included the measurements of apo AII, CIII, and AIV in either plasma or lipoprotein particles[61–64] showed increased apo CIII concentrations in the patients. Studies conducted in patients with peripheral vascular disease[8] are difficult to evaluate due to complications such as diabetes, hypertension, and obesity which are often present in such patients.

A further application domain for plasma apolipoprotein assays is monitoring and early detection of an hereditary risk for CHD. In one study, young adults whose fathers suffered a well-documented myocardial infarction below age 50 were compared to an age- and sex-matched controls who had no familial history of CHD.[65] Lower apo AI concentrations were the only significant difference in the lipid and apolipoprotein values between the cases and the controls. This difference in apo AI concentrations is not reflected in either the HDL-C or apo AII concentrations. In a comprehensive study based on the same design which was carried out in students of 12 European countries, results showed that apo B and LDL-C were the best discriminative parameters between students with and without a paternal history of CHD.[66]

An early screening of neonates, including assays for apolipoproteins and TC, can be carried out on a minimal quantity of blood[67] to detect increased apo B and decreased apo AI plasma concentrations. Other apolipoproteins, including AII, CII, CIII, and AIV, can also be measured in newborns.[68–70]

Apolipoproteins and Primary Dyslipidemias

Patients with Type I dyslipidemia are characterized by significantly decreased plasma concentrations of apo AI and apo B, but have normal apo AI/B ratios. When lipoprotein lipase deficiency is responsible for this disease, apo CII and CIII are increased about threefold compared to normals. Apo CII deficiency can also induce this severe form of dyslipidemia, in which case apo CII concentrations are undetectable even by the most sensitive ELISA or RIA assays.[9, 10]

In Type II patients, the plasma apo B concentrations are severely increased while the apo AI concentrations are slightly decreased. A strong positive correlation between plasma TG and apo CII and CIII concentrations exists in these hypercholesterolemic patients, so that a distinction between Type IIa and Type IIb patients can be made on the basis of the increased plasma concentrations of these apolipoproteins. In addition to a two- to threefold increase of plasma apo B concentrations, Type III patients are characterized by increased apo E concentrations, whereas apo CII and apo CIII increase about fourfold above normal values. In Type IV patients, the apo AI/B ratio is significantly reduced, whereas apo CII and apo CIII are increased by a factor of about two to three.

The most pronounced increase in plasma apolipoproteins is observed in Type V patients. In these patients, plasma TG are positively correlated with apo CIII and inversely correlated with apo CII.

In hypobetalipoproteinemic patients, plasma apo B concentrations are strongly reduced, whereas apo AI concentrations are normal. In addition, apo CII concentrations are reduced by about 50%, whereas apo CIII concentrations are decreased about fourfold compared to normals.

Plasma concentrations of the C apolipoproteins are thus increased most significantly in the hypertriglyceridemic state, whereas apo A-IV concentrations are not significantly different in dyslipoproteinemic patients compared to controls.[13, 14] Apo AIV concentrations seem sensitive to the state of alimentary lipemia of the patients,[71] and this apolipoprotein was reported to decrease food intake in rats.[72]

Apolipoproteins and Secondary Dyslipidemias

In addition to the primary dyslipidemias, which are mostly due to genetic factors, a number of pathologies and dietary factors are responsible for inducing secondary dyslipidemias. These have essentially the same features and lead to the same phenotypes as the primary dyslipidemias. These are, however, reversible and should normalize upon treatment of the underlying disease. Secondary dyslipidemias can occur in response to renal disease, alcohol consumption, liver disease, thyroid dysfunction, and diabetes mellitus.[7–10]

Chronic renal failure can be treated either by hemodialysis or by continuous ambulatory peritoneal dialysis (CAPD). Both types of treatment induce a decrease of the HDL fraction, which can be accompanied by an accumulation of VLDL and an increase of apo CII and CIII.[73] The secondary dyslipidemia is usually more severe in CAPD patients, in whom a Type IIb or even a Type V dyslipidemia can occur.

In patients with nephrotic syndrome, pronounced secondary dyslipidemia develops readily due to decreased LDL-receptor synthesis and to apo B overproduction. HDL and especially apo AI are lost through the kidneys, accounting for the low apo AI and AII concentrations in these patients.[74]

Alcohol consumption influences the lipoprotein patterns according to its frequency and volume of consumption. Acute consumption increases HDL concentrations, especially that of apo AII.

Secondary dyslipidemias can occur consecutive to diabetes mellitus. In Type I, insulin-dependent diabetes, Type IV hyperlipidemia is frequent, characterized by decreased apo AI and AII concentrations and increased apo B, CII, CIII concentrations.[7-10] In Type II, non-insulin dependent diabetes, a Type V hyperlipoproteinemia is frequently observed. In both cases, this secondary dyslipidemia results in CHD, which represents a major complication in diabetics.

Among the liver diseases, acute hepatitis decreases the apo AI and AII concentrations due to impaired HDL synthesis, while a decrease of hepatic lipase activity increases the apo CII and CIII concentrations. In chronic hepatitis, hepatic lipase activity is decreased while apo AI and AII synthesis is normal. In liver cirrhosis, protein synthesis is impaired and the concentrations of the liver-synthesized apolipoproteins—AI, AII, B, and the C apolipoproteins—are significantly decreased.

In thyroid disorders, hyperthyroidism is accompanied by severely decreased apo B concentrations due to increased LDL catabolism. In non-obese hypothyroid patients, apo B is increased and a IIa phenotype is most common due to reduced LDL catabolism.[75] In obese hypothyroid patients, apo CII and CIII can accumulate due to impaired IDL-to-LDL conversion, inducing a Type V dyslipidemia.

CONCLUSION

The apolipoprotein profiles can now be routinely assessed by a variety of immunological procedures and can be usefully combined with lipid quantitation for the characterization of primary and secondary hyperlipidemias. Since the apolipoproteins constitute the physiological carriers of the plasma lipids, their plasma concentration and lipoprotein distribution are extremely sensitive to hormonal, functional, and dietary regulation. As a consequence, different patterns arise from the disregulation. The combination of apolipoprotein and lipid assays provides a

more sensitive and accurate way of monitoring primary and secondary hyperlipidemias, as well as of evaluating a patient's risk of developing cardiovascular diseases. ❖

REFERENCES

1. Rifkind BM. The Lipid Research Clinics Primary Prevention Trial results. I. Reduction in incidence of coronary heart disease. J Amer Med Assoc 1984;251:351–64.
2. Rifkind BM. The Lipid Research Clinics Coronary Primary Prevention Trial results. II. The relationship of reduction in incidence of coronary heart disease to cholesterol lowering. J Amer Med Assoc 1984;251:365–74.
3. De Backer G, Rosseneu M, De Slypere JP. Discriminative value of lipids and apoproteins in coronary heart disease. Atherosclerosis 1982;42:197–203.
4. Steinberg KK, Cooper GR, Rosseneu M. Evaluation and standardization of apolipoprotein A-I immunoassays. Clin Chem 1983;29:415–26.
5. Rosseneu M, Vercaemst R, Steinberg KK, Cooper GR. Some considerations of methodology and standardization of apolipoprotein B immunoassays. Clin Chem 1983;29:427–33.
6. Naito HK. The Clinical significance of apolipoprotein measurements. J. Clin. Immunoassay 1986;9:11–20.
7. Rosseneu M, Bury J. Apolipoproteins assays for the diagnosis of hyperlipidemias. In: Naito H, ed. Recent aspects on diagnosis and treatment of lipid disorders: Impact on prevention of atherosclerotic diseases. New York: A. R. Liss, 1988:143–5.
8. Kostner GM. Apolipoproteins and lipoproteins of human plasma: significance in health and disease. Adv Lipid Res 1983;20:1–43.
9. Bury J, Rosseneu M. Apolipoprotein quantitation by ELISA: technical aspects and clinical applications. Rev. Immunoassay Techn 1988;1:1–25.
10. Labeur C, Shepherd J, Rosseneu M. Immunological assays of lipoproteins in plasma: Methods and instrumentation. Clin Chem 1990;36:591–7.
11. Bury J, Michiels G, Rosseneu M. Human apolipoprotein C-II quantitation by sandwich enzyme linked immunosorbent assay. J Clin Chem Clin Biochem 1986;24: 457–61.
12. Bury J, Rosseneu M. Enzyme immunosorbent assay for human apolipoprotein C-III. J Clin Chem Clin Biochem 1985;23:63–8.
13. Rosseneu M, Michiels G, De Keersgieter W, et al. Human apolipoprotein A-IV quantitation by sandwich enzyme linked immunosorbent assay. Clin Chem 1988; 34:739–43.
14. Weinberg RB, Scanu AM. Isolation and characterization of human apolipoprotein A-IV from lipoprotein-depleted serum. J Lipid Res 1983;24:54–9.
15. Lowry OH, Rosebrough NJ, Farr AL, Randall RJ. Protein measurement with the Folin reagent. J Biol Chem 1951;193:265–75.
16. Rosseneu MY, Labeur C. Apolipoprotein structure, function and measurement. Current Opinions on Lipidology 1990;1:508–13.
17. Mordasini RC, Riesen WF. Electroimmunoassay and radioimmunoassay for the quantitation of high-density apolipoproteins AI and AII. J Clin Chem Clin Biochem 1980;18:917–20.
18. Mao SJT, Gotto AM Jr, Jackson RL. Immunochemistry of human plasma high density lipoproteins: radioimmunoassay of apolipoprotein A-II. Biochemistry 1975;14:4127–31.

19. Schonfeld G, Chen J, McDonell WF, Jeng I. Apolipoprotein A-II content of plasma high density lipoproteins measured by radioimmunoassay. J Lipid Res 1977;18:645–54.
20. Goldberg RB, Karlin JB, Juhn DJ, Scanu AM, Edelstein C, Rubenstein AH. Characterization and measurement of human apolipoprotein A-II by radioimmunoassay. J Lipid Res 1980;21:902–12.
21. Ghiselli G, Krishnan S, Beigel Y, Gotto AM Jr. Plasma metabolism of apolipoprotein A-IV in humans. J Lip Res 1986;27:813–27.
22. Bisgaier CL, Sachdev OP, Lee ES, Williams KJ, Blum CB, Glickman RM. Effect of lecithin:cholesterol acyltransferase on distribution of apolipoprotein A-IV among lipoproteins of human plasma. J Lip Res 1987;28:693–703.
23. Bren ND, Rastogi A, Kottke BA. Quantification of human apolipoproteins C-I, C-II and C-III by radioimmunoassay. Mayo Clinic Proc 1993;68:657–64.
24. Polz E, Kotite L, Havel RJ, Kane JP, Sata T. Human apolipoprotein C-I: concentration in blood serum and lipoproteins. Biochem Med 1980;24:229–37.
25. Schonfeld G, George PK, Miller J, Reilly P, Witztum J. Apolipoprotein C-II and C-III levels in hyperlipoproteinemia. Metabolism 1979;28:1001–10.
26. Barr SI, Kottke BA, Chang JY, Mao SJT. Immunochemistry of human plasma apolipoprotein C-II as studied by radioimmunoassay. Biochem Biophys Acta 1981;663:491–505.
27. Dubois DY, Cantraine F, Malmendier CL. Comparison of different sandwich immunoassays for the quantitation of human apolipoproteins A-I and A-II. J Immunol Meth 1987;96:115–20.
28. Dufaux B, Ilsemann K, Assmann G. Competitive enzyme immunoassay for apolipoprotein A-II. J Clin Chem Clin Biochem 1983;21:39–43.
29. Stein EA, Dipersio L, Pesce AJ, et al. Enzyme-linked immunoabsorbent assay of apolipoprotein A-II in plasma with use of a monoclonal antibody. Clin Chem 1986;32:967–71.
30. Holmquist L. Quantitation of human serum very low density apolipoprotein B, C-I, C-II, C-III and E by enzyme immunoassay. J Immunol Meth 1980;34:243–51.
31. Carlson LA, Holmquist L. Concentrations of apolipoproteins B, C-I, C-II, C-III and E in sera from normal men and their relation to serum lipoprotein levels. Clin Chim Acta 1982;124:163–78.
32. Riesen WF, Sturzenegger E. Enzyme-linked immunosorbent assay for apolipoprotein C-I. J Clin Chem Clin Biochem 1986;24:723–7.
33. Curry MD, Alaupovic P, Suenram CA. Determination of apolipoprotein A and its constitutive A-I and A-II polypeptides by separate electroimmunoassays. Clin Chem 1976;22:315–22.
34. Alaupovic P, Curry MD, McConathy WJ. Quantitative determination of human plasma apolipoproteins by electroimmunoassay. Intern Confer Atheroscl 1978;109–15.
35. Utermann G, Beisiegel U. Apolipoprotein A-IV: a protein occurring in human mesenteric lymph chylomicrons and free in plasma. Eur J Biochem 1979;99:333–43.
36. Green PHR, Glickman RM, Riley JW, Quinet E. Human apolipoprotein A-IV. J Clin Invest 1980;65:911–9.
37. Bisgaier CL, Sachdev OP, Megna L, Glickman RM. Distribution of apolipoprotein A-IV in human plasma. J Lipid Res 1985;26:11–25.
38. Curry MD, McConathy WJ, Fesmire JD, Alaupovic P. Quantitative determination of apolipoproteins C-I and C-II in human plasma by separate electroimmunoassays. Clin Chem 1981;27:543–8.
39. Jauhiaianen M, Laitinen M, Penttilö I, Puhakainen E, Hietanen E. Determination of human apolipoprotein C-II by electroimmunoassay: studies on standardization

and determination before and after physical training. Int J Biochem 1983;15: 501–6.

40. Mörz W, Schenk G, Gross W. Apolipoproteins C-II and C-III in serum quantified by zone immunoelectrophoresis. Clin Chem 1987;33:664–9.
41. Curry MD, McConathy WJ, Fesmire JD, Alaupovic P. Quantitative determination of human apolipoprotein C-III by electroimmunoassay. Biochem Biophys Acta 1980;617:503–13.
42. Albers JJ, Cheung MC, Wahl PW. Effect of storage on the measurement of apolipoproteins A-I and A-II by radial immunodiffusion. J Lipid Res 1980;21:874–8.
43. Cheung MC, Albers JJ. The measurement of apolipoprotein A-I and A-II levels in men and women by immunoassay. J Clin Invest 1977;60:43–50.
44. Goto Y, Akanuma Y, Harano Y. Determination by the SRID method of normal values of serum apolipoproteins (A-I, A-II, B, CII, CIII and E) in normolipidemic healthy Japanese subjects. J Clin Biochem Nutr 1986;1:73–88.
45. Ishida BY, Paigen B. Silver-enhanced radial immunodiffusion assay of plasma apolipoproteins. J Lipid Res 1992;33:1073–8.
46. Kuusi T, Palosuo T, Ehnholm C. Immunonephelometric measurements of human plasma apolipoproteins A-I and A-II: an assay performed by consecutive addition of the two antisera to a single specimen. Scand J Clin Lab Invest 1985;45:245–53.
47. Rosseneu M, Vinaimont N, Musliner TA, Bernier D, Herbert PN, Belpaire F. Immunonephelometry of apolipoprotein A-II in plasma. Clin Chem 1984;30:234–7.
48. Bury J, De Keersgieter W, Rosseneu M, Belpaire F, Christophe J. Immunonephelometric quantitation of the apolipoprotein C-III in human plasma. Clin Chim Acta 1985;145:249–58.
49. Rifai N, King ME. Immunoturbidimetric assays of apolipoproteins A, AI, AII, and B in serum. Clin Chem 1986;32:957–61.
50. Noma A, Haras Y, Goto Y. Quantification of serum apolipoprotein A-I, A-II, B, C-II, C-III, and E by turbidimetric immunoassay: reference values and age- and sex-related differences. Clin Chim Acta 1991;199:147–57.
51. Rifai N, Silverman LM. Immunoturbidimetric techniques for quantifying apolipoproteins CII and CIII. Clin Chem 1986;32:1969–72.
52. Ikeda T, Shibuya Y, Senba U, et al. Automated immunoturbidimetric analysis of six plasma apolipoproteins: correlation with radial immunodiffusion assays. J Clin Lab Anal 1991;5:90–5.
53. Leitersdorf E, Gottehrer N, Fainaru M, et al. Analysis of risk factors in 532 survivors of first myocardial infarction hospitalized in Jerusalem. Atherosclerosis 1986;59:75–93.
54. De Backer G, Rosseneu M, Deslypere JP. Discriminative value of lipids and apoproteins in coronary heart disease. Atherosclerosis 1982;42:197–203.
55. Riesen WF, Mordasini R, Salzmann C, Theler A, Gurtner HP. Apolipoproteins and lipids as discriminators of severity of coronary heart disease. Atherosclerosis 1980;37:157–62.
56. Miller NE, Hammett F, Saltissi S, et al. Relation of angiographically defined coronary artery disease to plasma lipoprotein subfractions and apolipoproteins. British Med J 1981;282:1741–4.
57. Noma A, Yokosuka T, Kitmra K. Plasma lipids and apolipoproteins as discriminators for presence and severity of angiographically defined coronary artery disease. Atherosclerosis 1983;49:1–7.
58. Sedlis SP, Schechtman KB, Ludbrook PA, Sobel BE, Schonfeld G. Plasma apoproteins and the severity of coronary artery disease. Circulation 1986;73:978–86.
59. Kottke BA, Zinsmeister AR, Holmes DR, Kneller RW, Hallaway BJ, Mao ST. Apolipoproteins and coronary artery disease. Mayo Clin Proc 1986;61:313–20.

60. Labeur C, Vincke J, Muyldermans L, et al. Plasma lipoprotein (a) values and severity of coronary artery disease in a large population of patients undergoing coronary angiography. Clin Chem 1992;38:2261–6.
61. Genest JJ Jr, Bard JM, Fruchart JC, Ordovas JM, Wilson PF, Schaefer EJ. Plasma apolipoprotein A-I, A-II, B, E and CIII containing particles in men with premature coronary artery disease. Atherosclerosis 1991;90:149–57.
62. Kottke BA, Moll PP, Michels VV, Weidman WH. Levels of lipids, lipoproteins, and apolipoproteins in a defined population. Mayo Clin Proc 1991;66:1198–208.
63. Chivot L, Mainard F, Bigot E, et al. Logistic discriminant analysis of lipids and apolipoproteins in a population of coronary bypass patients and the significance of apolipoproteins C-III and E. Atherosclerosis 1990;82:205–11.
64. Ordovas JM, Civeira F, Genest J Jr, et al. Restriction fragment length polymorphisms of the apolipoprotein A-I, C-III, A-IV gene locus: relationships with lipids, apolipoproteins, and premature coronary artery disease. Atherosclerosis 1991;87:75–86.
65. De Backer G, Hulstaert F, De Munck K, Rosseneu M, Van Parijs L, Dramaix M. Serum lipids and apoproteins in students whose parents suffered prematurely from a myocardial infarction. Am Heart J 1986;112:478–84.
66. Rosseneu, M, Fruchart JC, Bard JM, Nicaud V, Vinaimont N, Cambien F. On behalf of the EARS group: plasma apolipoprotein concentrations in young adults with a parental history of premature coronary heart disease and in controls: The EARS Study. Circulation, in press.
67. Van Biervliet JP, Vinaimont N, Caster H, Rosseneu M. A screening procedure for dyslipoproteinemia in newborns: apoprotein quantitation on dried-blood spots. Clin Chim Acta 1982;120:191–200.
68. Averna MR, Barbagallo CM, Di Paola G. et al. Lipids, lipoproteins and apolipoproteins AI, AII, B, CII, CIII and E in newborns. Biol Neonate 1991;60:187–92.
69. Boediman D, Murakami R, Nakamura H. Relationship between plasma triglycerides and apolipoprotein CII in infants during the first year of life. J Pediatr Gastroenterol Nutr 1993;17:82–5.
70. Van Biervliet JP, Rosseneu M, Bury J, Caster H, Stul MS, Lamote R. Apolipoprotein and lipid composition of plasma lipoproteins in neonates during the first month of life. Pediatric Res 1986;20:324–8.
71. Annuzzi G, Holmquist L, Carlson LA. Concentrations of apolipoproteins B, C-I, C-II, C-III, E and lipids in serum and serum lipoproteins of normal subjects during alimentary lipaemia. Scand J Clin Lab Invest 1989;49:73–81.
72. Fujimoto K, Fukagawa K, Sakata T, Tso P. Suppression of food intake by apolipoprotein A-IV is mediated through the central nervous system in rats. J Clin Invest 1993;91:1830–3.
73. Alsayed N, Rebourcet R. Abnormal concentrations of CII, CIII, and E apolipoproteins among apolipoprotein B-containing, B-free, and A-I-containing lipoprotein particles in hemodialysis patients. Clin Chem 1991;37:387–93.
74. Muls E, Rosseneu M, Daneels R, Schurgers M, Boelaert J. Lipoprotein distribution and composition in the human nephrotic syndrome. Atherosclerosis 1985;54:225–37.
75. Muls E, Rosseneu M, Lamberigts G, De Moor P. Changes in the distribution and composition of high-density lipoproteins in primary hypothyroidism. Metabolism 1985;34:345–53.

CHAPTER 16

Apolipoprotein E: Laboratory Determination and Clinical Significance

Lily H. Wu, James T. Wu, and Paul N. Hopkins

INTRODUCTION

❖ Apolipoprotein E (apo E) is a 299 amino acid, arginine-rich glycoprotein with an M_r of 34,200 daltons. Apo E is synthesized by many different cell types throughout the body. The major site of synthesis and secretion is liver parenchymal cells, which secrete apo E in association with very-low-density lipoprotein (VLDL) particles. Besides being associated with VLDL, apo E is a protein component of several other classes of plasma lipoproteins, including chylomicrons, chylomicron remnants, intermediate-density lipoproteins (IDL), and some cholesterol-rich subclasses of high-density lipoproteins (HDL), including apo E-HDL and HDL_1 (in humans and some rodents), and HDL_c (in cholesterol-fed animals).[1] Only a minute amount of apo E is found in low-density lipoproteins (LDL). A significant amount of apo E is also produced in astrocytes, peripheral nerves, and mature macrophages. Particles containing only apo E, cholesterol, and cholesterol ester may also exist at least transiently in interstitial fluid. Apo E serves as a ligand for at least two lipoprotein receptors: the LDL (or apo B/E) receptor and the putative hepatic apo E receptor (or the LDL receptor related protein, LRP). As such, apo E plays an important role in lipoprotein metabolism.[2, 3]

The human apo E gene has been mapped to a single genetic locus on chromosome 19.[4] The structural gene for apo E is polymorphic, with three common alleles coding for three isoforms of the apo E protein: E2, E3, and E4. The three isoforms differ in their amino acid sequence at positions 112 and 158. Apo E3, the most prevalent allele, contains cysteine at position 112 and arginine at posi-

Figure 16–1 ❖ Specific Nucleotide Sequences and Their Coded Amino Acid Residues for the Three Common Apo E Alleles

Numbers at the bottom of the figure indicate the position of the amino acids in the protein sequence.

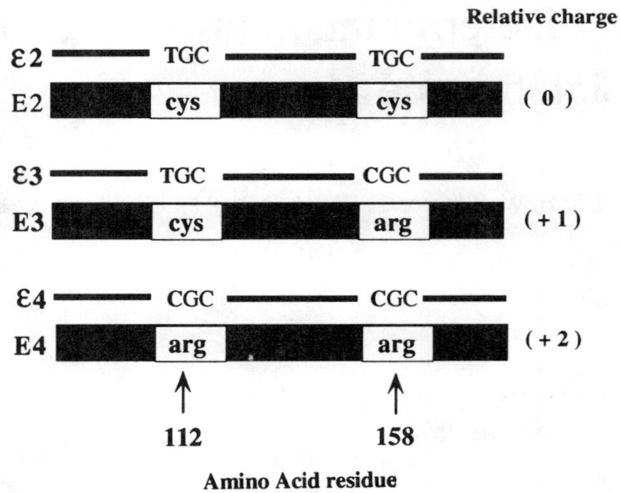

tion 158. In apo E2, a cysteine residue replaces the arginine at position 158 (Arg_{158}→Cys or R158C), while apo E4 contains arginine at both positions (Cys_{112}→Arg or C112R).[4] Individuals inherit one apo E allele from each parent in simple Mendelian fashion with resulting genotypes (or phenotypes) for apo E reported as E2/2, E3/2, E4/2, E3/3, E4/3, or E4/4 (see Figure 16–1).

Some investigators refer to the alleles which code for the E2, E3, and E4 isoforms as ε2, ε3, and ε4, respectively. However, only relatively recently have direct DNA methods been developed to determine apo E allele status. Such methods are referred to as "genotyping." Most studies have historically used isoelectric focusing of VLDL proteins to determine carrier state. Traditionally, this method has been called "phenotyping." Since only an error will cause a discrepancy between genotyping and phenotyping, we use the terms E2, E3, and E4 to signify E4 to signify either the common genotypes (alleles) or phenotypes (protein isoforms). Potential infrequent discrepancies between direct DNA methods and protein isoelectric focusing are discussed below.

Markedly different binding to the LDL receptor has been documented for the three common apo E isoforms. Apo E4 has a slightly higher affinity than apo E3, whereas apo E2 is clearly defective in its binding, having less than 2% of apo E3's affinity for the LDL receptor. Amino acid residues in the vicinity of 140–160 of the apo E molecule contain a high-affinity LDL receptor binding site. Surrounding residues may be responsible for the proper alignment of the receptor binding site and can also affect binding affinity. Loss of a positive charge as a result of amino acid substitutions in this region is thought to be the basis for the decrease in apo E2 binding. Homozygosity for apo E2 (an E2/2 phenotype or genotype) is the basis for Type III hyperlipidemia in most patients (see later discussion). In addition to the relatively common apo E2, there are a number of other rare isoforms with reduced binding of varying degrees to the LDL receptor, as shown in Table 16-1.[5–8] The potential differences in the tendency of apo E apolipoproteins to form homo- or mixed-disulfide protein dimers may also affect LDL receptor binding.[9, 10]

Consistent with the hyperlipidemia associated with defective or deficient apo E, a major role of apo E appears to be the clearance of apo B-containing remnants. Accelerated catabolism of such remnants occurs when rabbits are injected or infused with purified apo E.[11, 12] When either rat[13, 14] or normal human[15] apo E3 is overexpressed in transgenic mice, lower concentrations of VLDL result with more efficient clearance of apo B-containing lipoproteins. Conversely, apo E null mice[16, 17] and transgenic mice overexpressing dysfunctional apo E[18, 19] develop severe hyperlipidemia resembling Type III hyperlipidemia in humans.

The common polymorphisms of apo E influence plasma lipid and lipoprotein concentrations in human populations. Apo E4 imparts a tendency for modestly higher concentrations of total cholesterol (TC) and LDL cholesterol (LDL-C). Those with the E2 isoform tend to have lower LDL-C concentrations. Both apo E2 and E4 are associated with higher plasma triglyceride (TG) concentrations (see

Table 16–1 ❖ Apolipoprotein E Alleles and Associations with Type III Hyperlipidemia

Position on IEF	Amino Acid Change	Receptor Binding (% normal)	Comments
		COMMON	
E2	$Arg_{158} \rightarrow Cys$	< 2	Common cause of Type III (recessive). Also associated with reduced total and LDL cholesterol. Frequency 10%*
E3	None	100	Wild type allele. Frequency 75%*
E4	$Cys_{112} \rightarrow Arg$	> 100	Associated with increased total and LDL cholesterol. Frequency 15%*
		RARE	
E1	$Lys_{146} \rightarrow Glu$	Unknown	Dominant Type III
E2	$Arg_{136} \rightarrow Ser$	40	Unknown mode of inheritance.
E2	$Arg_{136} \rightarrow Cys$	Unknown	Subtle position difference on IEF compared to common E2. Recessive for Type III.
E3	$Arg_{142} \rightarrow Cys$	< 4	Dominant Type III. Expression at an early age.
E2	$Arg_{145} \rightarrow Cys$	45	Unknown mode of inheritance.
E2	$Lys_{146} \rightarrow Gln$	40	Dominant Type III.
E3	Insertion	25	Leiden variant. 7 amino acids inserted (121–127). Dominant Type III.

*Gene frequencies vary in different populations.

later section entitled "Clinical Significance"). There is no significant difference in the relative frequencies of the apo E alleles between men and women; however, the influence of apo E alleles on lipid variability appears to be greater in women.[20]

Apo E also appears to play an important role in HDL metabolism and reverse cholesterol transport, the process by which excess cholesterol is removed from peripheral tissues and transported to the liver. Apo E facilitates the expansion of the HDL core, enhancing its cholesterol-carrying capacity.[21, 22] Furthermore, apo E can mediate hepatic removal and catabolism of cholesteryl-ester-enriched HDL.[23] Apo E secretion by macrophages is stimulated three- to eightfold by cholesterol loading. The apo E secreted becomes associated with phospholipid in a disc having a density of < 1.21 Kg/L. While the fate of these nascent particles is not clear, they may facilitate cholesterol removal from cells, possibly in conjunction with HDL.[24] Indeed, intravenous administration of purified apo E into Watanabe heritable hyperlipidemic rabbits 3 times/w for 8.5 m greatly decreased the cholesterol deposition into the aortas of these atherosclerosis-prone animals.[25]

Physiological functions of apo E are summarized as follows:

1. As a ligand for the LDL (apo B/E) and apo E (LRP) receptors, apo E is a key mediator in the clearance of TG-rich lipoproteins, including VLDL, chylomicrons, and chylomicron remnants, and may be required for the conversion of VLDL to LDL. It also appears to be responsible for the removal of apo-E-containing HDL from the blood circulation.[1]
2. By increasing the affinity of cholesterol ester transfer protein (CETP) for lipoproteins, apo E further helps regulate HDL-mediated reverse cholesterol transport.[26]
3. Apo E secreted by macrophages inhibits the unregulated uptake of oxidized-VLDL via the scavenger receptor and prevents transformation of macrophages into foam cells.[14] Furthermore, apo E secretion by macrophages may provide an additional pathway for reverse cholesterol transport.
4. Apo E may be involved in nerve regeneration, modulation of immune response, and the growth and/or differentiation of neuron and smooth muscle cells.[27]

APO E CONCENTRATION

Clinical Significance

Since Type III hyperlipoproteinemia is usually associated with a large increase in apo E concentration, measurement of apo E in the plasma may be useful for detection of this disorder.[28, 29] Surprisingly, in plasma from normal, fasting adults, 58% of the apo E was found in the HDL fraction (primarily the less-dense HDL) with most of the remainder in TG-rich lipoproteins. An even higher percentage of apo E was associated with HDL in cord blood.[30] Normal concentrations range from 1.0 to 9.0 mg/dL. In the hypertriglyceridemic state, apo E is redistributed to TG-rich lipoproteins. The concentration of apo E in one Type III patient was 66.4

mg/dL.[31] Theoretically, the measurement of apo E may be useful for monitoring the therapeutic response of Type III patients taking lipid-lowering medications. However, in clinical practice, the plasma TC and TG concentrations generally suffice for this purpose.

Total plasma apo E concentration does not appear to predict risk of coronary heart disease (CHD). However, increased concentrations of apo B- and E-containing lipoproteins were seen in coronary patients, while apo E in the HDL fraction was probably lower.[32] These findings suggest a protective role for apo E associated with HDL, while high concentrations of apo E in TG-rich particles indicate accumulation of remnant particles and therefore increased risk for CHD.

Since apo E synthesis in the central nervous system increases with remyelination, measurement of apo E in cerebrospinal fluid may be useful in the assessment of certain neurological conditions. In one study, an apo E index (comparing apo E and albumin in cerebrospinal fluid and serum) was much higher in multiple sclerosis patients in remission (when remyelination was active) compared to normal controls or patients in exacerbation.[33]

Methods of Determination

Reference values of serum or plasma apo E in fasting adult subjects range from 1.0 to 9.0 mg/dL. Lack of standardization of the various methods has contributed to the wide range of normal values found in the literature. Various immunological methods have been employed successfully for the measurement of plasma apo E concentration, including radial immunodiffusion (RID), electroimmunodiffusion (EID), immunoturbidimetric assay, immunonephelometric assay, radioimmunoassay (RIA), and enzyme-linked immunosorbent assay (ELISA).

It should be noted that apolipoprotein concentrations in serum samples are expected to be 3–5% higher than those obtained from EDTA plasma samples. (Heparinized samples give similar results as serum samples.) Ultracentrifugation may cause artifactual loss of apolipoproteins as free proteins or result in the redistribution of apo E among lipoprotein particles. Discussed below are some available techniques and their major advantages and drawbacks.

Radial Immunodiffusion (RID)

Apo E concentration is estimated quantitatively by the diameter of the precipitation ring formed as the apo E present in a measured volume of sample placed in a central well diffuses outward and reacts with polyclonal antibody impregnated in an agarose gel.[34–38] This method is technically simple but can be time consuming. It may take as long as 72–96 h for apo E in VLDL to reach equilibrium and form a ring. In some cases, apoE never reaches equilibrium, resulting in spuriously low concentration estimates. The reading of the diameters can be somewhat subjective, and RID is less precise than other methods. The association of apo E with different sizes of lipoprotein particles may also alter the diffusion rate and the ring size. Finally, this method consumes rather large amounts of antibody.

Electroimmunodiffusion

Electroimmunodiffusion[39-41] is very similar to RID except that the height of a "rocket" formed as antigen-antibody complex after electrophoresis is used as a measure of concentration. As such, it shares most of the problems commonly encountered in RID. However, because of the use of electrophoresis to move the apo E antigen from the sample well into the gel in one direction, it is more sensitive and takes much less time to complete. Detergent is added to enhance the mobility of the VLDL fraction for apo E quantification.

Radioimmunoassay (RIA)

RIA[31, 42, 43] is very sensitive, capable of detecting as little as a few nanograms of protein. Either monoclonal or polyclonal antibody may be used. RIA uses either primary or secondary standards. Small amounts of antibody are used compared to other techniques. However, the large dilutions (> 10,000) employed can result in considerable imprecision and inaccuracy. No automated methods are available. RIA is not affected by turbidity—an important advantage. Competitive RIA requires the iodination of pure apolipoprotein, a procedure that might decrease the reactivity of apo E due to self-association and aggregation of the purified apolipoprotein.

Immunoprecipitation Assay

Immunoprecipitation assays[44-46] include immunonephelometry, which measures the light scattering of the particle, and immunoturbidimetry, which measures turbidity. Both assays are fast, precise, and can be readily automated. High turbidity in samples with increased TG (> 600–1000 mg/dL [6.78–11.30 mmol/L]) can lead to erroneous results. Immunoturbidity appears to be slightly less affected by the presence of lipemia than immunonephelometry. Secondary standards are used. Relatively large amounts of polyclonal or multiple monoclonal antibodies must be used.

Enzyme Immunoassay

Enzyme immunoassay[47-50] shares similar advantages and disadvantages with RIA. The sensitivity of this procedure is similar to that of RIA. There are several test formats for enzyme immunoassay, the most popular being the enzyme-linked immunosorbent assay (ELISA). With the microplate ELISA method, anti-apo E antibody is coated on polystyrene microplate wells for capturing apo E, and enzyme-conjugated anti-apo E antibody is then used to react with captured apo E for quantification.[51, 52] The microplate procedure can be semi-automated. Since the method is rapid, simple, and suitable for large population screening, the microplate ELISA format is the recommended method for measuring apo E concentration.

Apo E Standard

Purified apo E is recommended as the primary standard. However, pure apo E tends to aggregate over time, and the aggregate not only may have decreased immunoreactivity, but also migrates more slowly in a gel. Thus, using pure apo E as a standard may result in an overestimation of sample apo E concentration. To guard against this, plasma containing apo E calibrated against purified apo E should be used as a secondary standard for routine assays. The protein content of the highly purified apolipoproteins can be determined by the Lowry method or by amino acid analysis. Although in theory interaction of the proteins with lipids can mask antigenic determinants and affect immunoreactivity, in practice apo E epitopes in plasma lipoproteins appear to be fully exposed. Moreover, if a non-ionic detergent such as Tween 20 is included in the washing buffer with the ELISA method, delipidation of lipoprotein particles may occur without sample pretreatment and may help to expose all possible apo E antigenic sites.

PHENOTYPING AND GENOTYPING

Phenotyping

Because of their differences in charge, apo E isoforms can be conveniently separated by isoelectrofocusing (IEF). The IEF pattern of apo E phenotype is shown in Figure 16-2. Earlier procedures for apo E phenotyping required an isolation of VLDL from the plasma before the separation by IEF. VLDL was usually isolated either by ultracentrifugation or by a specific precipitation method followed by the delipidation step. After IEF, the apo E isoforms were visualized by a protein stain. Recently, the procedure of phenotyping by IEF has been simplified due to the availability of the anti-apo E antibodies. Plasma specimens can be subjected to IEF directly and transferred to a nitrocellulose paper. The apo E pattern is then revealed by immunoblotting. Below are several electrophoretic procedures that have been used by various investigators.

Polyacrylamide Gel–IEF (PAG-IEF)

In the PAG-IEF method,[53-56] IEF is carried out in polyacrylamide gels (5% or 7.5% gel) containing 2% ampholytes (pH 4–7). This method requires isolation of VLDL from the plasma specimen by ultracentrifugation and delipidation of the isolated VLDL. This procedure typically requires 4 mL of plasma (8 mL of whole blood) and takes several days to complete. Only about 40–50% of the total plasma apo E can be recovered in the VLDL fraction obtained by ultracentrifugation since the remainder is in the HDL fraction in normal subjects. Contamination by other proteins, such as serum amyloid protein A and apo AI in the isolated VLDL fraction, has been noted with this procedure, creating difficulties in interpretation of the phenotypes. The presence of sialylated apo E3 that focuses at the apo E2 position also can make reading difficult. Thus, the distinction between phenotypes E3/3 and E3/2 must be based on the relative intensities of the E2 and E3 bands, with

Figure 16–2 ❖ Polyacrylamide Gel Isoelectric Focusing Patterns for the Six Common Apo E Phenotypes Revealed by Coomassie Blue Stain

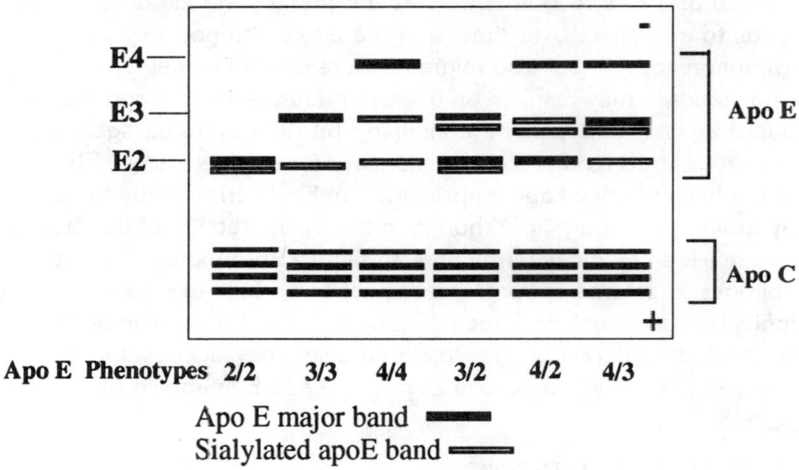

band E3 > E2 in the E3/3 phenotype and E2 > E3 in the E3/2 phenotype. Band staining may be quantified by densitometric scanning. PAG-IEF also lacks sensitivity for samples containing low concentrations of VLDL or TG (see Figure 16–2).

IEF-Immobiline

The major difference between IEF-immobiline[57, 58] and PAG-IEF is the use of immobiline, a different form of ampholyte. Because the buffering groups are polymerized to the acrylamide gel matrix, forming an immobilized pH gradient, the pH gradient will not drift during electrofocusing. Utilizing immobiline allows the separation of proteins with very small differences in isoelectric point. This technique allows the separation of a rare E2 variant (Lys146→Gln) from the common apo E2. The IEF-immobiline separation procedure can be combined with immunoblotting, thus eliminating the VLDL isolation procedure prior to IEF. Because of its high resolution, IEF-immobiline is a useful tool for the analysis of the structural integrity of apo E in various forms of dyslipoproteinemia.

SDS-PAGE

SDS-PAGE[59] accomplishes separation of common apo E isoforms and some rare mutants on polyacrylamide gel in the presence of sodium dodecyl sulfate (SDS).

PAG-IEF Immunoblotting

Availability of apo E antibodies and the need for a more rapid and reliable procedure for apo E phenotyping led to the application of immunoblotting to the analysis of apo E polymorphisms in unfractionated serum samples. Immunoblotting eliminates the need for VLDL isolation by ultracentrifugation. With the PAG-IEF method,[60-64] whole blood proteins are separated by IEF, transferred to nitrocellulose membranes, and then treated with anti-apo-E antibodies. An enzyme-conjugated second antibody is then added to visualize bands by substrate color development.

IEF-Agarose

With the IEF-agarose method,[65-67] PAG is replaced by agarose gel. After electrophoresis, apo E bands are visualized by immunoblotting or by the direct application of anti-apo E antibody to the gel without transferring apolipoproteins to nitrocellulose paper.

Additional Comments

Apo E variants with unusual amino acid composition or sequence may not be separated from common apo E isoforms by IEF if the isoelectric points of the variant proteins are similar to those of the three common isoforms of apo E.

Post-translational modifications, such as sialylation (enzymatic glycosylation) and glycation (nonenzymatic glycosylation), of apo E may alter the isoelectric point of the proteins and lead to erroneous interpretation of the phenotype.[56] For example, post-translational modifications can cause difficulties in the discrimination of phenotypes E4/3 from E4/4, and especially E3/2 from E3/3. Pretreatment of samples with neuraminidase to remove sialic acid residues or treatment with cysteinamine to add a positive charged group to cysteine helps to confirm the apo E phenotype.

Using "gel bond" for casting ultrathin gel plates for IEF allows up to 30 samples to be run at the same time. This minor modification makes it possible to screen phenotypes in a large population.

Combining IEF with immunoblotting reduces the sample requirement from 4 mL to a few microliters of plasma and also allows the entire procedure to be completed within a day.

Freezing serum samples at −70° C for as long as 15 y was found to have no effect on apo E phenotyping.[63]

Recommended Method: PAG-IEF with Immunoblotting

PAG-IEF

Blood should be collected in EDTA-containing tubes (1 mg/mL) and kept at 4° C. Serum is not suitable for this procedure. The plasma should be separated within 2 h and stored at −70° C if not used within 4 d.

Before electrophoresis, the plasma should be pretreated by incubating with dithiothreitol (5 mmol/L) containing Tween 20 (2.5 mmol/L) at 4° C for at least 15 min or overnight. Electrophoresis is carried out in 5% PAG containing 1.5% pharmalyte (a mixture of one-third pH 4.5–5.4 and two-thirds pH 5–8

Figure 16–3 ❖ Polyacrylamide Gel Isoelectric Focusing Patterns for the Six Common Apo E Phenotypes Identified by Immunoblotting Technique

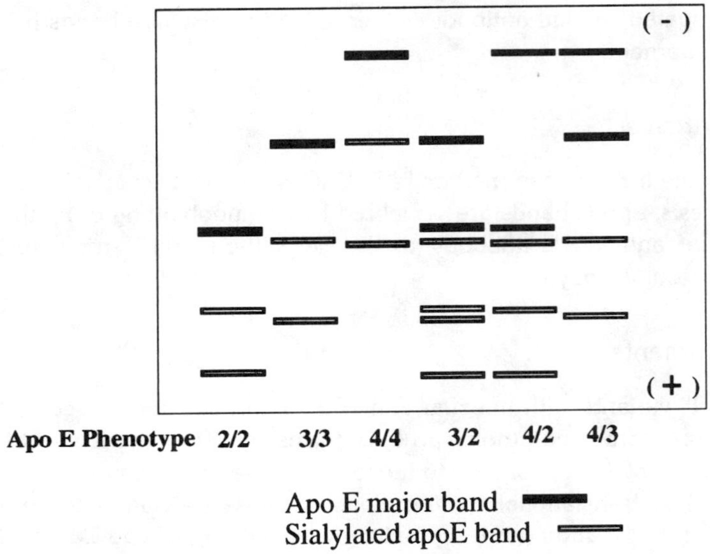

Adapted from Kataoka et al.[64]

pharmalyte) containing 3 mol/L urea. For details on IEF conditions, see Kataoka et al.[64]

Immunoblotting

Following IEF, the proteins should be transferred to nitrocellulose or (PVDF) membranes by passive diffusion. The transfer buffer is 0.03 mmol/L Tris-HCl containing 0.25 mmol/L NaCl, pH 8.0. After blocking the membrane with non-fat dry milk, the apo E bands on the membrane are stained by mouse monoclonal anti-apo E antibody and then by anti-mouse IgG-peroxidase conjugated antibody. Polyclonal anti-E antibodies are also available from several commercial sources. The pattern of the major bands and sialylated isoforms is shown in Figure 16–3.[64]

Comments

This method requires a very small amount of sample and does not require prior isolation of VLDL by ultracentrifugation or precipitation. The three common isoforms and their sialylated bands are well separated without having to be treated with neuraminidase or cysteinamine. This allows clear identification of phenotype by the band pattern rather than judging by the band intensities.

Genotyping

While DNA methods provide unambiguous results (assuming the amplification step is successful), they may not always be superior to protein separation (phenotyping). Genotyping is particularly helpful when the phenotype is difficult to interpret due to post-translational modification of apo E molecules, as in distinguishing phenotype E3/3 from E3/2. Genotyping methods may also complement phenotyping when certain structural variants of apo E have the same net charge as the three common isoforms. In this case, however, a genotyping method that specifically identifies rare variants must be employed. Indeed, most genotyping methods will omit new, rare variants that might be detected by IEF if a change in charge has occurred.

Apo E genotyping is based on the determination of nucleotide sequence variations. It requires less specimen than phenotyping. Direct determination of apo E genotype can be made with allele-specific oligonucleotides (ASO) or analysis of restriction enzyme digests after amplification of target sequences by the polymerase chain reaction (PCR). Various methods for genotyping are summarized below.

Southern Blotting and Hybridization with ASO Probes

Southern blotting was more widely used prior to the availability of PCR.[68-71] Relatively large amounts of genomic DNA is digested first with a restriction enzyme, separated by electrophoresis in agarose gels, and then transferred to nitrocellulose or nylon membranes. The DNA fragments are identified by hybridization with a radiolabeled probe. Different-sized DNA fragments (restriction fragment length polymorphisms [RFLP]) which hybridize with the probes are then visualized by autoradiography.

The introduction of PCR amplification allows for the use of much less specimen for genotyping. As little as 0.4–1.0 μg of DNA (10 μL of buffy coat, 100 μL of whole blood, or a 5 mm diameter spot of dried whole blood on filter paper) is sufficient for the identification of apo E phenotype. Methods based on PCR have essentially replaced Southern blotting.

PCR-RFLP

The PCR-RFLP[72-75] genotyping procedure for apo E takes advantage of Hha I RFLPs that are fortuitously introduced in the common apo E alleles. These alleles differ at amino acid residues 112 and 158 by the presence or absence of arginine. The Hha I enzyme cleaves at 5′...GCG|C...3′ sites while the Arg codon is CGC. Initially, the target apo E DNA sequence is amplified by PCR. The amplified product is then digested with Hha I and then separated by agarose gel or PAG electrophoresis. The bands are visualized by ethidium bromide staining, as shown in Figure 16–4.

Figure 16–4 ❖ Differentiation of apo E Alleles by Restriction Site Analysis

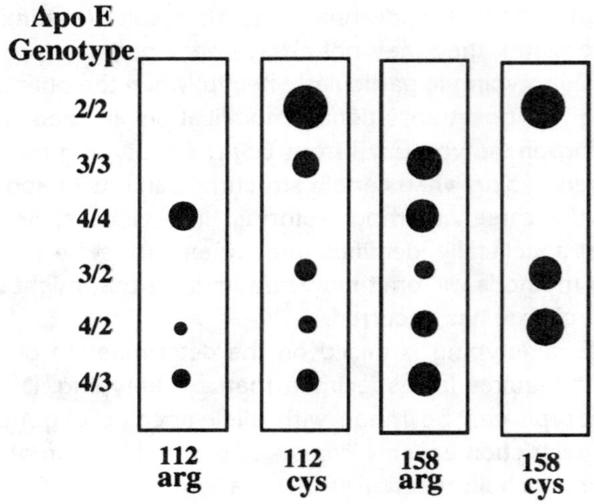

PCR-Dotting Method

The PCR-Dotting method[76] begins by spotting a PCR amplified apo E sequence on nitrocellulose or nylon membranes. The sequence on the membrane is then identified by hybridization with four different radioactively labeled ASO probes with sequences corresponding to the coding regions for cys 112, cys 158, arg 112, and arg 158 (see Figure 16–5). The major disadvantage of this otherwise excellent procedure is non-specific probe hybridization due to the high GC content of apo E. The procedure may be simplified by automating PCR and the hybridization procedure using non-radioisotopic, biotin-labeled probes.

PCR-DNA Melting Method

In the PCR-DNA melting method,[77, 78] a targeted apo E sequence is first amplified by PCR. Differences in DNA sequence are then detected by oligonucleotide "melting" or denaturing gradient gel. This procedure can detect single base substitutions and may facilitate the identification of new mutations. However, the procedure is technically very demanding.

Single Strand Conformational Polymorphism (PCR-SSCP)

In PCR-SSCP technique,[79] the PCR amplified products of the apo E gene are converted to single-strand DNA (SS DNA) by denaturation. The SS DNA is then separated by PAG electrophoresis (PAGE) under non-denaturing conditions. Differences in base pair sequence lead to changes in conformation of the folded molecule. After PAGE, these conformation polymorphisms are detected as discrete bands. This method is sensitive and can detect rare mutants. However, the particular conforma-

Figure 16–5 ❖ *Identification of Common apo E Genotypes by Dot Blotting Technique Using ASO Probes*

Apo E phenotypes of individual patients are shown on the left. A sample of the patient's amplified target DNA sequence is spotted onto each of four nitrocellulose strips. Each strip, in turn, is treated with a radiolabeled ASO which hybridizes to its matching sequence (shown at bottom). Dot patterns are revealed by autoradiography.

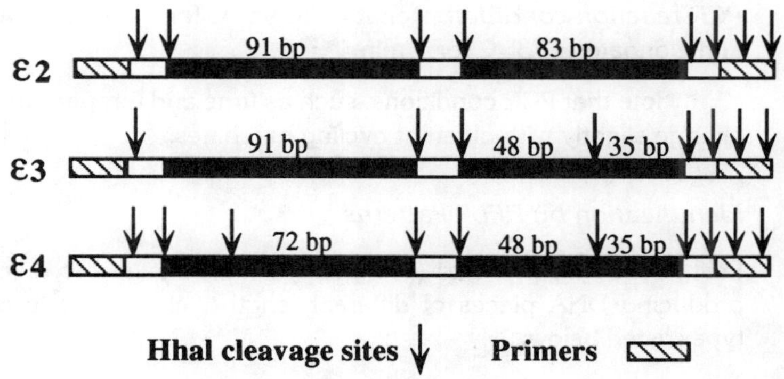

Adapted from Tsukamoto et al.[75]

tion that a given sequence will take (and hence its band pattern) is highly sensitive to experimental conditions such as salt concentrations and temperature.

Amplification Refractory Mutation Systems (ARMS)

The ARMS procedure[80-82] takes advantage of ASO primers that are specific for different alleles of the apo E gene. Multiple primers, each specific for a given mutation site, are added to the PCR reaction mixture. Depending on the presence of sequences complementary to the primers, products of differing length will be produced during the PCR reaction. The genotype is then determined by the pattern produced after agarose gel electrophoresis of the PCR products and ethidium bromide staining.

Recommended Method: PCR-RFLP

Amplification of Genomic DNA by PCR

This procedure requires about 1 µg of genomic DNA. The DNA source can be buffy coat, fresh whole blood, or whole blood on Whatman filter paper. Isolation of leukocyte DNA may be accomplished by the proteinase K/phenol method. Commercial kits for DNA isolation are available. Alternatively, DNA can be obtained by boiling the sample in deionized sterile water for 5 min. The following PCR primers, reagents, and conditions may be used:

Primer 1. 5′-GCCCCGGCCTGGTACAT-3′

Primer 2. 5′-GGCACGGCTGTCCAAGGA-3′

PCR reaction mixture: Mix 0.4 µg genomic DNA, 0.2 µmol/L of each primer, 1.5 units of Tag polymerase, 200 µmol/L of each deoxynucleotide triphosphate in 100 µL of 50 mmol/L KCl, 1.0 mmol/L Mgcl2, 10 mmol/L Tris, pH 8.4 buffer.

PCR reaction condition: Denature at 96° C for 2 min, anneal at 62° C for 1 min, and elongate at 73° C for 1 min.

Note that PCR conditions, such as time and temperature requirements, may change slightly with different cycling machines.

Identification by RFLP Patterns

Restriction enzyme Hha I (8 U/20µL) should be used to digest the PCR product, producing DNA pieces of different lengths, characteristic of the various genotypes listed below:

ε2: 91 and 83 base pairs

ε3: 91, 48, and 35 base pairs

ε4: 72, 48, and 35 base pairs

All these base pairs can be separated and identified by 20% PAGE (see Figure 16–6). For electrophoresis, 5 µg DNA should be applied in each trough. Bands representing the different-sized fragments are visualized by staining with ethidium bromide. For details on this method, see Tsukamoto et al.[75]

Figure 16–6 ❖ Restriction Fragment Length Polymorphisms (RFLPs) of PCR Products Revealed by Hha I Enzyme Digestion and PAGE.

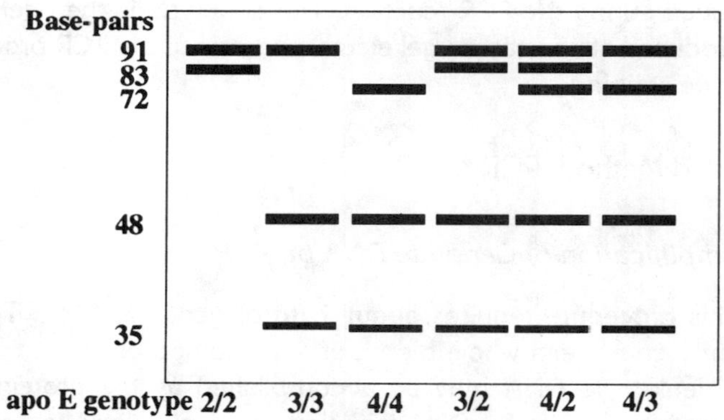

Comments

Due to the high GC content of the apo E gene, genotyping for apo E is a very demanding procedure. Unsuccessful amplification is a frequent problem if PCR conditions are not rigorously controlled. The hybridization step using ASO probes can also be problematic. The recommended method avoids ASO hybridization, does not require radiolabeled probes, and requires only one PCR amplification process with one pair of primers.[75]

CLINICAL SIGNIFICANCE OF APO E PHENOTYPES AND GENOTYPES

Lipoprotein Metabolism and Cardiovascular Disease

As a ligand for both the LDL (apo B/E) receptor and LRP, apo E plays an important role in the catabolism of lipoproteins. Because of the different affinities of various apo E isoforms for these receptors, apo E polymorphisms will have an impact on plasma lipoprotein metabolism. Resultant changes in plasma lipoprotein concentrations can have a significant impact on atherosclerotic vascular disease.

In a meta-analysis utilizing studies that included 45 population samples from 17 countries, Dallongeville et al.[83] analyzed the relationship between apo E phenotype and plasma TC, TG, and HDL cholesterol (HDL-C) concentrations. The mean plasma values of TC, TG, and HDL-C of subjects with the E2/2, E3/2, E4/2, E4/3, E4/4 phenotypes were compared those of subjects with the E3/3 phenotype. Results are illustrated in Figure 16–7. TC was clearly increased in subjects with the E4/3 and E4/4 phenotypes, while significantly lower values were seen in those with the E3/2 and E2/2 phenotypes. A number of investigators have reported similar results as to the effect of apo E genotype on TC and LDL-C.[20, 84–89]

With respect to serum TG concentrations, possession of the E2 allele in either homozygous or heterozygous form clearly increased the serum concentrations above those seen with the E3 allele. In addition, apo E4/3 heterozygotes had higher concentrations of TG than E3/3 homozygotes, but E4 homozygotes did not have significantly higher TG concentrations. For HDL-C, where E4/3 heterozygotes had slightly lower concentrations, E4/4 homozygotes had similar concentrations compared to those individuals who were E3/3 homozygotes. Apo E2 had no effect upon HDL-C concentrations. All of these effects were consistent across ethnic and patient groups.[83]

In several studies, patients with CHD[89–91] and stroke[92] had greater frequency of the E4 allele than was normally found in healthy controls; in contrast, the E2 allele was less frequently present in the same group and could therefore be considered generally protective. In most of these studies, the excess risk associated with apo E4 appeared to be due to higher plasma TC or LDL-C concentrations. Dallongeville et al.[83] suggested that the increased cardiovascular risk associated with the E4/3 allele may also be mediated by higher TG and lower HDL-C concentrations. More CHD cases in the population appear to be attribut-

Figure 16–7 ❖ Effect of Apo E Phenotype on the Plasma Concentrations of Cholesterol, Triglycerides, and HDL Cholesterol

Shown are mean levels for each phenotype and 95% confidence intervals. Mean lipids (for age 20+, both sexes averaged together) from the LRC Prevalence Study were taken as apo E3/3 levels. Deviations from these references levels were estimated as the *Z*-score deviations reported by Dallongeville[69] multiplied by the estimated LRC standard deviations (taken as 28.6 mg/dL [0.74 mmol/L] for total cholesterol, 49.1 mg/dL [0.55 mmol/L] for triglycerides, and 11.3 mg/dL [0.29 mmol/L] for HDL cholesterol).

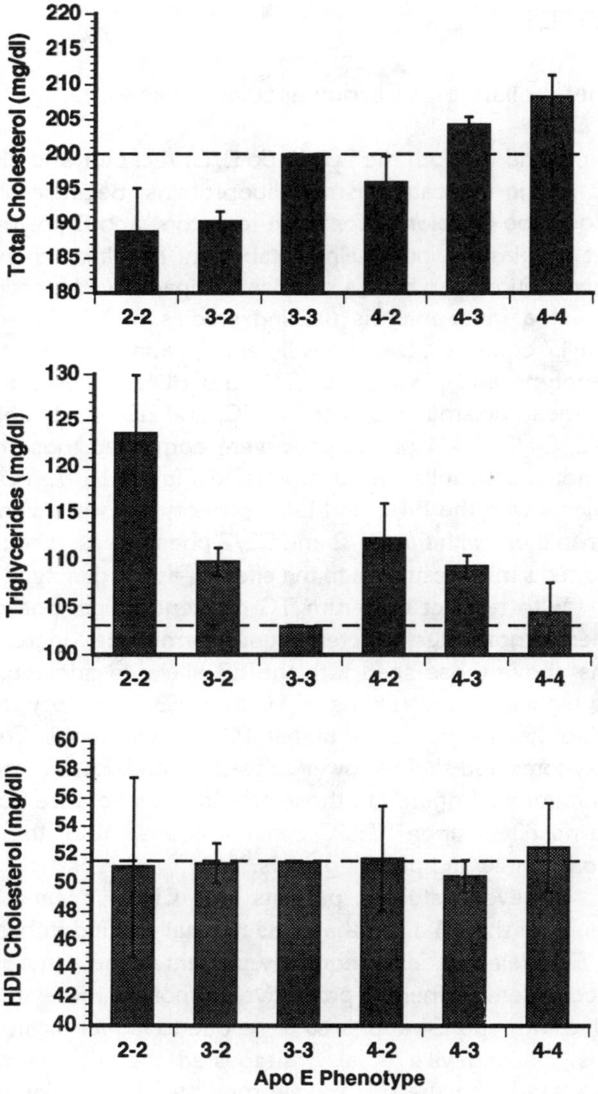

able to these modest effects on plasma lipids compared to the relatively rare occurrence of Type III hyperlipidemia in apo E2/2 individuals.

Differences in the production and removal rates of $VLDL_1$, $VLDL_2$, IDL, and LDL were examined in normolipidemic (TC 195–240 mg/dL [5.05–6.22 mmol/L]) subjects homozygous for E2, E3, or E4. Compared to E3/3 subjects, E2/2 subjects produced less $VLDL_1$ (the precursor of $VLDL_2$) and exhibited delayed conversion of the more dense $VLDL_2$ to LDL. This resulted in higher concentrations of $VLDL_2$ and IDL and more direct catabolism of these particles prior to conversion to LDL. These alterations led to low LDL. The LDL fractional catabolic rate was not significantly affected by E2. In contrast, the LDL fractional catabolic rate was depressed in E4/4 subjects. Conversion of more dense VLDL to LDL was also increased in E4/4 subjects, suggesting down-regulation of LDL receptors.[93] Down-regulation of LDL receptors in individuals bearing an E4 allele may be due to increased efficiency of intestinal cholesterol absorption (order of efficiency: E4 > E3 > E2)[94, 95] and more rapid clearance of chylomicron remnants (order of clearance: E4 > E3 > E2).[96–98]

Type III Hyperlipidemia

Type III hyperlipidemia (or familial dysbetalipoproteinemia) is characterized by increased plasma concentrations of TC and TG due to marked accumulation of abnormal chylomicron and VLDL remnants (collectively referred to as β-VLDL) and IDL. Serum TC concentrations are typically in the range of 300–600 mg/dL (7.77–15.54 mmol/L), with roughly equal increases in TG concentration (when reported as mg/dL). Individual patient lipid values may fluctuate widely, especially in response to body weight changes or effective treatment. Thus, an initial TC value of 490 mg/dL (12.7 mmol/L) with TG of 550 mg/dL (6.22 mmol/L) may decrease to 250 mg/dL (6.5 mmol/L) and 230 mg/dL (2.60 mmol/L), respectively, with a ten-pound weight loss. The usual clinical laboratory estimation of LDL-C is invalid in Type III since the accumulated β-VLDL particles have roughly equal weights of cholesterol and TG (unlike VLDL, which has a TG to cholesterol ratio of 5:1). Patients with Type III hyperlipidemia develop accelerated atherosclerosis involving both coronary and peripheral arteries. Type III hyperlipidemia may account for as much as 5% of the early familial CHD.[99]

Over 95% of patients with Type III hyperlipidemia are found to have the apo E2/2 phenotype. Nevertheless, only 5% of E2/2 individuals in the general population will develop Type III hyperlipidemia. Apparently, manifestation of Type III hyperlipoproteinemia requires additional inciting factors such as familial combined hyperlipidemia, familial hypercholesterolemia, diabetes, hypothyroidism, obesity, estrogen deficiency, high-fat diet, or certain drugs.[59, 100, 101] Other rare apo E mutations that cause defective receptor binding may also cause accumulation of atherogenic β-VLDL and Type III hyperlipoproteinemia (reviewed in Table 16–1). Some of these mutations appear to be dominant, with only a single copy required for manifestation of hyperlipidemia.

Diagnosis of Type III hyperlipidemia can usually be confirmed by the identification of the apo E2/2 phenotype or a clear demonstration of a broad-beta band by electrophoresis of the isolated VLDL fraction (which normally shows only a pre-beta band). A measured ratio of VLDL cholesterol (VLDL-C) to total TG ≥ 0.30 (when measured in mg/dL; 0.68 when using mmol/L) is a specific but somewhat insensitive indicator of Type III hyperlipidemia.[100] Recognition of this relatively rare disease (1 to 10 individuals in 10,000) is important not only because of its association with a strong predisposition to coronary and peripheral vascular disease, but also because patients typically have an excellent response to appropriate diet and drug (niacin or fibrates) therapy.

Interaction of Apo E with Other Genetic and Environmental Factors

Interaction of apo E genotype with a defective receptor or other genetic or environmental factors can lead to measurable effects on lipoprotein metabolism. A few examples have been reported.[102, 103]

Gene-Gene Interaction

Typically, patients with Type III hyperlipidemia are homozygous for apo E2. However, about 26% of heterozygous familial hypercholesterolemic patients with a single copy of E2 (phenotypes E3/2 and E4/2) develop a Type III pattern as defined by a measured ratio of VLDL-C to total TG ≥ 0.30 (0.68 when using mmol/L) and TG > 150 mg/dL (1.70 mmol/L). Furthermore, estimated concentrations of β-VLDL-C in these patients increased markedly with age. Estimated β-VLDL-C concentrations were lower and did not rise with age in close relatives who did not have an apo E2 allele or in those without familial hypercholesterolemia.[104] Presumably, the inherited LDL receptor defect together with a diminished receptor binding (imparted by the presence of even a single apo E2 allele) resulted in defective clearance and accumulation of abnormal remnant particles. Familial hypercholesterolemia patients with the superimposed Type III pattern may be at increased risk for premature CHD.

Diabetes

Plasma concentrations of TG and VLDL were increased in diabetic patients with the E2 allele.[56, 105–108] Thus, the E2 allele may be a marker for genetic susceptibility to hypertriglyceridemia in non-insulin-dependent diabetes mellitus patients.

Drug Response

In familial hypercholesterolemia, patients with an apo E4 allele may respond better to probucol treatment than those without the E4 allele.[109] Plasma LDL response to other drugs in familial hypercholesterolemic patients appears not to be affected by apo E phenotype.[110–112]

Dietary Manipulation

Some recent studies have reported an association between apo E phenotype and the response to dietary manipulation. Persons with phenotypes E4/3 and E4/4 exhibited a significantly greater increase in plasma LDL-C concentrations when dietary cholesterol intake was increased, whereas patients with phenotypes E2/3 and E2/2 tended to be resistant to changes. There may be considerable heterogeneity with respect to the magnitude of the apo E allelic effects in different populations, since not all investigators have found a significant relationship between response to diet and apo E genotype.[113, 114] Nevertheless, there may be a metabolic basis for a differential dietary response. In other studies (reviewed in Abey[113] and Boerwinkle et al.[114]) persons with an apo E4 allele were observed to have enhanced cholesterol absorption and chylomicron remnant removal rates. As a result, more dietary cholesterol is presumably delivered to the liver by apo-E4-mediated uptake, which should result in down-regulation of hepatic LDL (apo B/E) receptors and cause increased LDL in plasma. Interestingly, familial hypercholesterolemia patients with the E3/2 phenotype responded poorly to soy-protein diets, while those with E3/3 or E4/3 experienced a mean 26% reduction in LDL-C.[115]

Alzheimer's Disease

Alzheimer's disease (AD), a devastating neurological disorder, is the leading cause of dementia in the elderly. It affects millions of individuals of all races and ethnic backgrounds. There are two major types of AD: sporadic (85%) and familial (15%). In sporadic AD, unlike the familial form, a well-defined hereditary cause cannot be established. Both versions of the disease follow the same progression, with memory loss followed by loss of cognitive abilities. AD can also be differentiated by age of onset, with early-onset AD starting before age 65 and late-onset AD starting after age 65.

The apo E4 allele has been found to be a risk factor in both sporadic and familial late-onset AD and appears to transmit risk as an autosomal co-dominant trait. A total of 80% of cases with familial AD and 45–64% of sporadic late-onset AD have at least one apo E4 allele compared to controls, in whom frequency of an apo E4 allele is about 15%.[116–118] The proportion of affected individuals increases with the number of E4 alleles. For example, among individuals from AD-prone families with no E4 allele, only 20% developed AD by age 75. The risk for AD increased to 47% in those having one copy (genotypes E2/4 or E3/4) and to 91% in those having two copies (genotype E4/4).[119] Additional apo E4 alleles also shift the time of onset of AD to younger ages. The mean age of onset was found to be 84.3 for individuals without an E4 allele, 75.5 for those with one copy, and 68.4 for those with two copies. Almost all individuals homozygous for apo E4 developed AD by age 80.[119] Thus, the apo E gene in patients' blood samples can be used to predict whether an individual is likely to develop late-onset Alzheimer's disease and at what age the first symptoms might appear.[120]

Presence of amyloid β-peptide (Aβ) deposition in senile plaques and cerebral vessels is a neuropathological feature of AD. Neurofibrillary tangles are made of bundles of abnormal fibers (modified protein) in the cytoplasm of certain neurons. Several additional proteins are associated with Aβ, including APP, α_1-antichymotrypsin, complement factors, and immunoglobulins. Using immunochemical methods, apo E4 was recently found also in senile plaques in association with Alzheimer's Aβ deposits.[121] Apo E could also be seen in the vascular amyloid and neurofibrillary tangles of AD.

Evidence supporting the role played by apo E in the pathogenesis of AD is outlined below.[104, 110, 111]

1. Apo E plays an important role in nerve cell injury and regeneration.
2. The apo E gene is located on chromosome 19, within a region previously linked to late-onset AD.
3. Apo E can be detected in the senile plaques and the vascular and neurofibrillary tangles of AD.
4. The intensity of the immunohistochemical stain for apo E4 increases with the severity of AD.
5. Apo E was found to bind Aβ peptide in cerebrospinal fluid *in vitro*.
6. Apo E4 has a much higher affinity than apo E3 for Aβ.
7. Patients with one or two copies of apo E4 (phenotypes E2/4, E3/4, or E4/4) have greatly increased plaque amyloid deposits compared to individuals with the E3/3 phenotype.

The greatly increased risk of AD and the moderately increased risk of CHD in persons carrying an apo E4 allele may help explain the decreased survival rate of older carriers which was found in one study that compared apo E allele frequencies of older and younger women.[122] ❖

REFERENCES

1. Mahley RW, Innerarity TL. Apolipoprotein E: Cholesterol transport protein with expanding roles in cell biology. Science 1988;240:622–30.
2. Zannis VI, Breslow JL. Genetic mutations affect human lipoprotein metabolism. Adv Hum Gene 1985;14:125–215.
3. Getzs GS, Mazzone T, Soltys P, Bates SR. Atherosclerosis and apoprotein E. Arch Pathol Lab Med 1988;112:1048–55.
4. Rall SC, Weisgraber KH, Mahley RW. Human apolipoprotein E: the complete amino acid sequence. J Biochem 1982;247:4171–8.
5. Talmud P. Detection and physiological relevance of mutations in the apoprotein E, C-II, and B genes. In: Rosseneu M, ed. Structure and function of apolipoproteins. Boca Raton, FL: CRC Press, 1992:123–58.
6. Bhatnagar D, Durrington PN. Does measurement of apolipoproteins add to the clinical diagnosis and management of dislipidaemias? Curr Opin Lipidol 1993; 4:288–304.

7. Mahley RW, Innerarity TL, Rall SC, Weisgraber KH. Apolipoprotein E: genetic variants provides insights into its structure and function. Curr Opin Lipidol 1990; 1:87–95.
8. Walden CC, Huff MW, Leiter L, Connelly PW, Hegele RA. Variable expression of type III hyperlipidemia in compound heterozygotes for apolipoprotein E2 and a new mutant, R136C [abstract]. Circulation 1993;88(suppl I):461.
9. Weisgraber KH, Shinto LH. Identification of the disulfide-linked homodimer of apolipoprotein E3 in plasma. J Biol Chem 1991;266:12029–34.
10. Tozuka M, Yoshida Y, Tanigami J, Miyachi M, Katsuyama T, Masamitsu K. Development of an enzyme-linked immunosorbent assay of apolipoprotein E-AII complex in plasma. Clin Chem 1991;37:1645–8.
11. Yamada H, Shimano H, Mokuno H, et al. Increased clearance of plasma cholesterol after injection of apolipoprotein E into Watanabe heritable hyperlipidemic rabbits. Proc Natl Acad Sci USA 1989;86:665–9.
12. Mahley RW, Weisgraber KH, Hussain MM, Greenman B, Fisher M, Vobel T, Gorecki M. Intravenous infusion of apolipoprotein E accelerates clearance of plasma lipoproteins in rabbits. J Clin Invest 1989;83:2125–30.
13. Shimano H, Yamada N, Katsuki M, et al. Overexpression of apolipoprotein E in transgenic mice: marked reduction in plasma lipoproteins except high density lipoprotein and resistance against diet-induced hypercholesterolemia. Proc Natl Acad Sci USA 1992;89:1750–4.
14. Shimano H, Yamada N, Katsuki M, et al. Plasma lipoprotein metabolism in transgenic mice over-expressing apolipoprotein E. Accelerated clearance of lipoproteins containing apolipoprotein B. J Clin Invest 1992;90:2084.
15. de Silva HV, Lauer SJ, Wang J, et al. Opposing effects of human apolipoproteins E and C-II in lipoprotein remnant clearance in transgenic mice [abstract]. Circulation 1992;86(suppl I):471.
16. Zhang SH, Reddick RL, Piedrahita JA, Maeda N. Spontaneous hyperchoesterolemia and arterial lesions in mice lacking apolipoprotein E. Science 1992;258: 468–71.
17. Plump AS, Smith JD, Hayek T, et al. Severe hypercholesterolemia and atherosclerosis in apolipoprotein E-deficient mice created by homologous recombination in ES cells. Cell 1992;71:343–53.
18. Fazio S, Lee Y-L, Ji Z-S, Rall SCJ. Type III hyperlipoproteinemic phenotype in transgenic mice expressing dysfunctional apolipoprotein E. J Clin Invest 1993; 92:1497–1503.
19. Fazio S, Horie Y Simonet, WS, Weisgraber KH, Taylor JM, Rall SC. Altered lipoprotein metabolism in transgenic mice expressing low levels of a human receptor-binding-defective apolipoprotein E variant. J Lipid Res 1994;35:408–16.
20. Xhignesse M, Lussier-Cacan S, Sing CF, Kessling AM, Davignon J. Influences of common variants of apolipoprotein E on measures of lipid metabolism in a sample selected for health. Arterioscler Thromb 1991;11:1100–10.
21. Gordon V, Innerarity TL, Mahley RW. Formation of cholesterol- and apoprotein E-enriched high density lipoproteins in vitro. J Biol Chem 1983;258:6202–12.
22. Koo C, Innerarity TL, Mahley RW. Obligatory role of cholesterol and apolipoprotein E in the formation of large cholesterol-enriched and receptor-active high density lipoproteins. J Biol Chem 1985;260:11934–43.
23. Funke H, Boyles J, Weisgraber KH, Ludwig EH, Hui DY, Mahley RW. Uptake of apolipoprotein E-containing high density lipoproteins by hepatic parenchymal cells. Arteriosclerosis 1984;4:452–61.
24. Johnson WJ, Mahlberg FH, Rothblat GH, Phillips MC. Cholesterol transport between cells and high-density lipoproteins. Biochim Biophys Acta 1991;1085: 273–98.

25. Yamada N, Inoue I, Kawamura M, et al. Apolipoprotein E prevents the progression of atherosclerosis in Watanabe heritable hyperlipidemic rabbits. J Clin Invest 1992;89:706–11.
26. Kinoshita M, Arai H, Fukasawa T, et al. Apolipoprotein E enhances lipid exchange between lipoproteins mediated by cholesteryl ester transfer protein. J Lipid Res 1993;34:261–8.
27. Ishibashi S, Mori N, Shimada M, et al. Apolipoprotein E secreted by macrophage plays an important role in nerve injury and repair. Horm Metab Res 1993;25:82–7.
28. Gustafson S, Vessby B, Lindqvist OAM. Serum and interstitial fluid apolipoprotein E levels in the healthy and in hyperlipoproteinemia type III as studied by radioimmunoassay. Clin Chem Acta 1985;151:49–59.
29. Kushwaha RS, Hazzard WR, Wahl PW, Hoover JJ. Type III hyperlipoproteinemia: diagnosis in whole plasma by apolipoprotein E immunoassay. Ann Inter Med 1977;87:509–16.
30. Blum CB, Davis PA, Forte TM. Elevated levels of apolipoprotein E in the high density lipoproteins of human cord blood plasma. J Lipid Res 1985;26:755–60.
31. Blum CB, Aron L, Sciacca R. Radioimmunoassay studies of human apolipoprotein E. J Clin Invest 1980;66:1240–50.
32. Genest JJ, Bard JM, Fruchart JC, Ordovas JM, Wilson PF, Schaefer EJ. Plasma apolipoprotein A-I, A-II, B, E and C-III containing particles in men with premature coronary artery disease. Atherosclerosis 1991;90:149–57.
33. Rifai N, Christenson RH, Gelman BB, Silverman LM. Changes in cerebrospinal fluid IgG and apolipoprotein E indices in patients with multiple sclerosis during demyelination and remyelination. Clin Chem 1987;33:1155–7.
34. Carlson LA, Holmquist L. Concentrations of apolipoproteins B, C-I, C-II, C-III and E in sera from normal men and their relation to serum lipoprotein levels. Clin Chim Acta 1982;124:163–78.
35. Oikawa S, Suzuki N, Sakuma E, et al. Abnormal lipoprotein and apolipoprotein pattern in lipoprotein glomerulopathy. Am J Kid Dis 1991;18:553–8.
36. Averna MR, Barbagallo CM, Di Paola G, et al. Lipids, lipoproteins and apolipoproteins AI, AII, B, CII, CIII and E in newborns. Biol Neonate 1991;60:187–92.
37. Senti M, Romero R, Pedro-Botet J, Pelegri A, Nogues X, Rubies-Prat J. Lipoprotein abnormalities in hyperlipidemic and normolipidemic men on hemodialysis with chronic renal failure. Kidney International 1992;41:1394–9.
38. Catalano M, Aronica A, Carzaniga G, Seregni R, Libretti A. Serum lipids and apolipoproteins in patients with essential hypertension. Atherosclerosis 1991; 87:17–22.
39. Curry MD, McConathy WJ, Alaupovic P, Ledford JH, Popovic M. Determination of human apolipoprotein E by electroimmunoassay. Biochim Biophys Acta 1976; 439:413–25.
40. Bittolo BG, Cazzolato G, Saccardi M, Kostner GM, Avogaro P. Total plasma apo E and high density lipoprotein apo E in survivors of myocardial infarction. Atherosclerosis 1984;53:69–75.
41. Kuhl H, Marz W, Jung-Hoffman C, Weber J, Siekmeier R, Gross W. Effect on lipid metabolism of a biphasic desogestrel-containing oral contraceptive: divergent changes in apolipoprotein B and E and transitory decrease in Lp(a) levels. Contraception 1993;47:69–83.
42. Au YPT, Bren DN, Kottke BA. A rapid apolipoprotein E radioimmunoassay using solid phase Staphylococcus protein. Use of pooled plasma as a secondary standard. Biochim Biophys Res Comm 1986;138:455–62.
43. Mackie A, Caslake MJ, Packard CJ, Shepherd J. Concentration and distribution of human plasma apolipoprotein E. Clin Chim Acta 1981;116:35–45.

44. Weisweiler P, Schwandt P. Immunonephelometric quantitation of apolipoprotein E in human serum. J Clin Chem Clin Biochem 1983;21:227–30.
45. Noma A, Hata Y, Goto Y. Quantitation of serum apolipoprotein A-I, A-II, B, C-II, C-III and E in healthy Japanese by turbidimetric immunoassay: reference values, and age- and sex-related differences. Clin Chim Acta 1991;199:147–58.
46. Rifai N, Silverman LM. A simple immunotechnique for the determination of serum concentration of apolipoprotein E. Clin Chim Acta 1987;163:207–13.
47. Holmquist L. Quantitation of human serum very low density apolipoprotein C-I, C-II, C-III and E by enzyme immunoassay. J Immunol Methods 1980;34:243–51.
48. Bury J, Vercaemst R, Rosseneu M, Belpaire F. Apolipoprotein E quantified by enzyme-linked immunosorbent assay. Clin Chem 1986;32:265–70.
49. Koffigan M, Kora I, Clavey V, Bard JM, Chapman J. Quantification of human apolipoprotein E in plasma and lipoprotein subfractions by a non-competitive enzyme immunoassay. Clin Chim Acta 1987;163:245–56.
50. Alsayed N, Rebourcet R, Chapman J. Concentrations of apoprotein CII, CIII, and E in total serum and in the apoprotein B-containing lipoproteins, determined by a new enzyme-linked immunosorbent assay. Clin Chem 1990;36:2047–52.
51. Leroy A, Vu-dac N, Koffigan M, Clavey V, Fruchart J-C. Characterization of a monoclonal antibody that binds to apolipoprotein E and to lipoproteins of human plasma containing apo E. Applications of ELISA quantification of plasma apo E. J Immunoassay 1988;9:309–34.
52. Labeur C, van Huffel X, Rosseneu M. Evaluation of enzyme-linked immunoassays for the quantification of apolipoprotein E and B. In: Lentant C, et al., eds. Biotechnology of dyslipoproteinemias: application in diagnosis and control. New York: Raven Press, 1990:207–16.
53. Warnick GR, Mayfield C, Albers JJ, Hazzard WR. Gel isoelectric focusing method for specific diagnosis of familial hyperlipoproteinemia type 3. Clin Chem 1979;1979:279–84.
54. Eto M, Watanabe K, Ishii K. A rapid flat gel isoelectric focusing method for the determination of apolipoprotein E phenotypes and its application. Clin Chim Acta 1985;149:21–8.
55. Wu LL. Apo E phenotype by ultra-thin isoelectrofocusing method on polyacrylamide gel. In: Sunderman FW, ed. Manual of procedures for the applied seminar on six major advances in laboratory medicine and their clinical applications. Philadelphia: Institute for Clinical Science, 1990:71–5.
56. Salzer B, Stavljenic A, Jurgens G, Dumic M, Radica A. Polymorphism of apolipoprotein E, lipoprotein (a), and other lipoproteins in children with type I diabetes. Clin Chem 1993;39:1427–32.
57. Mailly F, Davignon J, Nestruck AC. Analytical isoelectric focusing with immobilized pH gradients of human apolipoprotein E from very low density lipoproteins and total plasma. J Lipid Res 1990;31:149–55.
58. Marz W, Cezanne S, Gross W. Phenotyping of apolipoprotein E by immunoblotting in immobilized pH gradients. Electrophoresis 1991;12:59–63.
59. Utermann G, Weisgraber KH, Weber W, Mahley RW. Genetic polymorphism of apolipoprotein E: a variant form of apolipoprotein E2 distinguished by sodium dodecylsulfate polyacrylamide gel electrophoresis. J Lipid Res 1984;25:378–82.
60. Menzel HJ, Utermann G. Apolipoprotein E phenotyping from serum by Western blotting. Electrophoresis 1986;7:492–95.
61. Havekes LM, de Knijff P, Beisiegel U, Havinga J, Smit M, Klasen E. A rapid micromethod for apolipoprotein E phenotyping directly in serum. J Lipid Res 1987; 28:455–63.
62. Hill JS, Pritchard PH. Improved phenotyping of apolipoprotein E: application to population frequency distribution. Clin Chem 1990;36:1871–74.

63. Eichner JE, Kuller LH, Ferrell RE, Kamboh ME. A simplified method for screening the apolipoprotein E polymorphism. Human Heredity 1991;41:61–4.

64. Kataoka S, Paidi M, Howard BV. Simplified isoelectric focusing/immunoblotting determination of apoprotein E phenotype. Clin Chem 1994;40:11–3.

65. Luley C, Baumstark MW, Wieland H. Rapid apolipoprotein E phenotyping by immunofixation in agarose. J Lipid Res 1991;32:880–3.

66. Luley CH, Haas B, Buhrer B, Wieland H. Improvement of apolipoprotein E phenotyping by isoelectric focusing/immunofixation. Clin Chem 1992;38:168.

67. McDowell IFW, Wisdom GB, Trimble ER. Apolipoprotein E phenotype determined by agarose gel electrofocusing and immunoblotting. Clin Chem 1989;35:2070–3.

68. Smeets B, Poddighe J, Brunner H, Ropers HH, Wieringa B. Tight linkage between myotonic dystrophy and apolipoprotein E revealed with allele specific oligonucleotides. Hum Genet 1988;80:49–52.

69. Weisgraber KH, Newhouse YM, Mahley RW. Apolipoprotein E genotyping using polymerase chain reaction and allele specific oligonucleotide probes. Biochem Biophys Res Commun 1988;157:1212–7.

70. Houlston RS, Wenham PR, Humphries SE. Detection of apolipoprotein E polymorphisms using PCR/ASO probes and Southern transfer: application for routine use. Clin Chim Acta 1990;189:153–8.

71. Richard P, Thomas G, de Zulueta MP, et al. Common and rare genotypes of human apolipoprotein E determined by specific restriction profiles of polymerase chain reaction-amplified DNA. Clin Chem 1994;40:24–9.

72. Nassar BA, McPherson R, Lamothe EMF, Al Sultan AI, Zhang ZJ, Rosenblatt DS. Identification of variant apolipoprotein E by isoelectric focusing and restriction isotyping in a patient with type III hyperlipoproteinemia. Clin Chem 1991;37:1308.

73. Hixson JE, Vernier DT. Restriction isotyping of human apolipoprotein E by gene amplification and cleavage with Hha I. J Lipid Res 1990;31:545–48.

74. Kontula K, Aalto-Setala K, Kuusi T, Hamalainen L, Syvanen AC. Apolipoprotein E polymorphism determined by restriction enzyme analysis of DNA amplified by polymerase chain reaction: convenient alternative to phenotyping by isoelectric focusing. Clin Chem 1990;36:2087–92.

75. Tsukamoto K, Watanabe T, Matsushima T, et al. Determination of PCR-RFLP of apo E genotype in a Japanese population. J Lab Clin Med 1993;121:598–602.

76. Emi M, Wu LL, Robertson MA, et al. Genotyping and sequence analysis of apolipoprotein E isoforms. Genomics 1988;3:373–9.

77. Funke H, Rust S, Assmann G. Detection of apolipoprotein E variants by an oligonucleotide "melting" procedure. Clin Chem 1986;32:1285–9.

78. Parker S, Angelico MC, Laffel L, Krolewski AS. Application of denaturing gradient gel electrophoresis to detect DNA sequence differences encoding apolipoprotein E isoforms. Genomics 1993;16:245–47.

79. Tsai MY, Suess P, Schwichtenberg K, et al. Determination of apolipoprotein E genotypes by single-strand conformational polymorphism. Clin Chem 1993;39:2121–4.

80. Main BF, Jones PJH, MacGillivray RTA, Banfield DK. Apolipoprotein E genotyping using the polymerase chain reaction and allele specific oligonucleotide primers. J Lipid Res 1991;32:183–7.

81. Green EK, Bain SC, Day PJR, et al. Detection of human apolipoprotein E3, E2 and E4 genotypes by an allele-specific oligonucleotide-primed polymerase chain reaction assay: development and validation. Clin Chem 1991;37:1263–8.

82. Wenham PR, Newton CR, Price WH. Analysis of apolipoprotein E genotypes by the amplification refractory mutation system. Clin Chem 1991;37:241–4.

83. Dallongeville J, Lussier-Cacan S, Davigon J. Modulation of plasma triglyceride levels by apo E phenotype: a meta analysis. J Lipid Res 1992;33:447–54.
84. Sing CF, Davignon J. Role of the apolipoprotein E polymorphism in determining normal plasma lipid and lipoprotein variation. Am J Hum Genet 1985;37:268–85.
85. Xu C-F, Talmud PJ, Angelico F, Ben MD, Savill J, Humphries SE. Apolipoprotein E polymorphism and plasma lipid, lipoprotein, and apolipoprotein levels in Italian children. Genetic Epidemiology 1991;8:389–98.
86. Hallman DM, Boerwinkle E, Saha N, et al. The apolipoprotein E polymorphism: a comparison of allele frequencies and effects in nine populations. Am J Hum Genet 1991;49:338–49.
87. Eichner JE, Kuller LH, Ferrell RE, Meilahn EN, Kamboh MI. Phenotypic effects of apolipoprotein structural variation on lipid profiles. III. Contribution of apolipoprotein E phenotype to prediction of total cholesterol, apolipoprotein B, and low density lipoprotein cholesterol in the Healthy Women Study. Arteriosclerosis 1990;10:379–85.
88. Eggertsen G, Tegelman R, Ericsson S, Angelin B, Berglund L. Apolipoprotein E polymorphism in a healthy Swedish population: variation of allele frequency with age and relation to serum lipid concentrations. Clin Chem 1993;39:2125–9.
89. Eto M, Watanabe K, Chonan N, Ishii K. Familial hypercholesterolemia and apolipoprotein E4. Atherosclerosis 1988;72:123–8.
90. Eichner JE, Kuller LH, Orchard TJ, et al. Relation of apolipoprotein E phenotype to myocardial infarction and mortality from coronary artery disease. Am J Cardiol 1993;71:160–5.
91. Davignon J, Gregg RE, Sing CF. Apolipoprotein polymorphism and atherosclerosis. Arteriosclerosis 1988;8:1–21.
92. Couderc R, Mahieux F, Bailleul S, Fenelon G, Mary R, Fermanian J. Prevalence of apolipoprotein E phenotypes in ischemic cerebrovascular disease: a case-control study. Stroke 1993;24:661–4.
93. Demant T, Bedford D, Packard CJ, Shepherd J. Influence of apolipoprotein E polymorphism on apolipoprotein B-100 metabolism in normolipemic subjects. J Clin Invest 1991;88:1490–501.
94. Kesaniemi YA, Ehnolm C, Miettinen TA. Intestinal cholesterol absorption efficiency in man is related to apoprotein E phenotype. J Clin Invest 1987;80:578–81.
95. Gylling H, Kuusi T, Vanhanen H, Miettinen TA. Apolipoprotein E phenotype and cholesterol metabolism in familial hypercholesterolemia. Atherosclerosis 1989; 80:27–32.
96. Weintraub MS, Eisenberg S, Breslow JL. Dietary fat clearance in normal subjects is regulated by genetic variation in apolipoprotein E. J Clin Invest 1987;80:1571–7.
97. Brenninkmeijer BJ, Stuyt PMJ, Demacker PNM, Stalenhoef AFH, van't Larr A. Catabolism of chylomicron remnants in normolipidaemic subjects in relation to the apoprotein E phenotype. J Lipid Res 1987;28:361–70.
98. Brown AJ, Roberts DCK. The effect of fasting triacylglyceride concentration and apolipoprotein E polymorphism on postprandial lipemia. Arterioscler Thromb 1991;11:1737–44.
99. Williams RR, Hopkins PN, Hunt SC, et al. Population-based frequency of dyslipidemia syndromes in coronary prone families in Utah. Arch Intern Med 1990;150:582–8.
100. Fredrickson DS, Morganroth J, Levy RI. Type III hyperlipoproteinemia: an analysis of two contemporary definitions. Ann Intern Med 1975;1975:150–7.
101. Mahley RW, Rall SC. Type III hyperlipoproteinemia (dysbetalipoproteinemia): the role of apolipoprotein E in normal and abnormal lipoprotein metabolism. In:

Scriver CR, Beaudet AL, Sly WS, Velle D, eds. The metabolic basis of inherited disease, 6th ed. New York: McGraw Hill, 1989: 1195–213.

102. Utermann G. Diabetes, obesity and hyperlipidemia III. In: Crepaldi G, et al, eds. New York: Elsevier Science Publishers, 1985:1–28.
103. Berglund L, Wiklund O, Eggertsen G, et al. Apolipoprotein E phenotypes in familial hypercholesterolaemia: importance for expression of disease and response to therapy. J Intern Med 1993;233:173–8.
104. Hopkins PN, Wu LL, Schumacher MC, et al. Type III hyperlipoproteinemia in patients heterozygous for familial hypercholesterol and apolipoprotein E2: evidence for a gene-gene interaction. Arterioscler Thromb 1991;11:1137–46.
105. Imari Y, Koga S, Ibayashi H. Phenotypes of apolipoprotein E and abnormalities in lipid metabolism in patients with non-insulin dependent diabetes mellitus. Metabolism 1988;37:1134–8.
106. Eto M, Watanabe K, Makino I, Ishii K. Apolipoprotein E allele frequencies in non-insulin-dependent diabetes mellitus with hypertriglyceridemia (type IIb, III, IV, and V hyperlipoproteinemia). Metabolism 1991;40:776–80.
107. Sakurai T, Oka T, Hasegawa H, Igaki N, Miki S, Goto T. Comparison of lipids, apoproteins and associated enzyme activities between diabetic and nondiabetic end-stage renal disease. Nephron 1992;61:409–14.
108. Despres JP, Verdon MF, Moorjani S, et al. Apolipoprotein E polymorphism modifies relation of hyperinsulinemia to hypertriglyceridemia. Diabetes 1993; 42:1474–81.
109. Nestruck AC, Bouthillier D, Sing CF, Davignon J. Apolipoprotein E polymorphism and plasma cholesterol response to probucol. Metabolism 1987;36:743–7.
110. Gylling H, Vanhanen H, Miettinen TA. Effects of acipimox and cholestyramine on serum lipoproteins, non-cholesterol sterols and cholesterol absorption and elimination. Eur J Clin Pharmacol 1989;37:111–5.
111. Knijff PD, Stalenhoef AFH, Mol MJTM, et al. Influence of apo E polymorphism on the response to simvastatin treatment in patients with heterozygous familial hypercholesterolemia. Atherosclerosis 1990;83:89–97.
112. O'Malley JP, Illingworth DR. The influence of apolipoprotein E phenotype on the response to lovastatin therapy in patients with heterozygous familial hypercholesterolemia. Metabolism 1990;39:150–4.
113. Abey M. The influence of apolipoprotein polymorphism on the response to dietary fat and cholesterol. Curr Opin Lipidol 1992;3:12–16.
114. Boerwinkle E, Brown SA, Rohrbach K, Gotto AMJ, Patsch W. Role of apolipoprotein E and B gene variation in determining response of lipid, lipoprotein, and apolipoprotein levels to increased dietary cholesterol. Am J Hum Genet 1991;49:1145–54.
115. Gaddi A, Ciarrocchi A, Matteucci A, et al. Dietary treatment for familial hypercholesterolemia: differential effects of dietary soy protein according to the apolipoprotein E phenotypes. Am J Clin Nutr 1991;53:1191–6.
116. Strittmatter WJ, Saunders AM, Schmechel D, et al. Apolipoprotein E: high-avidity binding to β-amyloid and increased frequency of type 4 allele in late-onset familial Alzheimer's disease. Proc Natl Acad Sci USA 1993;90:1977–81.
117. Saunders AM, Strittmatter WJ, Schmechel DS, et al. Association of apolipoprotein E allele 4 with late-onset familial and sporadic Alzheimer's disease. Neurology 1993;43:1467–72.
118. Saunders AM, Schmader K, Breitner JCS, et al. Apolipoprotein E ε4 allele distributions in late-onset Alzheimer's disease and in other amyloid-forming diseases. Lancet 1993;342:710–1.

119. Corder EH, Saunders AM, Strittmatter WJ, et al. Gene dose of apolipoprotein E type 4 allele and the risk of Alzheimer's disease in late onset families. Science 1993;261:921–3.
120. Poirier J, Davignon J, Bouthillier D, Kogan S, Bertrand P, Gauthier S. Apolipoprotein E polymorphism and Alzheimer's disease. Lancet 1993;342:697–9.
121. Wisniewski T, Golabek A, Matsubara E, Ghiso J, Frangione B. Apolipoprotein E: binding to soluble Alzheimer's β-amyloid. Biochem Biophys Res Comm 1993;192:359–65.
122. Cauley JA, Eichner JE, Kamboh MI, Ferrell RE, Kuller LH. Apo E allele frequencies in younger (age 42–50) vs. older (age 65–90) women. Genet Epidemiol 1993;10:27–34.

CHAPTER 17

Laboratory Assessment of Lipoprotein Oxidation

Ishwarlal Jialal and Cristina Scaccini

ROLE OF OXIDIZED LDL IN ATHEROSCLEROSIS

❖ Data continues to accrue supporting the hypothesis that the oxidative modification of low-density lipoprotein (LDL) is a key step in the genesis of the atherosclerotic lesion.

Figure 17–1 depicts the role of oxidatively modified LDL (OX-LDL) in foam cell formation. As shown in the figure, mildly oxidized LDL or minimally modified LDL (MM-LDL) is initially formed in the subendothelial space.[1, 2] MM-LDL can induce the endothelium to express adhesion molecules for monocytes and to secrete monocyte chemotactic protein (MCP-1) and macrophage-colony stimulating factor (M-CSF).[3] This sequence of molecular events results in monocytes binding to the endothelium and their subsequent migration into the subendothelial space, where M-CSF promotes differentiation into tissue macrophages. Macrophages can further modify MM-LDL to a more oxidized form through a mechanism involving reactive oxygen species. OX-LDL then can be taken up by the scavenger receptor of monocyte-derived macrophages; this uptake is not regulated by the cellular cholesterol content.[4] The processing of OX-LDL by macrophages results in appreciable cellular accumulation of cholesterol and possible foam cell formation.

LDL can be oxidatively modified by incubation with cells—endothelial cells, smooth muscle cells, macrophages, activated monocytes, and neutrophils—or by incubation with metal ions, copper and iron in particular.[5] Several biological effects ascribed to OX-LDL may contribute to the initiation and progression of the atherosclerotic process.[5, 6] The cytotoxicity of OX-LDL may be important in inducing dysfunction of endothelial cells, and/or in promoting the evolution of the

Figure 17–1 ❖ *The Role of OX-LDL in Foam Cel Formation*

Adapted from: Jialal I, Grundy SM. Influence of antioxidant vitamins on LDL oxidation. Annals NY Acad Sci 1992;669:237–48.

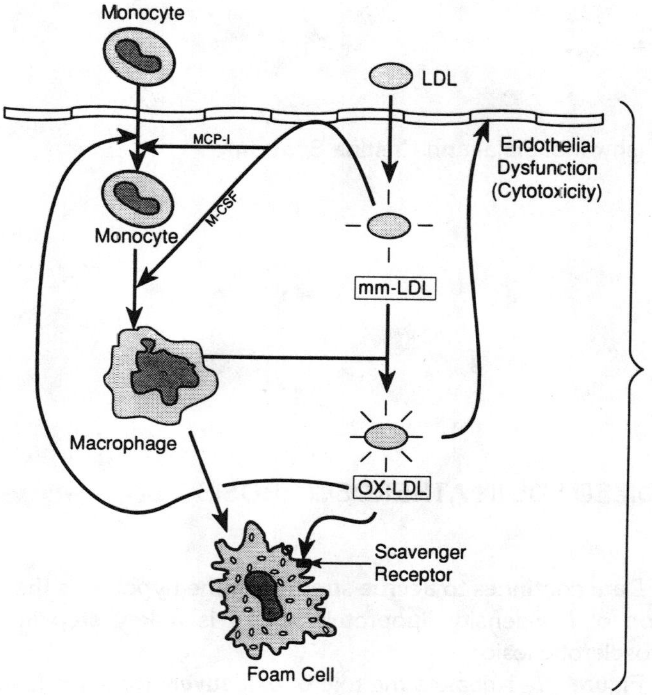

fatty streak to a more complex and advanced lesion which, in part, may be due to cytotoxicity of OX-LDL toward foam cells. OX-LDL is a potent chemo-attractant for the circulating monocytes. Since OX-LDL is recognized by the scavenger receptor, this could lead to cholesterol accumulation and foam cell formation. Because OX-LDL is a potent inhibitor of macrophage motility, it can also promote retention of macrophages in the arterial wall.

Furthermore, several lines of evidence support the *in vivo* existence of OX-LDL.[5, 6] Data has been presented for the occurrence of a modified form of LDL with many physical, chemical, and biological properties of OX-LDL in arterial lesions. Also, antibodies against epitopes on OX-LDL recognized material in atherosclerotic lesions, but not normal arteries, and circulating antibodies against epitopes of OX-LDL have been demonstrated in the plasma of humans. In fact, the titer of these antibodies correlates independently with the progression of atherosclerosis (carotid artery stenosis).[7] Additional support for the role of OX-LDL in atherosclerosis is found in the observation that antioxidants such as probucol, butylated hydroxytoluene, and vitamin E (α-tocopherol) can inhibit the development of atherosclerotic lesions in animal models.[8–10] Also, it was recently shown that the susceptibility of LDL to oxidation (the lag phase of oxidation) correlated with the severity of coronary atherosclerosis evidenced by angiography.[11]

The precise mechanism(s) involved in the oxidative modification of LDL *in vivo* and its exact locality are unclear.

Human LDL is the population of lipoproteins (density range of 1.019–1.063 Kg/L with diameter of 19–25 nm and molecular weight of 1.8–2.8 million Kd. The central lipophilic core consists mainly of cholesteryl esters and triglycerides surrounded by a monolayer of phospholipids and free cholesterol. The solubility in aqueous solutions of LDL is due to the location of the polar head of phospholipids at the surface of the particle. The apolipoprotein (apo) B-100, a large protein with molecular weight of 550 Kd, enfolds the whole surface of the LDL. About half of the fatty acids (FA) bound to the different lipids in LDL are polyunsaturated FA (PUFA), but the content and distribution of the FA can vary widely, depending on dietary habits. The composition of FA modulates the susceptibility of LDL to oxidative modification. In fact, PUFA are much susceptible to oxidation (due to the presence of double bonds) than mono-unsaturated and saturated FA.

PUFA in LDL are protected from attack by free radicals by several antioxidant compounds. On a molar basis, the major antioxidant is vitamin E, which is present in the LDL in a ratio of six molecules per particle.[12] Other antioxidants—γ-tocopherol, carotenoids, and ubiquinol-10–are present in smaller amounts.

EVALUATION OF LDL OXIDATIVE MODIFICATION

A number of problems make the evaluation of oxidative modification of LDL *in vivo* very difficult. First, lipoprotein oxidation is likely to take place in the subendothelial space, outside the circulation. Second, the residence time in plasma of OX-LDL is short because of their affinity and rapid uptake by the scavenger receptor mechanism. Finally, the sensitivity of the methods available makes it difficult to detect the oxidative modification of lipoproteins at very low concentrations.

While it has been suggested that there exists in plasma a more electronegative LDL which is more susceptible to oxidation, this methodology is tedious and is not easily applicable to routine use.[13] Thus, the most practical approach is to attempt to mimic the *in vivo* situation by subjecting LDL to an oxidative stress and monitoring its susceptibility to oxidation.

The LDL are attacked by free radicals in series of steps, but the final results are the peroxidation of lipids and the oxidative modification of the major protein (apo B-100).[12] This latter phenomenon is caused by either the reaction with final products of lipid peroxidation or the direct effect of free radicals. Some of the important properties of OX-LDL are depicted in Table 17–1. The various steps in oxidative modification of LDL can be monitored using different methodologies with varying degrees of complexity.

The initiation by oxygen free radicals able to abstract an H• from a double bond in PUFA is followed by a molecular rearrangement, leading to the formation of conjugated double bonds referred to as conjugated dienes (CD). During this initial phase of lipid peroxidation of LDL, the rate of oxidation is suppressed by the endogenous antioxidants in LDL, resulting in a lag phase preceding the

Table 17–1 ❖ Properties of Oxidized LDL*
● Decreased content of PUFA
● Loss of antioxidants
● Increased negative charge and density
● Increased content of lysolecithin
● Increased cholesterol oxidation products (7-keto-cholesterol)
● Fragmentation of apo B-100
● Increased fluorescence of apo B-100
● Increased content of aldehydes (MDA, HNE, etc.)
● Decreased uptake by LDL receptors
● Increased uptake by scavenger receptors
*LDL = low-density lipoprotein, PUFA = polyunsaturated fatty acids, apo B-100 = apolipoprotein B-100, MDA = malondialdehyde, HNE = hydroxynonenal

rapid propagation of lipid peroxidation. The length of this initial phase is largely determined by the antioxidant content and can be measured easily by spectrophotometry. The propagation phase starts only when antioxidants are depleted and results in the abstraction of $H^{\bullet}$ by PUFA-peroxyl radical ($LOO^{\bullet}$) from another PUFA. This reaction results in the formation of lipid peroxides (LPO). Both CD and hydroperoxides are labile species. A typical kinetic of formation is shown in Figure 17–2. The propagation and degradation phases overlap and the cleavage of carbon bonds during lipid peroxidation results in the formation of aldehydes. The major aldehydes produced include malondialdehyde (MDA), 4-hydroxynonenal (HNE), and hexanal. These can cross-link amino groups on apo B-100. MDA and other aldehydes are measured with the widely used thiobarbituric acid reactive substances method (TBARS). HNE can be measured with high-performance liquid chromatography (HPLC) or with gas chromatography (GC).

Another characteristic of LDL following oxidation is an increase in the negative charge on the particle (measured by electrophoresis on agarose gel), which is possibly due to the derivatization of positively charged amino groups through Schiff's base formation with aldehydes. The derivatization of apo B-100 by reactive aldehydes can also result in an increased fluorescence of OX-LDL. This increased fluorescence can be easily measured at 430 nm, with excitation set at 360 nm. The attack of apo B-100 by reactive oxygen species results in fragmentation.

These changes in the lipid and protein moiety of OX-LDL result in certain newer biological properties such as cytotoxicity and processing by the scavenger receptor mechanism.

Figure 17-2 ❖ *The Peroxidation of Polyunsaturated Lipid*

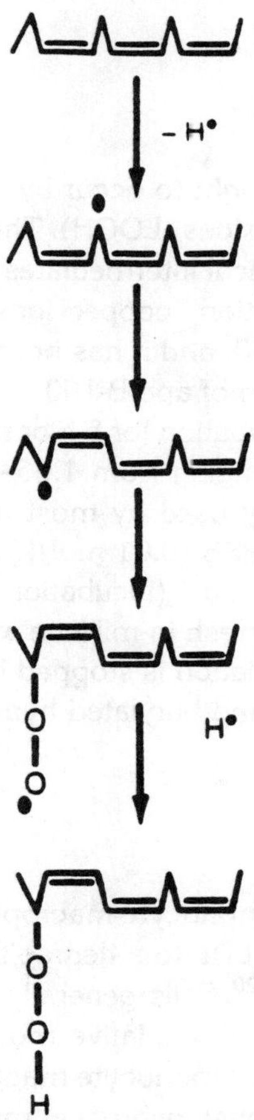

Unsaturated Fatty Acid

Hydrogen abstraction

Molecular rearrangement

Conjugated Diene

Oxygen Update

Peroxy Radical:
abstracts H• from another fatty acid, initiating an autocatalytic chain reaction

Hydroperoxide:
fragments to aldehydes and polymerization products

Adapted from: Sinclair AJ, Barnett AH, Lunec J. Free radicals and antioxidants in health disease. Br J Hosp Med 1990;43:334-44.

METHODS USED TO ASSAY LDL OXIDATION

In plasma, LDL is efficiently protected by the extracellular antioxidant defense system. In the course of isolation, a series of steps must be taken to minimize artifactual lipid peroxidation and maintain the integrity of LDL.

LDL is isolated from human plasma in EDTA (1 g/L) either by sequential ultracentrifugation in NaBr salt solutions as described by Havel,[14] or by rapid vertical spin gradient ultracentrifugation.[15] The isolated LDL is extensively dialyzed against a solution containing 150 mmol/L NaCl and 1 mmol/L EDTA (pH 7.4), filtered and stored at 4° C, under N_2. LDL is used within 7–10 d from the time of isolation. The EDTA is present in excess to chelate transition metals and thus prevent LDL oxidation during isolation. Protein content is measured by the method of Lowry et al.[16] using bovine serum albumin as a standard.

To eliminate EDTA before oxidation experiments, LDL is dialyzed in the dark for 24 h at 4° C against two changes (1000-fold volume) of 0.01 mol/L phosphate-buffered saline (PBS) at pH 7.4.

Cell-Free System of LDL Oxidation

The oxidative modification of LDL by copper is thought to occur by the reaction of Cu^{++} with preexisting endogenous lipid hydroperoxides (LOOH). The decomposition of LPO results in the production of several radical intermediates of lipid peroxidation. Besides the initiation of lipid peroxidation, copper ions can form complexes with the ligand-binding sites of apo B-100, and it has been suggested that this binding is linked with structural modification of apo B-100.

The procedure for preparing OX-LDL is an incubation for 5 h or more in PBS supplemented with Cu^{++} ions at a concentration ranging from 1.66–10 µmol/L. We describe here the standard procedure generally used by most laboratories. LDL (200 µg protein/mL) is incubated at 37° C in PBS (0.01 mol/L) for a 6–8 h time-course experiment in the presence of 5 µmol/L Cu^{++} (incubation volume 1.0 mL).[17] A stock solution of 1 mmol/L Cu^{++} is made fresh in millicue water and diluted to give a final concentration of 5 µmol/L. Oxidation is stopped by refrigeration and addition of 0.01 mL EDTA (200 µmol/L) and butylated hydroxytoluene (40 µmol/L).

Cellular System of LDL Oxidation

A number of cells, including endothelial cells, monocyte-macrophages, and smooth muscle cells are able to oxidatively modify LDL to a degree that permits uptake by scavenger receptor of macrophages.[18-20] Cells generally require the presence of metal ions in the medium to induce LDL oxidative modification. A procedure for oxidatively modifying LDL by human monocyte-macrophages is described below.[21] It should be pointed out that mouse peritoneal macrophages or human macrophages can be used.

Human monocytes are isolated by density-gradient centrifugation from blood derived from fasting normolipidemic subjects. 20 mL of blood (anticoagulated with 10 IU/mL heparin) is layered over 15 mL of Ficoll-Paque and centrifuged at 1500 rpm for 30 min at room temperature. The mononuclear cell band is removed by aspiration. Cells are washed twice in RPMI-1640 media (GIBCO, BRL, Grand Island, New York), supplemented with 100 U/mL penicillin, 100 µg/mL streptomycin, and 2 mmol/L L-glutamine. Cells are diluted at the concentration of 5×10^6 cell/mL and 1 mL of the cell suspension is plated in a 35 mm dish in the same medium. After a 2 h incubation at 37° C in 5% CO_2/95% air, nonadherent cells are removed by three washes with serum-free medium. Finally, cells are incubated in RPMI 1640 containing 20% autologous serum and the cells are fed twice a week with the same medium. Monocyte-derived macrophages are used within 10–14 d of plating. Before using, cells are washed three times with serum-free RPMI 1640 medium. For oxidation experiments, LDL (100 µg protein/mL) is incubated with monocyte-derived macrophages in Ham's F-10 medium for 24 h at

37° C in 5% CO_2/95% air (incubation volume 1.0 mL). Cell-free controls (LDL in Hams F-10 and RPMI-1640 media) are also undertaken. After incubation, medium is aspirated into tubes containing 200 μmol/L EDTA and 40 μmol/L BHT. The LDL solution is centrifuged at 1000 × g for 10 min and the supernatant is collected.

INDICES USED TO MONITOR LDL OXIDATION

Conjugated Dienes

Oxidation of PUFA side-chains is accompanied by formation of CD that absorb ultraviolet light in the wavelength range 230–235 nm (maximum at 234 nm). Measurement of this UV absorbance is a useful index of lipid peroxidation.[12] The CD peak after 3–4 h of incubation, depending on the incubation conditions. The second increase of absorbance does not correspond to newly formed CD but is due to the appearance of degradation products absorbing at the same wavelength. Since LDL is fully soluble in buffer, the generation of CD during LDL oxidation can be directly and continuously measured by recording the absorbance at 234 nm. This procedure has been extensively described by Esterbauer et al.[22] The authors follow the kinetics of CD formation by incubating LDL (50 μg protein/mL) at 37° C in PBS with 1.66 μmol/L Cu^{++}. Usually the instrument is blanked with native LDL, the Cu^{++} is added, and the time course of oxidation is monitored. If other indices of LDL oxidation are also desired, then the reaction is stopped at selected times, an aliquot is diluted, and its absorbance is measured at 234 nm. CD formation generally cannot be used to assess cell-mediated LDL oxidation since ingredients in media can interfere with the assay.

Spectrophotometric Assay for Lipid Peroxides

This is a quick and easy method for measuring lipid peroxides in lipoprotein using a commercially available test kit for cholesterol determination. The method is based on the oxidation of iodide to iodine by lipid peroxides.[23] The problem with this assay is that its lower detection limit is similar to the reagent blank. Its advantage is the rapid determination of the total amount of lipid peroxides in LDL following LDL oxidation (e.g., 5 μmol/L Cu^{++}). LDL (100 μL) is mixed on a vortex with 1 mL of the color reagent. (The constituents of the color reagent are detailed by El-Saadani et al.[23]) The sample is kept in the dark at room temperature for 30–60 min. Absorbance is measured at 365 nm against the reagent blank. The concentration of lipid peroxides is calculated using the molar absorption of iodide at 365 nm ($\varepsilon = 2.46 \times 10^4 mol/L^{-1}$).[23]

Gas chromatography–mass spectrometry (GC/MS) methods are available to characterize single classes of lipid hydroperoxides (i.e., phosphatidylcholine hydroperoxides), but these methods are cumbersome and are not suitable for routine use.[24] A very sensitive and specific method for measuring total lipid hydroperoxides which utilizes HPLC with chemiluminescence detection is also available.[25] However, these methods require additional instrumentation which is not always available to the average laboratory or to most investigators.

TBARS Assay

The thiobarbituric acid assay (TBA) is the most popular and easiest method used as an indicator of MDA formation during oxidation in biological samples. However, this test is not specific for MDA; in fact, many substances not related to lipid peroxidation—sugars, amino acids, etc.—may also form TBA adducts. Also, peroxides are formed during the heating phase of the assay. Thus, since this test does not only measure MDA formed in the peroxidation system, it is more appropriately referred to as TBA reactive substances, or TBARS. The chromogen is formed by reaction of one molecule of MDA with two molecules of TBA. Basically, the sample is heated with TBA under acidic conditions and the amount of chromogen formed (MDA-TBA adduct) is measured at 532 nm. The absorbance is expressed in MDA equivalents/mg of LDL protein, using a standard curve for 1,1,3,3 tetramethoxypropane. A number of variations of the original TBA assay, differing in sample pretreatment, acid concentration, heating time, absence or presence of antioxidants, have been described.[26]

A spectrophotometric test (adapted for LDL samples) and a fluorometric test are described below.

Spectrophotometric Test

This method is a modification of the original procedure and includes the precipitation of protein (which interferes with the assay) using trichloracetic acid (TCA).[27] The assay is conducted in the presence of BHT and EDTA to minimize the formation of TBARS activity during heating.[28]

LDL (500 µL, equivalent 100 µg and 50 µg LDL protein for copper-induced and cell-mediated oxidation, respectively) is combined with 1 mL of the TCA-TBA-HCl mixture (15% w/v TCA, 0.375% TBA, 25 mol/L HCl) and vortexed. The solution is heated for 20 min in a water bath at 100° C to precipitate protein. After cooling, samples are centrifuged at 2000 rpm for 20 min. The supernatant is carefully withdrawn and absorbance is determined at 532 nm against a blank containing all the reagents except LDL.

Fluorimetric Test

In this procedure, protein is precipitated using phosphotungstic acid prior to carrying out the TBARS reaction.[29] Butylated hydroxytoluene is added in the TBA reagent to prevent formation of new peroxides during the assay. For increased sensitivity, after reaction with TBA, the complex is extracted into butanol.

To 50 µL of LDL solution is added 0.5 mL phosphotungstic acid (10%) and 3.5 mL of water. Then 1 mL of TBA reagent (0.01% BHT in a mixture of equal volumes of 0.67% TBA and glacial acetic acid) is added to the sample. The solution is heated for 60 min in water bath at 95° C. After cooling, the sample is combined with 5 mL of n-butanol and vortexed. The solution is centrifuged at 3000 rpm for 5 min. Fluorescence of the butanol layer is measured at 535 nm with 515 nm excitation against a blank containing all the reagents except LDL.

Relative Electrophoretic Mobility (REM)

This assay is based on the modification of LDL due to the derivatization of lysine residues of apo B-100 by some reactive aldehydes (end products of lipid peroxidation) which lead to the increase in the negative charge of OX-LDL.

Commercially available reagents for electrophoretic separation of lipoproteins can be used. Electrophoresis is performed at pH 8.6 in 0.05 mol/L barbital buffer on 0.5% agarose gel. LDL (2–10 μg) is applied in each slot of a prepared agarose gel. After 5 min the gel is placed into the electrophoresis cell in barbital buffer and electrophoretic separation is performed for 30 min at 100 volts. The gel is then fixed in 1:3:6 solution of glacial acetic acid, water, and methanol and dried. The dried gel is placed for 10–15 min in the stain solution (Sudan Black B, 7%). After destaining in 45% methanol, the gel is rinsed with water and dried. The change in electrophoretic mobility is expressed relative to the mobility of native LDL. Figure 17-3 depicts the REM during a typical time-course experiment of copper-catalyzed LDL oxidation.

Apo B-100 Fluorescence

The same process that alters the electrophoretic mobility of OX-LDL also generates fluorophors with a very strong fluorescence at 430 nm with excitation at 360 nm. The fluorescence has been attributed to the derivatization of apo B-100 lysine residues by reactive aldehydes. A rapid method has recently been proposed in which the lipid extraction phase is omitted. This quick and simplified assay is useful to monitor apo B modification in time-course experiments,[30] as shown in Figure 17-3.

LDL solution is diluted to a concentration of 25 μg protein/mL in PBS. The emission is measured at 430 nm with excitation at 360 nm against a PBS blank. The fluorescence is corrected for protein concentration.

Loss of Endogenous Antioxidants and Fatty Acids

Measurements of the endogenous antioxidants and FA in LDL are important to have a better insight into the mechanism of LDL oxidative modification; however, they are difficult and need more complex instrumentation than the methods discussed above. The kinetic of disappearance of endogenous antioxidant during copper-catalyzed LDL oxidation can be monitored. As reported by Esterbauer et al.,[31] α-tocopherol is consumed first, followed by carotenoids. Measurement of α-tocopherol can be performed by HPLC with a reversed-phase column together with retinol.[32] Retinol and five carotenoids (lutein cryptoxanthin, lycopene, α- and β-carotene) can be measured simultaneously by HPLC (reversed-phase column) with an isocratic solvent system.[33]

About half of FA in LDL are PUFA, mainly linoleic acid with minor amounts of arachidonic and docosahexaenoic acid. There is a high inter-individual variation in LDL FA composition, probably due to dietary habits. Such variation in PUFA content can lead to changes in the resistance of LDL toward oxidative modification. Usually the decrease in linoleic acid, arachidonic acid, and docosa-

Figure 17-3 ❖ A Typical Time Course of LDL Oxidation

LDL (200 µg protein/mL) was incubated with 5 µmol/L Cu^{++} in PBS for 8 h.

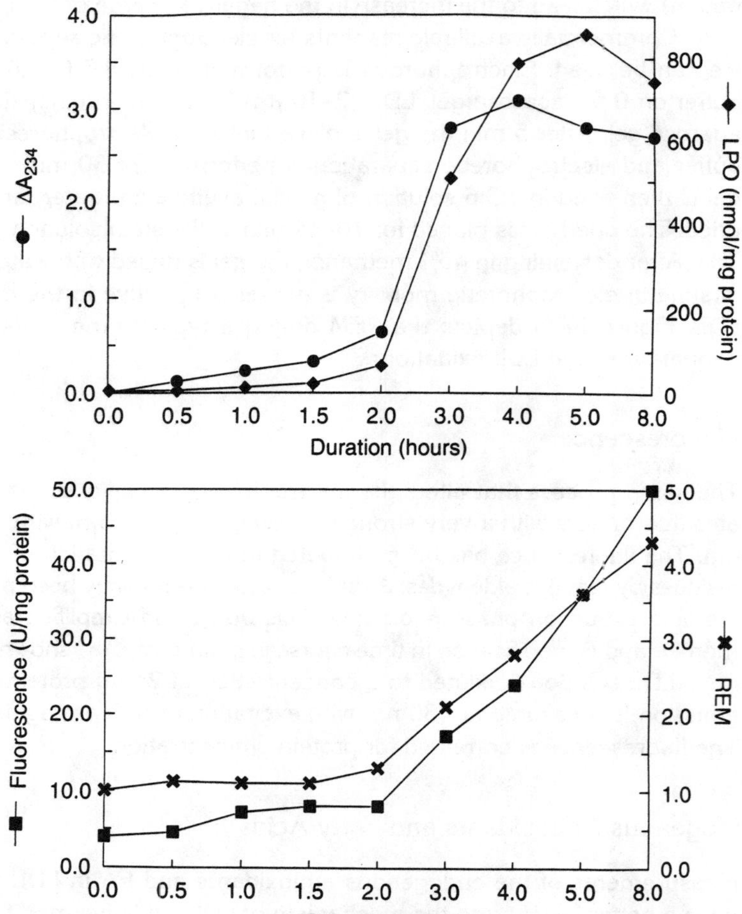

hexaenoic acid is monitored. The FA content of LDL is determined by GC following extraction and transmethylation.[34]

Fragmentation of APO B-100

Cell-induced and metal-catalyzed oxidative modification of LDL is accompanied by extensive breakdown of the major LDL protein (apo B-100) to small peptides.[35] The non-enzymatic oxidative cleavage of peptide bonds in apo B-100 leads to fragmentation, which is detectable by sodium dodecyl sulphate (SDS) polyacrylamide gradient gel electrophoresis.[36] Briefly, 20–30 µg of LDL protein is dissolved (1:3 v/v) with sample buffer containing 3% SDS, 10% glycerol, and 5% 2-mercaptoethanol in 0.5 mol/L Tris HCl at pH 6.8. The sample is incubated in boiling water for 5 min. Vertical gel electrophoresis is performed using 3–14% polyacrylamide gradient gel. The gel is fixed in 50% TCA and stained with Coomassie Blue R-250. Following LDL oxidation, the band of intact apo B-100 disappears with the formation of fragments.[35]

Oxysterols

It is also possible to evaluate the degree of modification of LDL by measuring of oxidation products of cholesterol. Several oxysterols have been reported following LDL oxidation. However, 7-ketocholesterol has been identified by most laboratories as the main oxysterol produced during Cu^{++} catalyzed and cell-mediated LDL oxidative modification.[28, 37] Since GC and GC-MS methods are used, the availability of instrumentation confines oxysterol measurements to research laboratories.

Assays for the Biological Effects of OX-LDL

As discussed at the beginning of this chapter, OX-LDL has numerous biological effects. For the sake of brevity, only the two most common biological assays, ^{125}I-OX-LDL degradation by macrophages and cytotoxicity, are discussed in this review.

Degradation by Macrophages

When incubated at 37° C with macrophages, LDL is rapidly internalized and then delivered to lysosomes, where protein and cholesteryl esters are hydrolyzed. To measure ^{125}I-LDL degradation by scavenger receptor on macrophages, ^{125}I-LDL is incubated with macrophages, usually for 5 h at 37° C in RPMI-1640 medium. ^{125}I-labeled LDL can be prepared by a modification of the iodine monochloride method.[38] The appearance of TCA-soluble radioactivity (non-iodide) formed by the cells and excreted in the media is determined by using a series of additions including albumin, TCA, H_2O_2, potassium iodide, and finally chloroform.[39] Degradation rates are corrected for cell-free controls incubated in parallel. Results are expressed as amount of protein degraded per mg cell protein.

Cytotoxicity

OX-LDL is a cytotoxin. The cytotoxicity of OX-LDL can be evaluated by incubating with cells (endothelial cells, embryonal fibroblasts, etc.) and monitoring the release of labeled chromium and lactate dehydrogenase, or determining cell numbers or hexosaminidase activity of remaining target cells.[40, 41]

Other Approaches for Assessing Lipid Peroxidation

These measures are not specific for lipoprotein oxidation and cannot delineate the tissue of origin of lipid peroxidation.

Breakdown products of lipid hydroperoxides can be measured in exhaled breath as volatile hydrocarbons. This is a very sensitive yet noninvasive measure of lipid peroxidation in man. Ethane and Pentane derived from hydroperoxides of W-3 and W-6 PUFA respectively can be quantitated by GC.[42, 43] It is essential to purify the inspired air because of the risk of hydrocarbon contamination arising from exogenous sources (motor vehicles, cigarette smoke, etc.) which create a high background. Thus, although this method is noninvasive, it is tedious and technically demanding because of its numerous pitfalls.[43] Its role in studying lipid peroxidation clearly needs to be established.

The total peroxyl radical trapping antioxidant parameter (TRAP) assay can be used as a measure of the antioxidant status of plasma.[44] Assessment of the relative contribution of the specific antioxidants (ascorbate, urate, protein sulphydryls, and α-tocopherol) requires that each be measured individually. Thus, while the TRAP assay may provide an index of antioxidant status, it only conveys part of the picture since the antioxidant deficiency is not defined. In fact, measurement of individual antioxidants such as ascorbate, α-tocopherol, and β-carotene in plasma or LDL may, in their own right, provide very useful information with respect to antioxidant status. If the oxidation hypothesis is proved, these antioxidants assays could easily be incorporated into the screening and management repertoire for atherosclerosis.

CONCLUSION

The methods that are presently available to measure LDL oxidation were reviewed in this chapter. It is clear that no single method conveys the entire picture, and that each has its limitations. Thus, there is an urgent need for newer techniques that afford increased sensitivity and specificity. While the focus of this review has been on LDL oxidation, other lipoproteins also undergo lipid peroxidation. In fact, it has been shown that HDL can be oxidized, and that this impairs its ability to mediate cholesterol efflux.[45] It would appear that for the clinical laboratory, the ideal will be a standardized model system of plasma oxidation that parallels lipoprotein oxidation, yet is simple and rapid and thus available for routine use. Much further work is needed before measurement of lipoprotein oxidation becomes a routine clinical chemistry test. ❖

REFERENCES

1. Cushing S, Berliner J, Valente A, et al. Minimally modified LDL induces monocyte chemotactic protein I in human endothelial cells and smooth muscle cells. Proc Natl Acad Sci USA 1990;87:5134–8.
2. Berliner J, Territo M, Sevanian A, et al. Minimally modified LDL stimulates monocyte endothelial interaction. J Clin Invest 1990;85:1260–6.
3. Rajavashisth TB, Andalibi A, Territo MC, et al. Modified low density lipoproteins induce endothelial cell expression on granulocyte and macrophage colony stimulating factors. Nature 1990;344:254.

4. Sparrow CP, Parthasarathy S, Steinberg D. A macrophage receptor that recognizes oxidized low density lipoprotein but not acetylated low density lipoprotein. J Biol Chem 1989;264:2599–604.
5. Witztum JL, Steinberg D. Role of oxidized LDL in atherogenesis. J Clin Invest 1991;88:1785–92.
6. Steinberg D, Parthasarathy S, Carew TE, Khoo JC, Witztum JL. Beyond cholesterol: modifications of low density lipoprotein that increase its atherogenicity. N Eng J Med 1989;320:915–24.
7. Salonen JT, Yla-Herttuala S, Yamamoto R, et al. Autoantibodies against oxidized LDL and progression of carotid atherosclerosis. Lancet 1992;339:883–7.
8. Kita T, Nagano Y, Yokode M, et al. Probucol prevents the progression of atherosclerosis in Watanabe heritable hyperlipidemic rabbit, an animal model for familial hypercholesterolemia. Proc Natl Acad Sci USA 1987;5928–31.
9. Björkhem I, Henriksson-Freyschuss A, Breuer O, Dicfalusy U, Berglund L, Henriksson P. The antioxidant butylated hydroxytoluene protects against atherosclerosis. Arteriosclerosis and Thrombosis 1991;11:15–22.
10. Verlangeri AJ, Bush M. Effects of d-alpha-tocopherol supplementation on experimentally induced primate atherosclerosis. J Am Coll Nutr 1992;11:131–8.
11. Regnstrom J, Nilsson J, Tornvall P, Landou C, Hamsten A. Susceptibility to low-density lipoprotein oxidation and coronary atherosclerosis in man. Lancet 1992;339:1183–6.
12. Esterbauer H, Gebicki J, Puhl H, Jürgens G. The role of lipid peroxidation and antioxidants in oxidative modification of LDL. Free Radic Biol Med 1992;13:341–90.
13. Cazzolato G, Avogaro P, Bittolo-Bon G. Characterization of a more electronegatively charged LDL subfraction by ion exchange HPLC. Free Rad Biol Med 1991;11:247–53.
14. Havel RJ, Eder HA, Bragdon JH. The distribution and chemical composition of ultracentrifugally separated lipoproteins in human serum. J Clin Invest 1955;34:1345–53.
15. Chung BH, Wilkinson T, Geer JC, Segrest JP. Preparative and quantitative isolation of plasma lipoproteins: rapid single discontinuous density gradient ultracentrifugation in a vertical rotor. J Lipid Res 1980;21:284–91.
16. Lowry OH, Rosebrough NJ, Farr AL, Randall RJ. Protein measurements with the Folin phenol reagent. J Biol Chem 1951;193:265–75.
17. Jialal I, Grundy SM. Effect of dietary supplementation with alpha-tocopherol on the oxidative modification of low density lipoprotein. J Lipid Res 1992;33:899–906.
18. Henriksen T, Mahoney EM, Steinberg D. Enhanced macrophages degradation of low density lipoprotein previously incubated with cultured endothelial cells: recognition by receptor of acetylated low density lipoprotein. Proc Natl Acad Sci USA 1981;78:6499–503.
19. Parthasarathy S, Printz DJ, Boyd D, Joy L, Steinberg D. Macrophage oxidation of low density lipoprotein generates a modified form recognized by the scavenger receptor. Arteriosclerosis 1986;6:505–10.
20. Heinecke JW, Rosen H, Chait A. Iron and copper promote modification of low density lipoprotein by human arterial smooth muscle cells in culture. J Clin Invest 1984;84:1890–94.
21. Jialal I, Grundy SM. Preservation of the endogenous antioxidants in low density lipoprotein by ascorbate but not probucol during oxidative modification. J Clin Invest 1991;87:597–601.
22. Esterbauer H, Striegl G, Puhl H, Rotheneder M. Continuous monitoring of in vitro oxidation of human low density lipoprotein. Free Rad Res Comms 1989;6:67–75.

23. El-Saadani M, Esterbauer H, El-Sayed M, Goher M, Nasser AY, Jürgens G. A spectrophotometric assay for lipid peroxides in serum lipoproteins using a commercially available reagent. J Lipid Res 1989;30:627–30.
24. Wang T, Yu W, Powell WS. Formation of monohydroxy derivative of arachidonic acid, linoleic acid, and oleic acid during oxidation of low density lipoprotein by copper ions and endothelial cells. J Lipid Res 1992;33:525–37.
25. Frei B, Yamamoto Y, Niclas D, Ames B. Evaluation of an isoluminol chemiluminescence assay for the detection of hydroperoxides in human blood plasma. Anal Biochem 1988;175:120–30.
26. Janero DR. Malondialdehyde and thiobarbituric acid-reactivity as diagnostic indices of lipid peroxidation and peroxidative tissue injury. Free Rad Biol Med 1990;9:515–40.
27. Buege JA, Aust SD. Microsomal lipid peroxidation. Methods Enzmol 1978;52: 302–10.
28. Jialal I, Freeman D, Grundy SM. Varying susceptibility of different LDLs to oxidative modification. Arteriosclerosis and Thrombosis 1991;11:482–8.
29. Maseki M, Nishigaki I, Hagihara M, Tomoda Y, Yagi K. Lipid peroxide levels and lipid content of serum lipoprotein fractions of pregnant subjects with and without pre-eclampsia. Clin Chim Acta 1981;115:155–61.
30. Cominacini L, Garbin U, Davoli A, et al. A simple test for predisposition to LDL oxidation based on the fluorescence development during copper-catalyzed oxidative modification. J Lipid Res 1991;32:349–58.
31. Esterbauer H, Puhl H, Waeg G, Krebs A, Dieber-Rotheneder M. The role of vitamin E in lipoprotein oxidation. In Packer L, Fuchs J, eds. Vitamin E: biochemistry and clinical application. New York: Marcel Dekker, 1992:649–71.
32. Bieri JG, Talliver GT, Catignani LG. Simultaneous determination of alpha-tocopherol and retinol in plasma or red cells by high pressure liquid chromatography. Am J Clin Nutr 1979;32:2143–9.
33. Stacewicz-Sapuntzakis M, Bowen PE, Kikendal JQW, Burgess M. Simultaneous determination of serum retinol and various carotenoids: their distribution in middle-aged men and women. J Micronut Anal 1987;3:27–45.
34. Esterbauer H, Jürgens G, Quehenberger O, Koller E. Autoxidation of human low density lipoprotein: loss of polyunsaturated fatty acids and vitamin E and generation of aldehydes. J Lipid Res 1987;28:495–509.
35. Fong LG, Parthasarathy S, Witztum JL, Steinberg D. Nonenzymatic oxidative cleavage of peptide bonds in apoprotein B-100. J Lipid Res 1987;28:1466–77.
36. Laemmli UK. Cleavage of structural proteins during the assembly of the head of bacteriophage T4. Nature 1970;227:680–85.
37. Zhang H, Basra HJK, Steinbrecher U. Effects of oxidatively modified LDL on cholesterol esterification in cultured macrophages. J Lipid Res 1990;31:1361–9.
38. Grundy SM, Vega GL. Influence of mevinolin on metabolism of low sensitivity lipoproteins in primary moderate hypercholesterolemia. J Lipid Res 1985;26: 1464–70.
39. Goldstein JL, Basu SK, Brown MS. Receptor-mediated endocytosis of low density lipoprotein in cultured cells. Meth Enzymol 1983;98:241–60.
40. Morel DW, Hessler JR, Chisolm GM. Low density lipoprotein cytotoxicity induced by free radical peroxidation of lipid. J Lipid Res 1983;24:1070–76.
41. Cathcart MK, McNally AK, Morel DW, Chisolm III, GM. Superoxide anion participation in human monocyte-mediated oxidation of low-density lipoprotein and conversion of low-density lipoprotein to a cytotoxin. J Immunol 1989;142(6): 1963–9.
42. Jeejeebhoy KN. In vivo breath alkane as an index of lipid peroxidation. Free Rad Biol Med 1991;10:191–3.

43. Kneepkens CM, Ferreira C, Lepage G, Roy CC. The hydrocarbon breath test in the study of lipid peroxidation: principles and practice. Clin Invest Med 1991; 15:163–86.

44. Wayner DDM, Burton GW, Ingold KU, Barklay LRG, Locke SJ. The relative contribution of vitamin E, urate, ascorbate, and proteins to the total peroxyl radical-trapping antioxidant activity of human blood plasma. Biochem Biophys Acta 1987;984:408.

45. Nagano Y, Arai H, Kita T. High density lipoprotein loses its effect to stimulate efflux of cholesterol from foam cells after oxidative modification. Proc Natl Acad Sci USA 1991;88:6457–61.

CHAPTER 18

Alternative Approaches to Lipoprotein Analysis

Pierre N. M. Demacker, James D. Otvos, Gerd Schmitz, Christoph Möllers, Paola Lombardi, Ben Cost, and Louis M. Havekes

INTRODUCTION

❖ To those who have followed the history of research into the role of lipoproteins in the development of atherosclerosis and coronary heart disease (CHD), it is clear that the availability of a new technique always promotes new fundamental and clinical studies directed toward the relevance of newly developed parameters based on that technique. Results of these studies, in turn, promote the development of new techniques that are simpler or more informative. Thus, there is a fruitful interaction between the development of a new technique and its application in clinical and fundamental studies.

A useful example is the progress in the diagnosis of Type III hyperlipoproteinemia. Over a period of 25 years, the sequential development of a spectrum of different techniques, such as electrophoresis, ultracentrifugation, isoelectric focusing, and, finally, molecular biological techniques, made it possible to improve the sensitivity and specificity of the diagnosis. This fruitful interaction between the development of diagnostic methods and their application led to an enormous increase in the fundamental knowledge of lipoprotein metabolism and ultimately to a definitive diagnostic test for Type III hyperlipoproteinemia.

In the previous chapters, the clinical significance of various lipoprotein or apolipoprotein parameters have been discussed by a number of experts. The clinical significance of these parameters became clear after case-control or epidemiological studies. For the more experienced lipidologist, it is clear that many of the parameters discussed are metabolically interrelated. As such, deviating concentrations are simply the multiple expression of a metabolic error that is directly associated with the risk for atherosclerosis. Simply speaking, there are only two metabolic processes that determine excessive atherosclerosis as far as the lipoprotein metabolism is concerned: the infiltration rate of cholesterol-rich lipoproteins into the blood vessel wall, and the removal rate of cholesterol from the vessel wall with subsequent excretion from the body. In this balanced system, increased concentrations of chylomicron remnants and very-low-density lipoprotein (VLDL) remnants, low-density lipoprotein (LDL), and lipoprotein(a) [Lp(a)], as well as heavy LDL and increased oxidizability of LDL, are predictors of excessive deposition of cholesterol into the vessel wall. One the other hand, increased concentrations of high-densitity lipoproteins (HDL) and HDL_2 promote increased removal of cholesterol from atherosclerotic plaques. Because of this abundance of possible risk factors, there is a need for simple, rapid, and inexpensive methods that allow quantitative information on a maximum number of parameters in a single run.

From the fundamental research we have selected a number of methods that meet these requirements. In this chapter, the possibilities of these potentially important methods are outlined briefly by pioneering scientists who evaluated the techniques in their laboratories. It is hoped that wider application of these methods ultimately will result in a more accurate delineation of lipoprotein parameters associated with atherosclerosis. Because of the diagnostic importance of molecular biological methods to identify molecular defects in lipoprotein metabolism associated with atherosclerosis, the potential of these methods is also reviewed. Clinical chemists and their co-workers are invited to explore the possibilities of these innovative techniques in case control and in epidemiological studies.

SINGLE SPIN DENSITY GRADIENT ULTRACENTRIFUGATION, GRADIENT GEL ELECTROPHORESIS, AND DENATURING SDS-PAGE

Pierre N. M. Demacker

Single Spin Density Gradient Ultracentrifugation in a Fixed Angle Rotor

Lipoproteins are commonly prepared by sequential ultracentrifugation[1] or density gradient ultracentrifugation[2, 3], which requires 20–36 h of ultracentrifugation. Both lipoprotein isolation procedures, especially when using high salt concentrations, may result in partial denaturation of the lipoproteins with loss of some apolipoproteins, especially apolipoprotein (apo) AI.[4, 5] For future studies, there is a need for lipoprotein isolation methods that are nondestructive and therefore result in more native particles. One potentially suitable method is vertical spin density gradient ultracentrifugation which, using a discontinuous density gradient, is able to almost completely separate LDL, VLDL, and HDL.[6] Results are comparable to those obtained using a density gradient in swinging bucket rotors.

Depending on the purpose, several rotor types are available in which 4.5–92 mL of total plasma can be fractionated over a range of 25–330 min of ultracentrifugation; 13 separate vertical spin procedures were described.[6] After ultracentrifugation, the gradients are analyzed by downward fractionation with on-line cholesterol determination.[6]

For preparative purposes, sequential aspiration of small fractions with a simple Pasteur pipette is also possible, and probably results in less cross-contamination. After fractionation of the gradient, characteristic patterns can be achieved for plasma of patients with Type IIa, Type III, Type IV, and Type V hyperlipoproteinemia. Also, Lp(a), when present in above-normal concentrations, can be detected as a separate peak between the LDL and HDL. Using this procedure, the separation of the main lipoproteins shows some overlap; in particular, the separation between HDL and the serum proteins is incomplete, as can be concluded from sodium dodecylsulphate-polyacrylamide gradient gel electrophoresis (SDS-PAGE) of the isolated fractions.[6] However, in this way, lipoproteins, especially certain HDL subfractions with an increased albumin binding, can be identified without the risk of excessive degradation of an eventually present albumin-HDL complex. Also, studies of the real antioxidant capacity of a given LDL are possible.[7, 8]

In our laboratory we use an adaptation of the single spin density ultracentrifugation procedure for the isolation of LDL. The vertical rotor is exchanged by a fixed angle rotor with a capacity of 40 tubes. The LDL preparation isolated in this way is more native than when isolated with sequential ultracentrifugation.[8] Results obtained for VLDL plus intermediate-density lipoprotein cholesterol (IDL-C), LDL cholesterol (LDL-C), and HDL cholesterol (HDL-C) agreed well with the values determined by sequential ultracentrifugation ($r > 0.92$ for all three parameters), except in plasma from subjects with Type III hyperlipoproteinemia, which could be identified by their increased content of VLDL-remnants in the top of the density gradient.

Non-denaturing Gradient Gel Electrophoresis for the Analysis of Native Lipoproteins and their Subfractions

Another method for analyzing lipoproteins under non-denaturing conditions makes use of 2–16% PAGE. With these gels, which are commercially available (Isolab, Akron, OH), it is possible to analyze subfractions of LDL and HDL by electrophoresis of plasma followed by lipid staining, as well as the various proteins in plasma, whether or not complexed with lipids.

The electrophoresis is performed according to the directions of the manufacturer.[9-11] (See also Chapters 12 and 13.) Depending on the purpose, in our laboratory we use variable electrophoresis conditions. We first employ a standard pre-electrophoretic run for 20 min at 70 volts. For identification of LDL or HDL-heterogeneity, we apply 10 µL of sample containing plasma enriched with bromphenol blue and glycerol, electrophorese for 30 min at 70 volts and 330 min at 400 volts, followed by staining with Sudan Black.[10] For analysis of the apolipoprotein composition of lipoproteins, we apply 2–10 µL of the isolated lipoprotein fractions or 10 µL of tenfold diluted plasma enriched with bromphenol blue and glycerol, and electrophorese for 30 min at 70 volts and subsequently at 150 volts until the bromphenol blue stain has reached the basis of the gel. The protein bands are stained with Coomassie Brilliant Blue R using standard conditions (discussed below). If desired, the lipoproteins can be transferred quantitatively to nitrocellulose membranes by electrophoresis preceding immunological detection by an enzyme linked immunoabsorbent assay (ELISA) technique.[9]

The possibilities of this method for the identification of a complex present in the very-high-density fraction of a patient with hyperalphalipoproteinemia are shown in Figure 18–1.

Denaturing SDS-PAGE for the Analysis of Apolipoproteins

Clinical laboratories involved in lipoprotein research always need a rapid and sensitive method that gives insight into the apolipoprotein content and composition of a given lipoprotein fraction. For this purpose, in our laboratory we use the conventional denaturing SDS electrophoresis method, a standard method in biochemistry laboratories.[12] This method is more rapid and at least as informative as a number of immunological methods for the analysis of apolipoproteins apo AI, apo AII, apo B-100 and apo C, all of which all require specific, antibody-based as-

Figure 18–1 ❖ Continuous 2–16% Non-denaturing Polyacrylamide Gradient Gel Electrophoresis of Native Lipoproteins

The following samples were applied: (1) whole plasma of a normal subject, (2) whole plasma with hyperalphalipoproteinemia with an albumin-apo AI complex, (3–4) the respective VLDL fractions of both plasma samples, (5-6) the respective LDL samples, (7, 12) HDL_2 of plasma 1, (8) HDL_2 of plasma 2, (9) HDL_3 of plasma 1, (10) HDL_3 of plasma 2, (11) heavy HDL fraction of plasma 2.

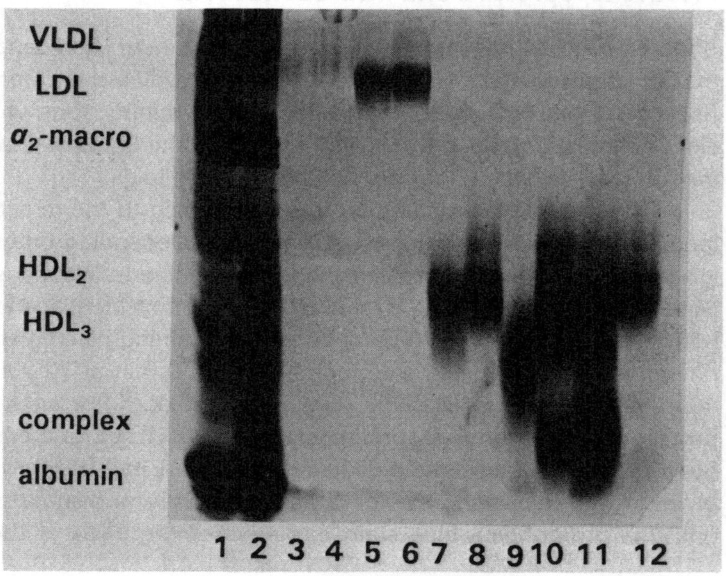

says. Using special denaturing conditions and a rather low acrylamide concentration in the top of the gel, this method also allows semi-quantitative estimation of the apo(a), apo B-48, and apo B-100, which carry Lp(a), chylomicrons, and VLDL+LDL, respectively. Apo(a), apo B-48, and apo B-100 are very large proteins which become insoluble after delipidation with organic solvents. By boiling the samples for a short time preceding the electrophoresis, these proteins form complexes with the dodecylsulphate molecules. Therefore, they remain in solution when the complexes penetrate the large pores of the low component polyacrylamide gel (PAG), while the lipids are lost. By applying similar volumes instead of similar protein loads, it is possible, within certain limits, to compare apolipoprotein compositions and concentrations to a control fraction. Because of this necessary quantitative delipidation procedure, as well as the rather large sample volume and protein content applied to the gel, it is difficult to obtain satisfactory results with commercial gels. Samples can be analyzed without prior desalting. The procedure is as follows:

The electrophoresis unit is Hoefer SE 500 dual vertical slab gel unit, or any other similar vertical slab gel unit. The slab gels that can be poured measure 180 × 160 × 0.75 mm and consist of two separating gels and a stacking gel as a discontinuous PAG gradient (from bottom to top of the gel sequentially: 10 cm 11% PAG, 3 cm 9%, and 3 cm 4.5% of PAG). Prior to polymerization, the spacer comb for the application of 20 samples is inserted into the stacking gel. The composition of the various gel solutions is as follows:

	11% gel	9% gel	stacking gel
acryl/bis 375 g/L	5.9	2.4	1.2
1.5 mol/L Tris, pH 8.8	5	2.5	2.5 mL, 0.5 mol/L, pH 6.8
SDS 100 g/L	0.2	0.1	0.1
dist. water	8.8	5.0	6.1
TEMED 100 g/L	0.1	0.05	0.01

After de-aeration of these solutions with a vacuum pump, 0.05 mL (100 g/L) of freshly prepared ammonium peroxodisulphate is added, followed by careful mixing. A solution containing per liter 3 g Tris, 14.4 g glycine, and 1.0 g SDS (pH = 8.4–8.6) is used as an electrophoresis buffer. The samples are incubated for 10 min at room temperature and then boiled for exactly 45 sec in a boiling water bath in a solution with final concentrations of 4% (weight volume) SDS, 4% saccharose, 0.8% dithiothreitol, and 5% (by volume) bromphenol blue (1 g/L). After cooling, 20 μL of the sample is rapidly applied into the slots, followed by electrophoresis initially at 10 mA and, after the blue-colored band has reached the separating gel, at 17 mA until the blue-colored front has reached the lower end of the gel. In this procedure, total plasma can also be analyzed if diluted five- to tenfold with the sample buffer. After electrophoresis, the protein bands are stained with Coomassie Brilliant Blue R (2 g/L of methanol, water, and acetic acid, 36/36/8 by volume) at 60° C, followed by destaining in the same solution without the stain twice for 24 h at 60° C. Other, more sensitive protein staining methods using colloidal gold are also possible.

The protein bands can eventually be scanned using a suitable laser densitometer. The proteins can also be transferred to a membrane for immunological analysis, for which we advise the method described by Havekes et al.[13] However, the transfer of the high-molecular-weight apolipoproteins is not quantitative. An example of the protein bands visible after electrophoresis of VLDL and of total plasma is shown in Figure 18–2. All relevant apolipoproteins are detectable.

Figure 18–2 ❖ SDS-PAGE Discontinuous Gel Showing the Typical Apolipoprotein Patterns of VLDL and of Whole Plasma.

The gel consists of three layers with acrylamide concentrations of 11%, 9%, and 4.5%, respectively. Samples were diluted with sample buffer, boiled for 45 sec, and after cooling applied to the slots in the stacking gel. The following samples were applied: (1) VLDL of a non-fasting normolipidemic subject, (2) VLDL of a patient with turbid plasma, (3) VLDL of a patient with increased VLDL concentrations, (4) human albumin, (5–7) whole plasma of three different subjects. All samples were applied in duplicate to give insight into the reproducibility of the method. Molecular weight markers are given in kilodaltons (K) at the left; most apolipoproteins and some main plasma protein bands are identified at the right. Note that by application of similar volumes of the VLDL-fractions or of whole plasma, it is possible to compare the relative concentrations of the proteins.

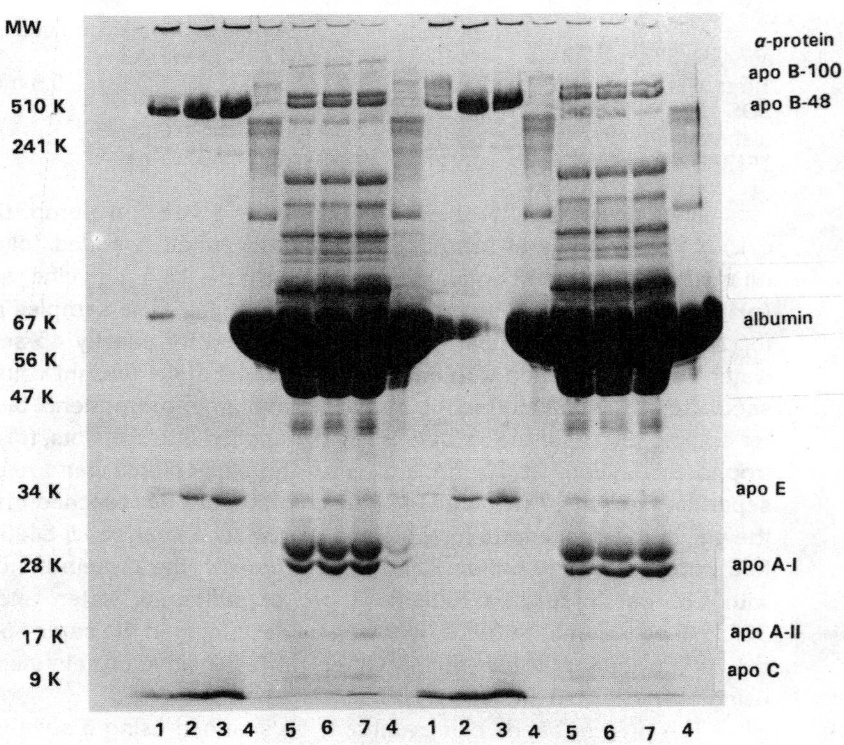

DETERMINATION OF LIPOPROTEINS AND THEIR SUBFRACTIONS BY NUCLEAR MAGNETIC RESONANCE

James D. Otvos

As knowledge grows about the complex associations between CHD and the concentrations of various plasma lipoproteins and their subclasses, it becomes increasingly important for accurate risk assessment and patient monitoring that laboratories have available automated analytical methods capable of producing detailed lipoprotein profiles at low cost. Ideally, such profiles would include total cholesterol (TC), triglycerides (TG), VLDL-TG, LDL-C, and HDL-C, as well as the concentrations of the different-sized subspecies within the VLDL, LDL, and HDL classes. With current technology, a profile of this kind is impractical for routine clinical use, owing to the expense of physically separating the lipoprotein (sub)fractions using methods such as ultracentrifugation, electrophoresis, or chromatography. A promising alternative is offered by a recently developed proton nuclear magnetic resonance (NMR) spectroscopic method.[14, 15] This method requires no physical fractionation of the plasma nor the use of any reagents, yet provides rapid, simultaneous quantification of all lipoprotein classes and their subclasses (including chylomicrons, but excluding Lp(a)).

This new method exploits the fact that the NMR signals from the terminal methyl groups of the various lipids in plasma, which might be expected to be indistinguishable, have characteristic frequencies and line shapes that differ according to the size of the lipoprotein particles with which they are associated. Thus, as seen in Figure 18–3, subspecies of VLDL, LDL, and HDL fractionated by agarose gel filtration chromatography all display spectrally distinct methyl lipid signals. For reasons that are not yet fully understood, lipoprotein particles of larger diameters always give rise to lipid signals that appear downfield (to the left) of those of smaller particles (except Lp(a), which cannot be distinguished from LDL by NMR).

These size-dependent spectral differences among lipoprotein subfractions are completely reproducible and apparently unaffected by lipid compositional variability. Thus, since each lipoprotein subspecies present in plasma makes a spectral contribution to the overall plasma spectrum proportional to its concentration, it is possible to work backwards from the line shape of the plasma methyl signal envelope to deduce the entire lipoprotein profile.[14, 15]

Computer analysis of the plasma signal line shape uses a linear least-squares fitting procedure that currently employs as reference standards the spectra of the 13 isolated subspecies shown in Figure 18–3. The spectrum of each plasma sample (0.8 mL) is acquired automatically in 2 min on a dedicated 360 MHz spectrometer (Analogic Corp., Peabody, MA). The line shape analysis is performed in about 2 sec on a model 486 personal computer. As shown by the example in Figure 18–4, excellent agreement is routinely obtained between experimental and calculated plasma spectra, and the calculated amplitudes of the reference signals yield lipid and lipoprotein concentrations that agree closely

Figure 18–3 ❖ Methyl Lipid Resonances of Lipoprotein Subfractions Isolated by Agarose Gel Filtration Chromatography.

Each subfraction was analyzed chemically for cholesterol and triglycerides, and mean particle diameters were determined for LDL and HDL subfractions by gradient gel electrophoresis (GGE). Larger numbers identify subfractions of larger particle size. (To convert cholesterol and triglyceride values in mg/dL to mmol/L, multiply by 0.0259 and 0.0113, respectively.)

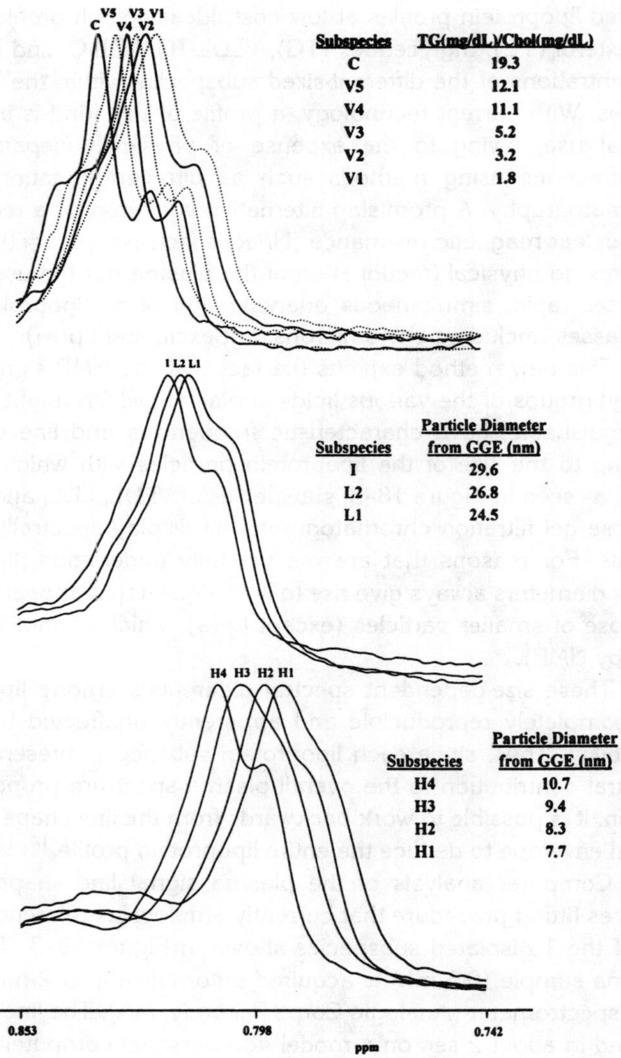

C = chylomicrons V = VLDL I = IDL L = LDL H = HDL.

Figure 18-4 ❖ Computer Line Shape Analysis of a Representative Plasma Spectrum Acquired at 360 MHz and 47°C.

Beneath the experimental (solid line) and calculated (dashed line) plasma spectra are the calculated amplitudes of the protein andlipoprotein subfraction reference spectra. On the right are the derived lipid, lipoprotein, and subspecies concentrations. (To convert cholesterol and triglyceride values in mg/dL to mmol/L, multiply by 0.0259 and 0.0113, respectively.)

Total Plasma Cholesterol = 234 mg/dL
Total Plasma Triglyceride = 347 mg/dL

		Triglyceride (mg/dL)
VLDL		318
	V5	103.3
	V4	135.1
	V3	60.5
	V2	17.5
	V1	2.0
		Cholesterol (mg/dL)
LDL		153
	I	0.5
	L2	60.0
	L1	92.1
HDL		43
	H4	6.4
	H3	10.6
	H2	24.4
	H1	1.2

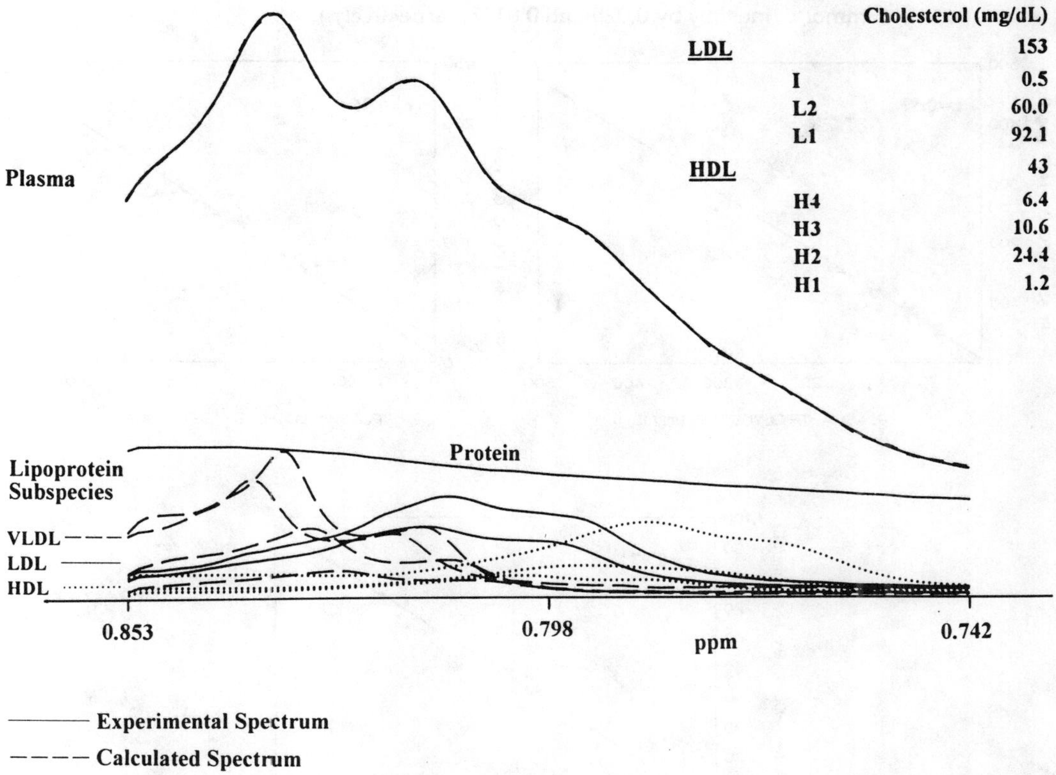

with those determined by chemical analysis (Figure 18–5). The small discrepancies between LDL-C and HDL-C values measured by NMR and chemical methods stem mainly from the fact that the NMR signal arises from bulk lipoprotein lipid rather than cholesterol alone. For that reason, NMR lipoprotein concentrations more closely reflect true particle concentrations and are relatively insensitive to triglyceride-cholesterol ester exchange reactions.

Direct quantitative comparisons between subspecies distributions determined by NMR and methods such as gradient gel electrophoresis (GGE) are hindered by differences in the ways the subspecies are quantified. For example, HDL subspecies separated by GGE are usually quantified by protein staining,[16] whereas NMR measures lipid concentrations. Nevertheless, some conclusions can be made based on qualitative comparative studies of more than 100 plasma samples.

Figure 18–5 ❖ Correlations between Lipid and Lipoprotein Concentrations

Shown are concentrations of 54 fasting plasma samples determined by NMR (*y*) and chemical analysis (*x*). (To convert cholesterol values in mg/L to mmol/L, multiply by 0.025 and 0.0113, respectively.)

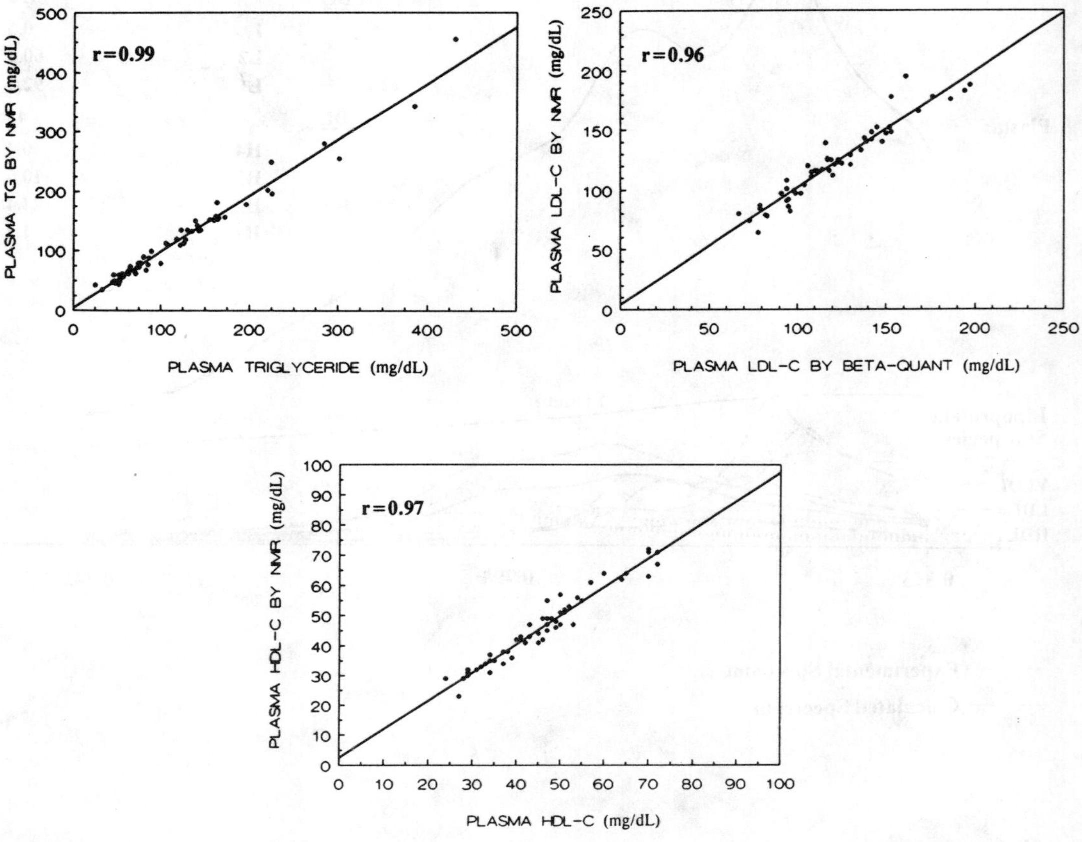

First, NMR clearly does not have the same resolving power as GGE, yet it reliably distinguishes individuals with predominantly large LDL (pattern A) from those with the reportedly more atherogenic, small, dense LDL (pattern B).[17, 18] The more pronounced spectral differences among HDL subspecies as compared to those of LDL (Figure 18–3) make NMR particularly sensitive to HDL subclass composition. This allows the large HDL_{2b} and small $HDL_{3b/3c}$ subclasses that reportedly have opposite associations with CHD risk[19, 20] to be measured separately. Studies of fasting and postprandial plasma indicate that at least four or five size ranges of chylomicron/VLDL particles are quantifiable by NMR. Since these TG-rich subfractions are not easily studied by other methods and consequently little is known about their associations with CHD, NMR may provide a valuable research tool for such investigations.

In summary, NMR lipoprotein analysis offers significant advantages over analytical methods currently employed in both clinical and research settings. Foremost among these is the extensive amount of information on lipoprotein status produced by a single, rapid, completely automated measurement made on fasting or nonfasting plasma specimens. Although the clinical utility of much of the subspecies information is not yet established, the availability of a method that provides subspecies concentrations at no incremental cost should greatly aid research that seeks to define their metabolic interrelationships and associations with CHD. The only significant drawback of the method is the relatively high cost of the instrumentation required, but this is offset when laboratories can analyze several hundred samples per day.

Compact, turn-key NMR analyzers that can be operated by technicians with no background in NMR spectroscopy are under development and should become commercially available in the near future.

ANALYSIS OF PLASMA LIPOPROTEINS BY FAST PROTEIN LIQUID CHROMATOGRAPHY

Ben Cost and Louis M. Havekes

The various lipoprotein classes show typical size differences, which allows their separation by gel permeation chromatography. In the past, conventional agarose gels have been used successfully, mainly for preparative purposes.[21, 22] For analytical purposes, high-performance liquid chromatography (HPLC) using TSK columns was used.[23, 24] Neither technique is suitable for serial analyses because of the low flow rates in the use of agarose gel columns, or because the interaction of the TSK column material with the lipoproteins to be separated results in clogging.

Recently, a number of investigators applied Superose 6 columns connected to a fast protein liquid chromatography (FPLC) equipment.[25, 26] These systems are usually combined with post-column on-line reaction with cholesterol reagent.[27] Based on the analysis of 20 μL of serum, it was possible to achieve complete and reproducible separations of lipoprotein classes VLDL, LDL, Lp(a), and HDL, even in the analysis of lipemic plasma, with minimal degradation of the lipoproteins. Quantitatively, the results agree well with a combined ultracentrifugation-precipitation method.[27] The number of analyses is limited to 10–20 samples in a working day. With adaptations, the method can be made preparative.

Below we present some lipoprotein profiles that our laboratory obtained for preparative purposes using FPLC-Superose 6. With the application of 100–200 μL of plasma, complete separation of the lipoproteins is achievable with a tandem of two Superose columns (30 mL each). Figure 18–6 shows the cholesterol and triglyceride profiles of a normolipidemic subject and of patients with either familial hypercholesterolemia, combined hyperlipidemia, or hypertriglyceridemia. In each sample, three separate peaks were obtained, corresponding to VLDL, LDL, and HDL, respectively. The profiles are in full accordance with the profiles to be expected from these different types of hyperlipidemic subjects.

We also analyzed plasma of a normal subject before and 3 h after a conventional fat load, and found a dramatic increase in VLDL-sized TG-rich particles, suggesting incomplete separation of the VLDL and chylomicrons. To improve the separation between these two lipoprotein fractions, we applied Biogel A 150m (operating range 10^6 to 150×10^6 mol. weight). Using two Biogel A 150m columns in tandem, VLDL, LDL, and HDL could not be separated. However, 3 h after a fat load, a peak of TG-rich lipoproteins was found in the very-large-size fraction preceding those characteristic for VLDL (Figure 18–7). Thus, using Biogel A 150m FPLC, chylomicron and VLDL particles can be separated clearly, at least at a reduced flow rate of 25% from that used with the Superose 6 gel columns.

Employing both procedures in parallel, we were able to obtain characteristic lipoprotein profiles of small experimental animals such as mice. Since the lipoprotein fractions supposedly do not lose apolipoproteins, the procedure described allows the accurate determination of the chemical composition of the different fractions. In addition, it provides a profile of the lipoprotein size distributions based on cholesterol or triglyceride measurement. If only the cholesterol profile is desired, then we prefer on-line measurement of cholesterol; if both the cholesterol and triglyceride profiles are desired, we sample fractions of 0.5 mL for sequential measurement of both lipids. We expect that this FPLC technique will become a widely used method for plasma lipoprotein analysis in the very near future.

CAPILLARY ISOTACHOPHORESIS OF LIPOPROTEINS IN HUMAN BODY FLUIDS

Gerd Schmitz and Christoph Möllers

In clinical practice, analysis of lipoproteins and description of lipoprotein profiles are needed to recognize and classify disorders of lipoprotein metabolism and to relate them to risk of CHD. Beyond the analysis of TG, VLDL-C, LDL-C, and HDL-C, there has been an enormous increase in analytical efforts (and associated costs) to elucidate the underlying causes of lipoprotein abnormalities, with special emphasis on the risk of atherosclerosis.

Therefore, we have developed an analytical free-flow capillary isotachophoresis (ITP) procedure for the detailed analysis of lipoproteins on commercially

Figure 18-6 ❖ Lipoprotein Profiles Obtained by Superose 6 FPLC of Normal and Hyperlipoproteinemic Plasmas

Two Superose 6 columns (30 mL each) were connected in series; 200 μL of plasma were added; flow rate was 0.5 mL/min. Fractions of 0.5 mL were collected. The fractions representing VLDL, LDL, and HDL are indicated. (To convert cholesterol and triglyceride values in mg/dL to mmol/L multiply by 0.259 and 0.0133, respectively.)

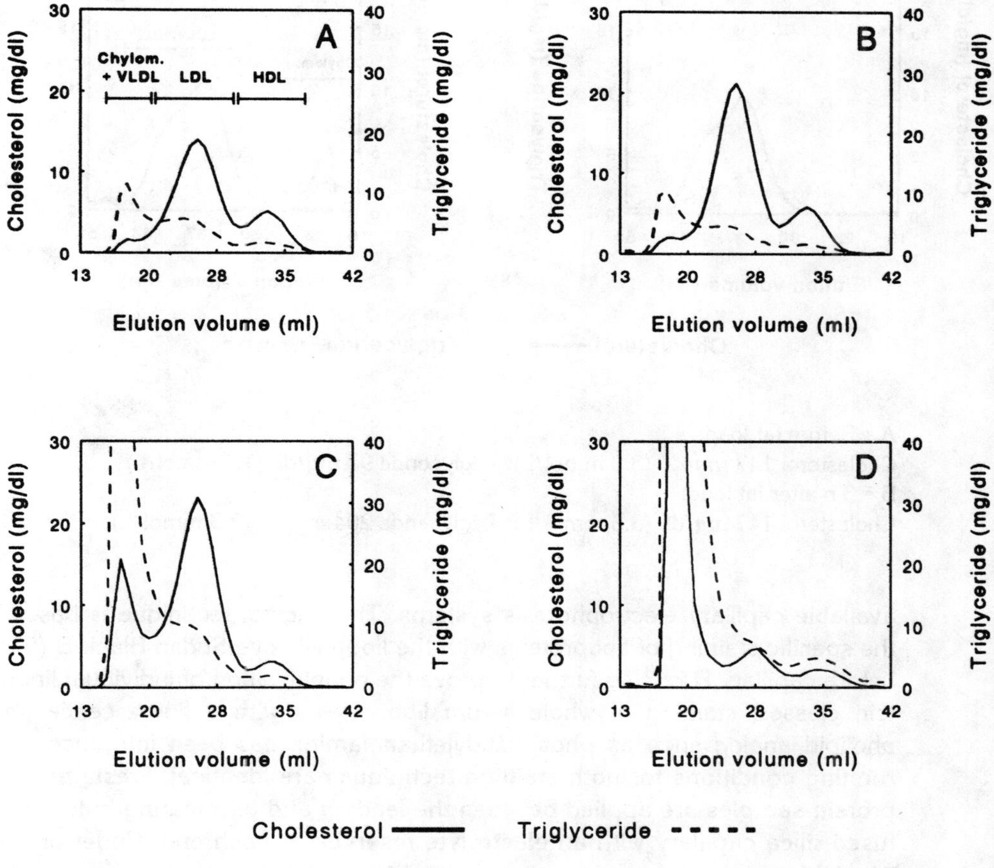

A = Normolipidemic subject: Cholesterol 174 mg/dL (4.5 mmol/L); Triglyceride 150 mg/dL (1.7 mmol/L)
B = Familial hypercholesterolemic patient: Cholesterol 266 mg/dL (6.9 mmol/L); Triglyceride 159 mg/dL (1.8 mmol/L)
C = Combined hyperlipidemic patient: Cholesterol 320 mg/dL (8.3 mmol/L); Triglyceride 584 mg/dL (6.6 mmol/L)
D = Hypertriglyceridemic subject: Cholesterol 320 mg/dL (8.3 mmol/L); Triglyceride 1,867 mg/dL (21.1 mmol/L)

Figure 18–7 ❖ Lipoprotein Profiles Obtained by Biogel A 150m of Plasma from a Subject Before and 3 h After a Fat Load.

Two Biogel A 150m columns (30 mL each) were connected in series. The flow rate was 0.1 mL/min. The fractions representing chylomicrons and VLDL, LDL, and HDL are indicated, respectively. (To convert cholesterol and triglyceride values in mg/dL to mmol/L, multiply by 0.0259 and 0.0113, respectively.)

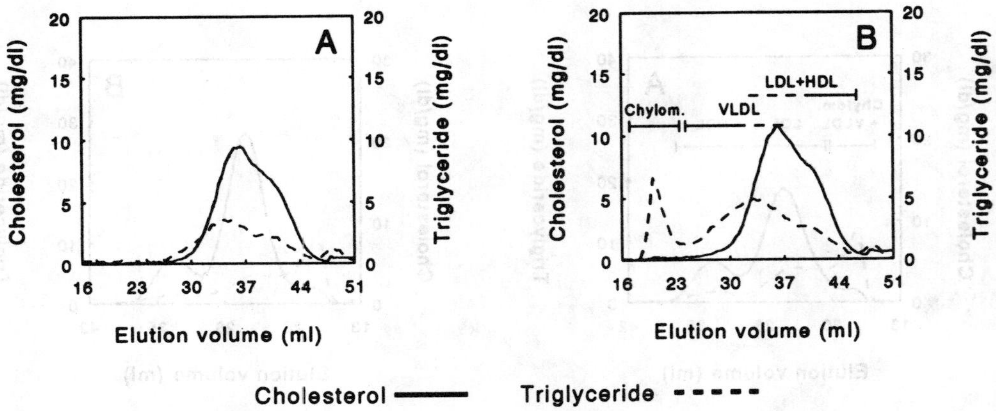

A = Before fat load:
Cholesterol 147 mg/dL (3.8 mmol/L); Triglyceride 97 mg/dL (1.1 mmol/L)
B = 3 h after fat load:
Cholesterol 147 mg/dL (3.8 mmol/L); Triglyceride 204 mg/dL (2.3 mmol/L).

available capillary electrophoresis systems. The original technique is based on the specific staining of lipoproteins with the lipophilic dye Sudan Black B (SBB), before capillary ITP.[28] To further improve the quantification of individual lipoprotein classes, staining of whole serum lipoproteins with a fluorescence phospholipid analog such as phosphatidylethanolamine has been introduced. The running conditions for both staining techniques are identical. Prestained lipoprotein samples are applied between the leading and terminating buffer into a fused-silica capillary with an electrolyte reservoir at each end. Under an electric field, lipoproteins are separated into 14 well-characterized subfractions according to their electrophoretic mobility and are monitored either with UV-VIS detection at 570 nm[29, 30] or with laser-induced fluorescence detection (excitation 488 nm; emission 510 nm). (See Figures 18–8 and 18–9.) Advantages of the isotachophoretic separation method are the high resolution and the small sample volumes. Moreover, lipoprotein analysis can be done directly from native serum. In isotachophoresis, the concentration- and zone-sharpening effect allows a simultaneous separation of major and minor lipoprotein subclasses in the same run.

Figure 18-8 ❖ Characterization of Subpopulations of High-Density Lipoproteins Separated from Normal Serum by Analytical and Preparative Isotachophoresis

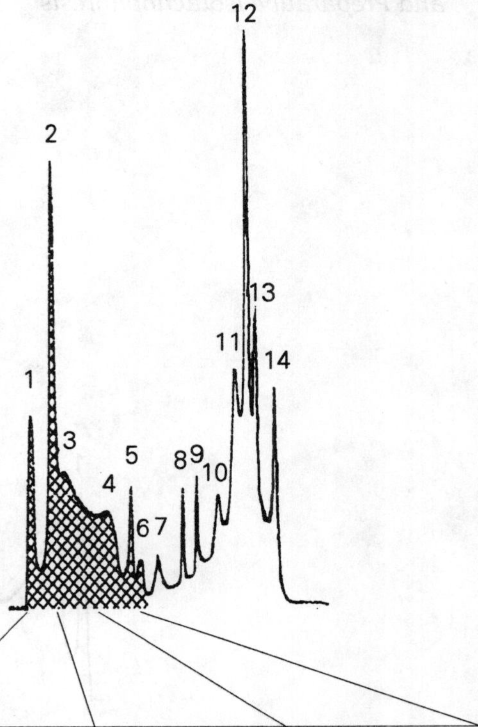

Composition	fast	intermediate	slow
Particle size	predominantly HDL2b, HDL3a	HDL2b, HDL2a HDL2a, HDL3b	predominantly HDL2b, HDL3a
Cholesterol (UC,EC)	poor in EC	rich in EC	poor in EC
Phospholipids (PC,SPM)	rich in PC	rich in SPM	rich in PC
Apolipoproteins	rich in apo A-I	rich in apo E, A-II C-II, C-III	rich in apo A-IV
LCAT activity	low	low	high
CETP associated	no	no	yes
Binding a) affinity (KD) b) nonspecific bind. c) Bmax	7.7 µg/ml low 245.6 ng	17.7 µg/ml low 198.4 ng	7.3 µg/ml high 95.9 ng
Internalization rate	high	low	low
Cholesterol efflux	effective	poor	effective

Figure 18–9 ❖ Characterization of Subpopulations of apo-B-containing Lipoproteins (VLDL, IDL, LDL) Separated from Normal Sera by Analytical and Preparative Isotachophoresis

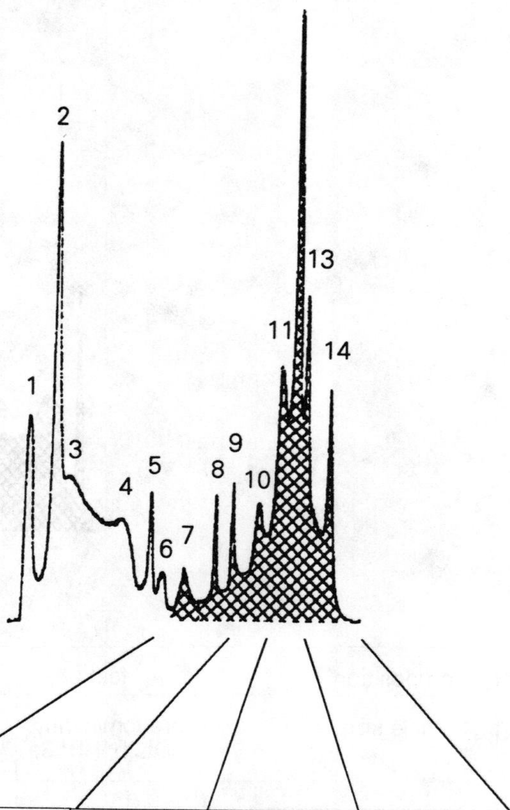

Composition	VLDL	IDL	LDL	LDL
Major lipids	TG, PC	TG, EC	EC	EC
Apolipoproteins	apo B, C, E	apo B, E	apo B ↑ EC/apo B	apo B ↓ EC/apo B
binding to fibroblast receptors KD (μg/ml) Bmax (ng)	nonspecific	7.8 54	4.4 75	2.3 58
binding to HepG2 receptors KD (μg/ml) Bmax (ng)	3.6 37	3.5 16.9 22 53	7.7 50 50 160	11.2 68 58 170

The HDL particles (fractions 1 to 6) are separated into three major subpopulations: the fast-migrating HDL subpopulation, containing mainly apo AI and phosphatidylcholine; the subpopulation with intermediate mobility, consisting of particles rich in apo AII, apo E, and apo C's; and the slow-migrating HDL subfraction, containing mainly particles rich in apo AI, apo AIV, and lecithin-cholesterol acyltransferase (LCAT) activity. The slow-migrating HDL population resembles the pre-β $HDL_{1,2,3}$ subfraction. The apo-B-containing lipoproteins (fractions 7 to 14) can be subdivided into four major functional groups. The first represents chylomicron-derived particles and large, TG-rich VLDL; the second group consists of small VLDL and IDL particles; the third and fourth groups represent the LDL. By means of these data, the precursor/product relationships for the major lipolytic enzymes can be generated with a single isotachophoretic analysis associated with a simple interpretation algorithm (see Figure 18–10). Thus, the ratio of slow-migrating HDL (peaks 4, 5, and 6) versus fast-migrating HDL (peaks 1 and 2) gives an estimate of LCAT activity. The ratio of peaks 7 and 8 versus peaks 9 and 10 provides a rough estimate of lipoprotein lipase (LPL) activity in the metabolic overload syndrome and in the genetic defects of TG metabolism. Furthermore, the ratio of peaks 10 and 11 versus peaks 12 to 14 provides a rough estimate of hepatic triglyceride lipase (HTGL) in mixed hyperlipidemias.

Figure 18–10 ❖ *Precursor/Product Relationships for Major Lipolytic Enzymes Shown by Analytical Capillary Isotachophoresis of Human Lipoproteins*

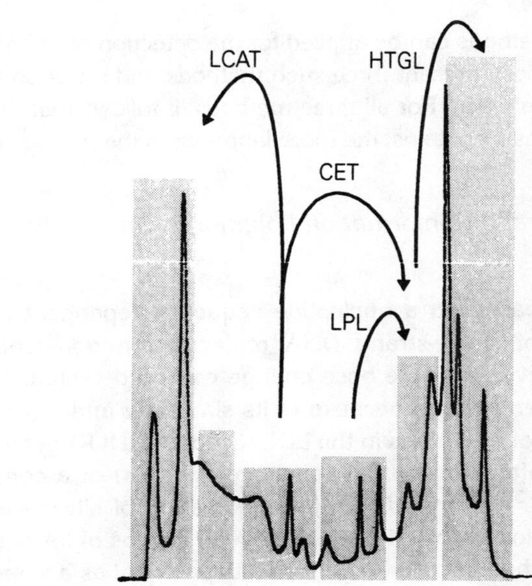

Results obtained by the isotachophoretic lipoprotein analysis revealed a good correlation in the range of HDL with routinely used techniques, such as lipoprotein electrophoresis, HDL-C analysis by a precipitation procedure, or turbidimetric determination of apo AI. Similar correlations with other analytical techniques were found for the quantitation of the apo-B-containing lipoproteins. With all these characteristics, analytical capillary isotachophoresis can become the method of choice for a fast and reliable automated analysis of lipoprotein subpopulations in biological fluids in the clinical laboratory.

MOLECULAR BIOLOGICAL METHODS FOR DIAGNOSIS OF CONGENITAL DISORDERS OF LIPOPROTEIN METABOLISM

Paola Lombardi and Louis Havekes

Methods to detect DNA sequence alterations have become increasingly important in all areas of biology.[31-33] Especially in clinical research and diagnosis, specific demand exists for linking mutations of identified genes into particular disease phenotypes. The introduction of polymerase chain reaction (PCR)[31] has enormously increased the efficiency of mutation detection, and PCR has been incorporated into virtually all molecular biological diagnostic methods. Methods can be subdivided into methods for detecting unknown mutations and methods for detecting previously characterized mutations.

Detection of Unknown Mutations

Several methods can be applied for the detection of unknown mutations. In this section we briefly present three such methods and focus on the advantages and disadvantages of each. For all three methods, it follows that the longer the gene and the more exons it contains, the more laborious is the process of detecting the mutation.

Single-Strand Conformation Polymorphism (SSCP)

SSCP is based on a nucleotide sequence-dependent change in electrophoretic mobility of single-strand DNA molecules in non-denaturing PAG.[34] With this method, even a single base change can be detected. This method is very rapid and is often favored because of its simplicity and low cost. Its application to the detection of mutations in the LDL receptor (LDLR) gene has resulted in the identification (after subsequent sequence analysis) of a considerable number of new mutations.[35-37] In addition, the application of silver staining has eliminated the use of radioactivity. At present, several groups of investigators are improving the SSCP method in order to increase its potential as a screening tool. It is likely that the SSCP method will soon be modified to enable screening for sequence alterations in hundreds of PCR fragments per day by one person.

SSCP does present some major drawbacks. It lacks 100% detection sensitivity, especially in the case of point mutations. Different rates of detection efficiency have been reported, with an average of 80%.[38] The sensitivity can be improved by using RNA molecules instead of DNA, since the RNA structure is more sensitive to a change in the sequence. However, even in this case the sensitivity remains lower than 100%. One reason for this may be that occasionally the bands are less sharp, requiring some further optimization of the electrophoretic conditions, such as gel porosity and temperature. Another disadvantage is that the optimal electrophoresis conditions need to be defined for each PCR fragment.

The SSCP is able to detect the presence of a sequence alteration. This implies that a DNA sequence analysis still needs to be done to fully characterize the mutation. With SSCP, the position of the mutation is located in a certain PCR fragment, thus narrowing down the region of the gene that requires subsequent sequence analysis.

Denaturing Gradient Gel Electrophoresis (DGGE)

DGGE is based on the sequence-dependent electrophoretic mobility of double-stranded DNA fragments in a PAG that contains a linear gradient of denaturing agents.[39] DNA molecules melt in discrete segments called domains, each of which is characterized by a distinctive melting temperature (Tm). For the best application of DGGE, the DNA fragment must consist of two melting domains, a high-melting domain and a low-melting domain. In a PAG containing a gradient of denaturant agent, DNA fragments migrate according to their size. When they reach a point in the gel at which the concentration of denaturant equals the Tm of the lowest melting domain, the molecule begins to denature and assumes a branched structure that prevents further migration of the molecule in the gel. Since the Tm for a certain domain is sequence-dependent, a single base change in the lowest melting domain will alter the maximal migration of the fragment in the gel, thereby allowing the separation of the mutant from the wild-type fragment. With the DGGE method, the base changes in the undenatured, highest melting domain cannot be detected.

It is essential that a branched-like structure is formed upon electrophoresis into the denaturant gradient. To ensure the formation of these structures for each PCR fragment, a 50 base pair containing GC-rich sequence is introduced on one or both sides of the PCR fragment (GC-clamp).[40] Such a clamp introduces an artificially high melting domain.

DGGE allows for the detection of nearly all, if not all, possible mutations. Usually, the electrophoretic patterns show sharp bands. Nucleotide sequence alterations are easily revealed by the presence of two or more DNA bands on the gel, as compared to the single band present in normal subjects. In principle, in the case of heterozygosity alterations are characterized by the typical four-band pattern for the mutation: two upper bands representing the heteroduplexes (formed between the two different alleles during the random reordering of strands that occurs during PCR), and two lower bands representing the homoduplexes of mutant and normal DNA strands, respectively. (For some examples of DGGE electrophoretic patterns, see Figure 18–11.) The formation of heteroduplexes increases the resolution of the method, especially in those cases in which the nucleotide change itself does not considerably alter the electrophoretic mobility of the DNA fragment.

Because of its high sensitivity, the DGGE method can also be readily used to detect and identify separately discovered mutations in a specific PCR fragment. Another important advantage of this method is the fact that the homoduplex mutant band can be easily isolated from the gel for subsequent sequencing analysis, although the presence of the GC-clamp presents some serious drawbacks in this respect. Furthermore, like SSCP, the DGGE technique can easily be done in less well-equipped laboratories and without the use of radioactive isotopes.

In our laboratory, DGGE has been successfully applied for the characterization of mutations in the apo E gene.[41] More recently, we have applied DGGE to the detection of mutations in the LDLR gene of patients with familial hypercholesterolemia (FH). A preliminary screening identified the presence of mutations in the coding sequences of the LDLR gene in 93% of all FH patients analyzed.

DGGE also exhibits several disadvantages. For each gene of interest, the initial setup of the method is rather complex and time consuming. Furthermore, this method is far more time consuming than the SSCP method for a specific gene study. As in SSCP, DGGE only detects the presence of mutations in a specific fragment of DNA. The full characterization by sequence analysis still needs to be done.

Figure 18–11 ❖ Typical DGGE Electrophoretic Patterns Representing the Presence of Mutations in Different Exons of the LDLR Gene

The numbers on top of each lane represent the exons amplified. In all lanes, three or four bands are visible instead of one, thus indicating the presence of a sequence alteration in one of both strands. The upper two bands are the heteroduplex bands and the lower two bands are the homoduplex bands. The lowest band is a constant band and represents the wild-type homoduplex.

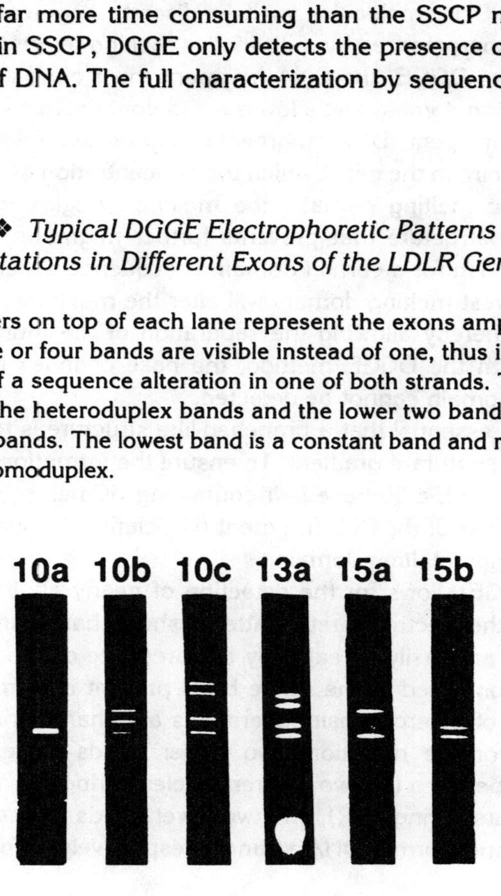

Sequencing

Sequencing represents the crucial step for all mutation detection methods, since it is the only method that describes the nature of the mutation. DNA sequencing used to be highly laborious, but the development of methods for direct sequencing of PCR products improved the process and made it useful for detection of unknown mutations. Direct sequencing enables the avoidance of the laborious and time-consuming process of cloning into a M13 sequencing vector. Furthermore, the rapid progress in the development of automated sequencing methods allows for the use of direct sequencing as a screening tool.[42, 43] At present, several groups routinely use automated direct sequencing as a routine method for detection of unknown mutations. The high cost of the equipment needed for this method is its major drawback. In addition, once a laboratory is equipped with the sequencing machine, it normally takes a long time and much effort to get the technique running on a routine basis.

Detection of Known Mutations

Once a sequence alteration has been characterized, a number of detection methods exist for screening for these mutations in relatively large populations.[42–44]

Small sequence alterations sometimes create or disrupt a cutting site for a restriction enzyme. These mutations can be easily identified by PCR followed by restriction enzyme cleavage analysis. For known mutations that do not alter a restriction enzyme cutting site, several less simple techniques for their identification are available. Three different but related methods are explained briefly below. For more details, the reader is referred to the review by Prosser.[45]

Allele-Specific Oligonucleotide Hybridization (ASO)

This is the oldest method for mutation detection. It relies on the differential hybridization of sequence-specific synthetic oligonucleotides (mutant or wild-type) to mutant PCR amplified DNA, due to the presence of mismatches between the mutant PCR product and the wild-type oligonucleotides. The application of colorimetric detection with horseradish peroxidase-labelled probes and of the dot-blot technique has increased the versatility of this method. The major drawback of this approach is the initial setup of the hybridization conditions for each individual mutation.

Allele-Specific Amplification (ASA)

This method is the basis for another rapid PCR-based technique. It is based on the selective amplification of the mutant or wild-type form when PCR is conducted in the presence of a primer that is placed at the site of a mutation and made either in the wild-type or in the mutant form. Several modifications of this method have been introduced. The most interesting one is described below.

Restriction-Site-Generating PCR (RG-PCR)

This method is based on the introduction of a restriction site by mismatched primers in the region of a mutation for either the wild-type or mutant sequence. The mismatch is usually introduced near the extreme 3′-nucleotide of one of the primers. The presence of the mutation is then simply detected after digestion with the appropriate restriction enzyme followed by electrophoresis. This method is used for the identification of carriers of familial defective apo B-100.[44]

SUMMARY

A number of parameters of lipoprotein metabolism are positively or negatively associated with the risk for atherosclerosis and CHD. Rather than determine the most important parameters by different methods, it is desirable to use simple methods that allow the quantification of a maximum of these parameters in a single run. The fundamental research has always been an important source of diagnostic methods.

In this chapter we presented several techniques commonly used in fundamental biochemistry that, in our opinion, are of potential importance for future case control or epidemiological studies on risk factors associated with the disturbance of lipoprotein metabolism. ❖

REFERENCES

1. Havel RJ, Eder HA, Bragdon JH. The distribution and chemical composition of ultracentrifugally separated lipoproteins in human serum. J Clin Invest 1955;34: 1345–53
2. Terpstra AHM, Woodward CJH, Sanchez-Muniz FJ. Improved techniques for the separation of serum lipoproteins by density gradient ultracentrifugation: visualization by prestaining and rapid separation of serum lipoproteins from small volumes of serum. Anal Biochem 1981;111:149–57
3. Demacker PNM, van Sommeren-Zondag DF, Stalenhoef AFH, Stuyt PMJ, van't Laar A. Ultracentrifugation in swinging-bucket and fixed angel rotors evaluated for isolation and determination of high-density lipoprotein subfractions: HDL_2 and HDL_3. Clin Chem 1983;29:656–63
4. Fainaru M, Glangeaud MC, Eisenberg S. Radioimmunoassay of human high density lipoprotein apoprotein A-1. Biochem Biophys Acta 1975;386:432–43.
5. Cheung MC, Wolf AC. Differential effect of ultracentrifugation on apolipoprotein A-1 containing lipoprotein subpopulations. J. Lipid Res 1988;29:15–26.
6. Chung BH, Segrest JP, Ray MJ, Brunzell JD, Hokanson JE, Krauss RM, Beaudrie K, Cone JT. Single vertical spin density gradient ultracentrifugation. In: Segrest JP, Albers JJ, eds. Plasma lipoproteins, part A. Methods in enzymology, vol. 128. Orlando, FL: Academic Press, 1986:181–209.
7. Mohr D, Bowry VW, Stocker R. Dietary supplementation with coenzyme Q10 results in increased levels of ubiquinol 10 within circulating lipoproteins and increased resistance of human low-density lipoprotein to the initiation of lipid peroxidation. Biochim Biophys Acta 1992;1126:247–54
8. Kleinveld HA, Hak-Lemmers HLM, Stalenhoef AFH, Demacker PNM. Improved measurement of low-density-lipoprotein susceptibility to copper-induced oxidation: application of a short procedure for isolating low-density lipoprotein. Clin Chem 1992;38:2066–72

9. Nichols AV, Krauss RM, Musliner TA. Non-denaturing polyacrylamide gradient gel electrophoresis. Meth Enzymol 1986;128;417–31
10. Dormans TPJ, Swinkels DW, de Graaf J, Hendriks JCM, Stalenhoef AFH, Demacker PNM. Single spin density gradient ultracentrifugation versus gradient gel electrophoresis: a comparison of two methods for detection of low density lipoprotein heterogeneity. Clin Chem 1991;37:853–8
11. Lefevre M, Goudey-Lefevre JC, Roheim PS. Gradient gel electrophoresis-immunoblot analysis (GGEI): a sensitive method for apolipoprotein profile determinations. J Lipid Res 1987;28:1495–507
12. Laemmli UK. Clearage of structural proteins during assembly of the head bacteriophage T4. Nature (London) 1970;227:680–5
13. Havekes LM, de Knijff P, Beisiegel U, Havinga J, Smit M, Klasen E. A rapid method for apolipoprotein E phenotyping directly in serum. J Lipid Res 1987;28:455–63
14. Otvos JD, Jeyarajah EJ, Bennett DW. Quantification of plasma lipoproteins by proton nuclear magnetic resonance spectroscopy. Clin Chem 1991;37:377–86.
15. Otvos JD, Jeyarajah EJ, Bennett DW, Krauss RM. Development of a proton nuclear magnetic resonance spectroscopic method for determining plasma lipoprotein concentrations and subspecies distributions from a single, rapid measurement. Clin Chem 1992;38:1632–38.
16. Blanche PJ, Gong EL, Forte TM, Nichols AV. Characterization of human high-density lipoproteins by gradient gel electrophoresis. Biochim Biophys Acta 1981; 665:408–19.
17. Austin MA, Breslow JL, Hennekens CH, Buring JE, Willet WC, Krauss RM. Low-density lipoprotein subclass patterns and risk of myocardial infarction. JAMA 1988;260:1917–21.
18. Crouse JR, Parks JS, Schey HM, Kahl FR. Studies of low density lipoprotein molecular weight in human beings with coronary artery disease. J Lipid Res 1985;26: 566–74.
19. Cheung MC, Brown BG, Wolf AC, Albers JJ. Altered particle size distribution of apolipoprotein A-1-containing lipoproteins in subjects with coronary artery disease. J Lipid Res 1991;32:383–94.
20. Williams PT, Krauss RM, Vranizan KM, Stefanik ML, Wood PDS, Lindgren FT. Associations of lipoproteins and apolipoproteins with gradient gel electrophoresis estimates of high density lipoprotein subfractions in men and women. Arterioscler Throm 1992;12:332–40.
21. Sata T, Havel RJ, Jones AL. Characterization of subfractions of triglyceride-rich lipoproteins separated by gel chromatography from blood plasma of normolipidemic and hyperlipidemic humans. J Lipid Res 1972;13:757–68.
22. Rudel LL, Marzetta CC, Johnson FL. Separation and analysis of lipoproteins by gel filtration. Methods Enzymol 1986;129:45–57.
23. Okazaki M, Itakura H, Shiraishi K, Hara I. Serum lipoprotein measurement. Liquid chromatography and sequential flotation (ultracentrifugation) compared. Clin Chem 1983;29:768–73.
24. Carroll RM, Rudel LL. Lipoprotein separation and low density lipoprotein molecular weight determination using high performance gel-filtration chromatography. J Lipid Res 1983;24:200–7.
25. Van Gent T, van Tol A. Automated gel permeation chromatography of plasma lipoproteins by preparative fast protein liquid chromatography. J Chromatography 1990;525:433–41.
26. Van den Maagdenberg AMJM, Hofker MH, Krimpenfort PJA, de Bruijn IG, van Vlijmen B, van der Boom H, Havekes LM, Frants RR. Transgenic mice carrying the apolipoprotein E3-Leiden gene exhibit hyperlipoproteinemia. J Biol Chem 1993;268:10540–5.

27. März W, Siekmeier R, Scharnagl H, Seiffert UB, Gross W. Fast liquid chromatography: new method of analysis for plasma lipoproteins. Clin Chem 1993;39:2276–81.
28. Schmitz G, Nowicka G, Möllers C. Capillary isotachophoresis in the analysis of lipoproteins. In: Christie WW, ed. Advances in lipid methodology. Dundee: Oily Press Ltd., 1993:157–93
29. Nowicka G, Brüning T, Grothaus B, Kahl G, Schmitz G. Characterization of apolipoprotein B-containing lipoproteins separated by preparative free flow isotachophoresis. J Lipid Res 1990;31:1173–86
30. Nowicka G, Brüning T, Böttcher A, Kahl G, Schmitz G. Macrophage interaction of HDL subclasses separated by free flow isotachophoresis. J Lipid Res 1990;31:1947–63.
31. Saiki RK, Walsh PS, Levenson CH, Erlich HA. Analysis of enzymatically amplified B-globin and HLA-DQ alpha DNA with allelic specific oligonucleotides probes. Nature 1986;324:163–6.
32. Leren TP, Rodningen OK, Rosby O, Solberg K, Berg K. Screening for point mutations by semi-automated DNA sequencing using Sequenase and magnetic beads. Biotechniques 1993;14:618–23.
33. Funke H, Assmann G. The rapid molecular diagnosis of genetic cardiovascular risk factors. Curr Opinion Lipid 1993;4:453–60.
34. Orita M, Iwahana H, Kanazawa H, Hayashi K, Sekiya T. Detection of polymorphisms of human DNA by gel electrophoresis as single-strand conformation polymorphisms. Proc Natl Acad Sci (USA) 1989;86:2766–70.
35. Hobbs HH, Brown MS, Goldstein JF. Molecular genetics of the LDL receptor gene in familial hypercholesterolemia. Hum Mut 1992;1:445–66.
36. Gudnason V, Mak Y-T, Betteridge J, McCarthy SN, Humphries S. Use of the single-strand conformational polymorphism method to detect recurrent and novel mutations in the low-density lipoprotein receptor gene in patients with familial hypercholesterolemia: detection of a novel mutation Asp200→Gly. Clin Invest 1993;71:331–7.
37. Loux N, Saint-Jore B, Collod G, Dairou F, Benlian P, Truffert J, Dastugue B, Douste-Blazy P, de Gennes JL, Junien C, Boileau C. Screening for new mutations in the LDL receptor gene in seven French familial hypercholesterolemia patients by the single strand conformation polymorphism method. Hum Mut 1992;1:325–32.
38. Cotton RGH. Current methods of mutation detection. Mut Res 1993;285:125–44.
39. Myers MR, Lumelsky N, Lerman LS, Maniatis T. Detection of single base substitutions in total genomic DNA. Nature 1985;313:495–8.
40. Sheffield VC, Cox DR, Lerman LS, Meyers RM. Attachment of a 40-base-pair G+C-rich sequence (GC-clamp) to genomic DNA fragments by the polymerase chain reaction results in improved detection of single-base changes. Proc Natl Acad Sci (USA) 1989;86:232–6.
41. van den Maagdenberg AMJM, Weng W, de Bruin I, de Knijff P, Funke H, Smelt AHM, Gevers Leuven JA, van't Hooft FM, Assmann G, Hofker MH, Havekes LM, Frants RR. Characterization of five new mutants in the carboxyl-terminal domain of human apolipoprotein E: no cosegregation with severe hyperlipidemia. Am J Hum Gen 1993;52:937–46.
42. Leitersdorf E, van der Westhuyzen DR, Coetzee GA, Hobbs HH. Two common low density lipoprotein receptor gene mutations cause familial hypercholesterolemia in Afrikaners. J Clin Invest 1989;84:954–61.
43. Soutar AK, Knight BL, Patel DD. Identification of a point mutation in growth factor repeat C of the low density lipoprotein receptor gene in a patient with ho-

mozygous familial hypercholesterolemia that affects ligand binding and intracellular movement of receptors. Proc Natl Acad Sci (USA) 1989;86:4166–70.
44. Motti C, Funke H, Rust S, Dergunov A, Assmann G. Using mutagenic polymerase chain reaction to detect carriers of familial defective apolipoprotein B100. Clin Chem 1991;37:1762–6.
45. Prosser J. Detecting single-base mutations. Tibtech 1993;11:238–46.

Lipid and Lipoprotein Values

Lipid and lipoprotein values for white and black males and females by age. Data presented as mean and 5–95 percentile values in mg/dL (Table I.A–IV-A) and in mmol/L (Tables I.B–IV.B).

From Hainline A, Karon J, Lippel K, eds., *Manual of Laboratory Operations,* Lipid Research Clinics Program, and *Lipid and Lipoprotein Analysis,* 2nd ed., Bethesda MD: U.S. Department of Health and Human Services, 1982.

TABLE I.A PLASMA TOTAL AND LIPOPROTEIN CHOLESTEROL AND TRIGLYCERIDE MEAN AND PERCENTILE VALUES (mg/dL) FOR WHITE MALES, BY AGE

Age in Years	N	Cholesterol (mg/dL) X (5–95 % tile)	Triglycerides (mg/dL) X (5–95 % tile)	Cholesterol in HDL (mg/dL) N (5–95 % tile)	HDL X	Cholesterol in LDL (mg/dL) N (5–95 % tile)	LDL X	Cholesterol in VLDL (mg/dL) N (5–95 % tile)	VLDL X
0	(10)	143	63						
1	(36)	150	73						
2	(49)	158	55						
3	66	151 (110–193)	53 (28–84)						
4	77	159 (115–221)	52 (29–89)						
5	72	156 (113–203)	56 (26–88)						
6	278	158 (120–201)	54 (31–96)	66	56.0 (38–72)	61	95.0 (69–129)	61	6.9 (0–16)
7	275	159 (122–195)	55 (29–107)						
8	332	162 (121–206)	56 (29–101)	68	55.6 (39–73)	65	90.5 (65–123)	64	9.6 (0–21)
9	294	161 (123–203)	58 (30–111)						
10	455	159 (127–196)	58 (29–105)	94	57.3 (38–76)	87	96.8 (64–131)	90	10.4 (2–23)
11	409	162 (124–211)	57 (29–106)						
12	572	160 (121–203)	68 (34–125)	144	55.9 (39–75)	139	95.4 (64–129)	138	9.0 (1–19)
13	411	153 (115–195)	71 (35–143)						
14	426	152 (114–195)	73 (35–134)	129	49.2 (34–69)	129	95.5 (57–130)	127	10.7 (0–25)
15	508	149 (111–194)	74 (34–134)						
16	639	149 (113–192)	77 (35–146)	160	45.6 (30–62)	160	93.2 (64–129)	160	13.5 (2–27)
17	417	150 (110–204)	79 (39–160)						
18	238	152 (113–199)	81 (39–155)	67	43.7 (32–60)	66	99.3 (62–142)	66	13.8 (2–25)
19	173	154 (114–196)	85 (44–172)						
20–24	882	166.5 (124–218)	100.3 (44–201)	118	45.4 (30–63)	118	103.3 (66–147)	118	13.7 (1–28)
25–29	2042	182.2 (133–244)	115.8 (46–249)	253	44.7 (31–63)	253	116.7 (70–165)	253	17.4 (3–36)
30–34	2444	192.2 (138–254)	128.3 (50–266)	403	45.5 (28–63)	403	126.4 (78–185)	403	21.3 (5–48)
35–39	2320	201.3 (146–270)	144.9 (54–321)	371	43.4 (29–62)	371	133.2 (81–189)	372	24.1 (3–56)
40–44	2428	206.5 (151–268)	151.4 (55–320)	383	44.3 (27–67)	385	135.6 (87–186)	384	25.5 (5–56)
45–49	2296	212.2 (158–276)	151.7 (58–327)	325	45.4 (30–64)	325	143.9 (98–202)	326	24.4 (5–51)
50–54	2138	212.7 (158–277)	151.8 (58–320)	340	44.1 (28–63)	340	142.3 (89–197)	340	26.8 (8–62)
55–59	1621	213.9 (156–276)	141.4 (58–286)	261	47.6 (28–71)	261	145.8 (88–203)	261	21.6 (3–49)
60–64	905	213.0 (159–276)	142.3 (58–291)	131	51.5 (30–74)	131	146.3 (83–210)	131	18.9 (3–44)
65–69	750	212.6 (158–274)	136.7 (57–267)	105	51.1 (30–78)	105	150.4 (98–210)	105	19.7 (0–45)
70+	850	206.8 (151–270)	129.8 (58–258)	119	50.5 (31–75)	119	142.9 (88–186)	119	17.0 (0–38)

TABLE II.A PLASMA TOTAL AND LIPOPROTEIN CHOLESTEROL AND TRIGLYCERIDE MEAN AND PERCENTILE VALUES (mg/dL) FOR WHITE FEMALES NOT TAKING HORMONES, BY AGE

Age in Years	N	Cholesterol (mg/ dL) X (5–95 % tile)	Triglycerides (mg/ dL) X (5–95 % tile)	Cholesterol in HDL (mg/ dL) N	X (5–95 % tile)	LDL (mg /dL) N	X (5–95 % tile)	VLDL (mg/ dL) N	X (5–95 % tile)
0	(6)	148	82						
1	(31)	156	70						
2	(33)	152	65						
3	57	160 (118–201)	63 (32–110)						
4	59	155 (108–196)	59 (34–94)						
5	86	162 (126–195)	56 (28–93)						
6	253	165 (129–205)	57 (32–95)	58	50.1 (28–68)	53	100.5 (66–124)	50	9.8 (1–23)
7	220	160 (124–202)	61 (32–112)						
8	297	166 (129–209)	61 (34–104)	60	55.7 (37–75)	55	100.0 (67–142)	54	10.6 (1–25)
9	261	163 (120–205)	63 (32–110)						
10	417	163 (127–205)	66 (35–116)	100	51.5 (34–72)	99	98.1 (70–140)	99	11.8 (2–23)
11	338	161 (126–200)	73 (36–126)						
12	536	161 (124–203)	79 (38–132)	102	53.0 (37–69)	101	97.7 (68–133)	101	10.6 (1–23)
13	344	155 (122–192)	79 (43–136)						
14	441	157 (120–198)	79 (38–138)	122	51.0 (34–70)	120	93.5 (60–129)	120	10.6 (2–23)
15	553	155 (119–199)	74 (39–127)						
16	675	155 (121–197)	71 (37–120)	165	52.8 (35–77)	165	95.2 (57–138)	164	12.3 (3–26)
17	351	158 (119–202)	70 (40–113)						
18	193	157 (116–199)	75 (39–126)	53	53.2 (37–74)	53	101.8 (58–143)	53	11.1 (0–24)
19	137	162 (123–212)	79 (40–135)						
20–24	778	164.1 (122–216)	72.4 (36–131)	96	52.2	96	98.1	96	11.9
25–29	1329	170.7 (128–222)	74.7 (37–145)	181	56.0 (37–81)	181	106.0 (70–151)	181	12.0 (2–24)
30–34	1569	175.4 (130–231)	78.5 (39–151)	233	55.4 (38–75)	232	108.9 (67–150)	233	10.8 (0–25)
35–39	1606	184.4 (140–242)	86.3 (40–176)	241	54.7 (34–82)	242	118.8 (76–172)	240	14.4 (1–35)
40–44	1583	193.8 (147–252)	98.4 (45–191)	244	57.1 (33–87)	244	125.1 (77–174)	244	13.9 (3–29)
45–49	1515	202.5 (152–265)	104.5 (46–214)	249	57.7 (33–86)	247	129.7 (80–187)	247	16.8 (2–38)
50–54	1257	212.7 (162–285)	114.8 (52–233)	177	60.1 (37–89)	177	146.1 (90–215)	177	16.4 (0–37)
55–59	1112	230.5 (173–300)	125.0 (55–262)	172	59.1 (36–86)	172	151.8 (95–213)	172	21.2 (2–51)
60–64	723	230.8 (172–297)	127.0 (56–239)	111	62.0 (36–91)	111	156.3 (100–234)	111	17.6 (0–40)
65–69	593	232.8 (171–303)	131.3 (60–243)	101	60.5 (34–89)	101	161.6 (97–223)	101	17.6 (0–40)
70+	748	228.1 (169–289)	132.4 (60–237)	127	60.1 (33–91)	127	148.9 (96–207)	127	16.2 (0–52)

TABLE III.A PLASMA TOTAL AND LIPOPROTEIN CHOLESTEROL AND TRIGLYCERIDE MEAN AND PERCENTILE VALUES (mg/dL) FOR WHITE FEMALES TAKING HORMONES, BY AGE

Age in Years	N	Cholesterol (mg/ dL) X (5–95 % tile)	Triglycerides (mg/ dL) X (5–95 % tile)	Cholesterol in HDL (mg/ dL) N	HDL X (5–95 % tile)	LDL (mg /dL) N	LDL X (5–95 % tile)	VLDL (mg/ dL) N	VLDL X (5–95 % tile)
15–19	167	169.0 (121–231)	106.3 (49–200)	19	-----	19	-----	20	-----
20–24	788	179.2 (131–236)	105.3 (55–176)	101	54.7 (34–79)	101	108.3 (62–163)	101	15.2 (1–30)
25–29	855	183.6 (141–236)	110.4 (57–191)	132	56.1 (35–83)	132	115.9 (72–169)	132	15.3 (2–36)
30–34	579	188.5 (139–246)	115.8 (58–206)	97	58.0	95	117.4	95	15.6
35–39	406	194.4 (146–249)	126.0 (56–241)	50	57.0	50	118.9	50	20.4
40–44	466	199.4 (152–258)	128.9 (58–238)	63	61.1	63	124.8	63	17.2
45–49	627	209.4 (155–276)	129.6 (53–260)	71	65.6	71	127.3	72	19.0
50–54	729	218.1 (164–283)	130.0 (62–238)	70	67.6	70	119.7	70	18.7
55–59	577	218.1 (164–283)	126.5 (62–238)	60	70.8	60	132.5	60	19.2
60–64	341	224.4 (172–285)	126.0 (57–240)	32	70.4	32	136.6	32	13.5
65–69	224	222.5 (173–279)	129.5 (60–234)	25	76.1	25	126.4	25	15.6
70+	132	215.9 (160–274)	121.1 (60–216)	10	-----	10	-----	10	-----

TABLE IV.A PLASMA TOTAL CHOLESTEROL AND TRIGLYCERIDE MEAN AND PERCENTILE VALUES (mg/dL) FOR BLACK MALES AND FEMALES, BY AGE

	MALES			FEMALES		
Age in Years	N	Cholesterol (mg/ dL) X (5–95 % tile)	Triglycerides (mg/ dL) X (5–95 % tile)	N	Cholesterol (mg/ dL) X (5–95 % tile)	Triglycerides (mg/ dL) X (5–95 % tile)
0	(0)			(2)	124	86
1	(2)	139	49	(6)	154	52
2	(4)	154	52	(5)	160	43
3	(8)	149	46	(7)	167	51
4	(3)	164	64	(4)	147	68
5	(8)	166	45	(8)	183	51
6	97	164 (128–213)	53 (33–94)	102	172 (138–216)	57 (29–108)
7	52	167 (116–208)	49 (33–75)	60	170 (126–216)	52 (27–82)
8	103	167 (129–218)	54 (28–96)	100	172 (124–226)	54 (31–90)
9	74	168 (135–200)	53 (33–81)	68	176 (133–235)	60 (33–100)
10	118	167 (126–206)	53 (29–95)	131	169 (128–211)	60 (39–97)
11	68	167 (133–191)	52 (30–81)	65	168 (131–213)	61 (33–92)
12	120	171 (127–216)	63 (32–100)	128	161 (121–203)	68 (37–109)
13	88	159 (124–195)	58 (33–90)	70	167 (120–219)	70 (37–112)
14	91	157 (115–195)	58 (32–98)	80	161 (120–205)	66 (38–110)
15	138	154 (110–200)	59 (28–97)	150	166 (129–211)	64 (34–121)
16	200	157 (120–199)	57 (31–102)	192	162 (120–206)	59 (32–95)
17	126	156 (118–208)	65 (32–122)	83	166 (124–208)	62 (36–107)
18	(38)	163	65	(23)	167	66
19	(12)	160	60	(6)	166	55
20–29	97	178.5	81.3	129	178.1 (125–234)	71.3 (36–127)
30–39	178	191.6 (138–253)	106.9 (42–224)	291	186.4 (132–243)	76.4 (38–140)
40–49	167	206.9 (148–267)	126.1 (52–294)	188	201.8 (145–267)	95.1 (43–175)
50–59	70	206.5	142.2	80	215.5	102.1
60+	46	220.7	108.5	48	233.6	115.4

TABLE I.B PLASMA TOTAL AND LIPOPROTEIN CHOLESTEROL AND TRIGLYCERIDE MEAN AND PERCENTILE VALUES (mmol/L) FOR WHITE MALES, BY AGE

				Cholesterol in					
Age in Years	N	Cholesterol (mmol/L) X (5–95 % tile)	Triglycerides (mmol/L) X (5–95 % tile)	HDL (mmol/L) N	HDL X (5–95 % tile)	LDL (mmol/L) N	LDL X (5–95 % tile)	VLDL (mmol/L) N	VLDL X (5–95 % tile)
0	(10)	3.70	0.71						
1	(36)	3.89	0.83						
2	(49)	4.09	0.62						
3	66	3.91 (2.85–5.00)	0.60 (0.32–0.95)						
4	77	4.12 (2.98–5.72)	0.59 (0.32–1.01)						
5	72	4.04 (2.93–5.26)	0.63 (0.29–0.99)						
6	278	4.09 (3.11–5.21)	0.61 (0.35–1.08)	66	1.45 (0.98–1.87)	61	2.46 (1.79–3.34)	61	0.18 (0–0.41)
7	275	4.12 (3.16–5.05)	0.62 (0.33–1.21)						
8	332	4.20 (3.13–5.34)	0.63 (0.33–1.14)	68	1.44 (1.01–1.89)	65	2.34 (1.68–3.19)	64	0.25 (0–0.54)
9	294	4.17 (3.19–5.26)	0.65 (0.34–1.25)						
10	455	4.12 (3.29–5.08)	0.65 (0.33–1.19)	94	1.48 (0.98–1.97)	87	2.51 (1.66–3.39)	90	0.27 (0.05–0.60)
11	409	4.20 (3.21–5.46)	0.64 (0.33–1.20)						
12	572	4.14 (3.13–5.26)	0.77 (0.38–1.41)	144	1.45 (1.01–1.94)	139	2.47 (1.66–3.34)	138	0.23 (0.03–0.49)
13	411	3.96 (2.98–5.05)	0.81 (0.40–1.62)						
14	426	3.94 (2.95–5.05)	0.83 (0.40–1.51)	129	1.27 (0.88–1.79)	129	2.47 (1.48–3.37)	127	0.28 (0–0.65)
15	508	3.86 (2.88–5.02)	0.84 (0.38–1.51)						
16	639	3.86 (2.93–4.97)	0.87 (0.40–1.65)	160	1.18 (0.78–1.61)	160	2.41 (1.66–3.34)	160	0.35 (0.05–0.70)
17	417	3.89 (2.85–5.28)	0.89 (0.44–1.81)						
18	238	3.94 (2.93–5.15)	0.92 (0.44–1.75)	67	1.13 (0.83–1.55)	66	2.57 (1.66–3.34)	66	0.36 (0.05–0.65)
19	173	3.99 (2.95–5.08)	0.96 (0.50–1.94)						
20–24	882	4.31 (3.21–5.65)	1.13 (0.50–2.27)	118	1.18 (0.78–1.63)	118	2.68 (1.71–3.81)	118	0.36 (0.03–0.73)
25–29	2042	4.72 (3.44–6.32)	1.31 (0.52–2.81)	253	1.16 (0.80–1.63)	253	3.02 (1.81–4.27)	253	0.45 (0.08–0.93)
30–34	2444	4.98 (3.57–6.58)	1.45 (0.57–3.01)	403	1.18 (0.73–1.63)	403	3.27 (2.02–4.79)	403	0.55 (0.13–1.24)
35–39	2320	5.21 (3.78–6.99)	1.64 (0.61–3.62)	371	1.12 (0.75–1.61)	371	3.45 (2.09–4.90)	372	0.62 (0.08–1.45)
40–44	2428	5.35 (3.91–6.94)	1.71 (0.62–3.62)	383	1.15 (0.70–1.74)	385	3.51 (2.25–4.82)	384	0.66 (0.13–1.45)
45–49	2296	5.50 (4.09–7.15)	1.71 (0.66–3.70)	325	1.18 (0.78–1.66)	325	3.73 (2.54–5.23)	326	0.63 (0.13–1.32)
50–54	2138	5.51 (4.09–7.17)	1.72 (0.66–3.62)	340	1.14 (0.73–1.63)	340	3.69 (2.31–5.10)	340	0.69 (0.21–1.61)
55–59	1621	5.54 (4.04–7.15)	1.60 (0.66–3.23)	261	1.23 (0.73–1.84)	261	3.78 (2.28–5.26)	261	0.56 (0.08–1.27)
60–64	905	5.52 (4.12–7.15)	1.61 (0.66–3.29)	131	1.33 (0.78–1.92)	131	3.79 (2.15–5.44)	131	0.49 (0.08–1.14)
65–69	750	5.51 (4.09–7.10)	1.54 (0.64–3.02)	105	1.32 (0.78–2.02)	105	3.90 (2.54–5.44)	105	0.51 (0–1.17)
70+	850	5.36 (3.91–6.99)	1.47 (0.66–2.92)	119	1.31 (0.80–1.94)	119	3.70 (2.28–4.82)	119	0.44 (0–0.98)

TABLE II.B PLASMA TOTAL AND LIPOPROTEIN CHOLESTEROL AND TRIGLYCERIDE MEAN AND PERCENTILE VALUES (mmol/L) FOR WHITE FEMALES NOT TAKING HORMONES, BY AGE

				Cholesterol in					
				HDL (mmol/L)		LDL (mmol/L)		VLDL (mmol/L)	
Age in Years	N	Cholesterol (mmol/L) X (5–95 % tile)	Triglycerides (mmol/L) X (5–95 % tile)	N	X (5–95 % tile)	N	X (5–95 % tile)	N	X (5–95 % tile)
0	(6)	3.83	0.93						
1	(31)	4.04	0.79						
2	(33)	3.94	0.74						
3	57	4.14 (3.06–5.21)	0.71 (0.36–1.24)						
4	59	4.02 (2.80–5.08)	0.67 (0.38–1.06)						
5	86	4.20 (3.26–5.05)	0.63 (0.32–1.05)						
6	253	4.27 (3.34–5.31)	0.64 (0.36–1.07)	58	1.30	53	2.60	50	0.25
7	220	4.14 (3.21–5.23)	0.69 (0.36–1.27)		(0.73–1.76)		(1.71–3.21)		(0.03–0.60)
8	297	4.30 (3.34–5.41)	0.69 (0.38–1.18)	60	1.44	55	2.59	54	0.28
9	261	4.22 (3.11–5.31)	0.71 (0.36–1.24)		(0.96–1.94)		(1.74–3.68)		(0.03–0.65)
10	417	4.22 (3.29–5.31)	0.75 (0.40–1.31)	100	1.33	99	2.54	99	0.31
11	338	4.17 (3.29–5.31)	0.83 (0.41–1.42)		(0.88–1.86)		(1.81–3.63)		(0.05–0.60)
12	536	4.17 (3.21–5.26)	0.89 (0.43–1.49)	102	1.37	101	2.53	101	0.28
13	344	4.02 (3.16–4.97)	0.89 (0.49–1.54)		(0.96–1.79)		(1.76–3.44)		(0.03–0.60)
14	441	4.07 (3.11–5.13)	0.89 (0.43–1.56)	122	1.32	120	2.42	120	0.28
15	553	4.02 (3.03–5.15)	0.84 (0.44–1.44)		(0.88–1.86)		(1.55–3.34)		(0.05–0.60)
16	675	4.02 (3.13–5.10)	0.80 (0.42–1.36)	165	1.37	165	2.47	164	0.32
17	351	4.09 (3.03–5.23)	0.79 (0.45–1.28)		(0.91–1.99)		(1.48–3.57)		(0.08–0.67)
18	193	4.07 (3.00–5.15)	0.85 (0.44–1.42)	53	1.38	53	2.64	53	0.29
19	137	4.20 (3.19–5.49)	0.89 (0.45–1.53)		(0.96–1.92)		(1.50–3.70)		(0–0.62)
20–24	778	4.25 (3.16–5.59)	0.82 (0.41–1.48)	96	1.35	96	2.54	96	0.31
25–29	1329	4.42 (3.32–5.75)	0.85 (0.42–1.64)	181	1.45 (0.96–2.10)	181	2.75 (1.81–3.91)	181	0.31 (0.05–0.62)
30–34	1569	4.54 (3.37–5.98)	0.89 (1.44–1.71)	233	1.43 (0.98–1.94)	232	2.82 (1.74–3.89)	233	0.28 (0–0.65)
35–39	1606	4.78 (3.63–6.27)	0.98 (0.45–1.99)	241	1.42 (0.88–2.12)	242	3.08 (1.97–4.46)	240	0.37 (0.03–0.91)
40–44	1583	5.02 (3.81–6.53)	1.11 (0.51–2.16)	244	1.48 (0.86–2.25)	244	3.24 (1.99–4.51)	244	0.36 (0.08–0.75)
45–49	1515	5.25 (3.94–6.86)	1.18 (0.52–2.42)	249	1.49 (0.86–2.23)	247	3.36 (2.07–4.84)	247	0.44 (0.05–0.98)
50–54	1257	5.64 (4.20–7.38)	1.30 (0.59–2.63)	177	1.56 (0.96–2.31)	177	3.78 (2.33–5.57)	177	0.42 (0–0.96)
55–59	1112	5.97 (4.43–7.77)	1.41 (0.62–2.96)	172	1.53 (0.93–2.23)	172	3.93 (2.46–5.52)	172	0.55 (0.05–1.32)
60–64	723	5.98 (4.46–7.69)	1.44 (0.63–2.70)	111	1.61 (0.93–2.36)	111	4.05 (2.59–6.06)	111	0.46 (0.03–1.17)
65–69	593	6.03 (4.43–7.85)	1.48 (0.68–2.75)	101	1.57 (0.88–2.31)	101	4.19 (2.51–5.78)	101	0.46 (0–1.04)
70+	748	5.91 (4.33–7.49)	1.50 (0.68–2.68)	127	1.56 (0.86–2.36)	127	3.86 (2.49–5.36)	127	0.42 (0–1.35)

TABLE III.B PLASMA TOTAL AND LIPOPROTEIN CHOLESTEROL AND TRIGLYCERIDE MEAN AND PERCENTILE VALUES (mmol/L) FOR WHITE FEMALES TAKING HORMONES, BY AGE

Age in Years	N	Cholesterol (mmol/L) X (5–95 % tile)	Triglycerides (mmol/L) X (5–95 % tile)	Cholesterol in HDL (mmol/L) N (5–95	HDL X % tile)	LDL (mmol/L) N (5–95	LDL X % tile)	VLDL (mmol/L) N (5–95	VLDL X % tile)
15–19	167	4.38 (3.13–5.98)	1.20 (0.55–2.26)	19	-----	19	-----	20	-----
20–24	788	4.64 (3.39–6.11)	1.19 (0.62–1.99)	101	1.42 (0.88–2.05)	101	2.81 (1.61–4.22)	101	0.39 (0.03–0.78)
25–29	855	4.76 (3.65–6.11)	1.24 (0.64–2.19)	132	1.45 (0.91–2.15)	132	3.00 (1.86–4.38)	132	0.39 (0.05–0.93)
30–34	579	4.87 (3.60–6.37)	1.31 (0.65–2.33)	97	1.50	95	3.04	95	0.40
35–39	406	5.04 (3.78–6.45)	1.43 (0.63–2.72)	50	1.48	50	3.08	50	0.53
40–44	466	5.16 (3.94–6.68)	1.46 (0.65–2.72)	63	1.58	63	3.23	63	0.45
45–49	627	5.42 (4.01–7.15)	1.47 (0.60–2.94)	71	1.70	71	3.30	72	0.49
50–54	729	5.65 (4.32–7.23)	1.47 (0.64–2.80)	70	1.75	70	3.10	70	0.48
55–59	577	5.65 (4.25–7.33)	1.44 (0.70–2.69)	60	1.83	60	3.43	60	0.50
60–64	341	5.81 (4.46–7.38)	1.43 (0.64–2.71)	32	1.82	32	3.54	32	0.35
65–69	224	5.76 (4.48–7.23)	1.47 (0.68–2.64)	25	1.97	25	3.27	25	0.40
70+	132	5.59 (4.14–7.10)	1.37 (0.68–2.44)	10	-----	10	-----	10	-----

TABLE IV.B PLASMA TOTAL CHOLESTEROL AND TRIGLYCERIDE MEAN AND PERCENTILE VALUES (mmol/L) FOR BLACK MALES AND FEMALES, BY AGE

	MALES			FEMALES		
Age in Years	N	Cholesterol (mmol/L) X (5–95 % tile)	Triglycerides (mmol/L) X (5–95 % tile)	N	Cholesterol (mmol/L) X (5–95 % tile)	Triglycerides (mmol/L) X (5–95 % tile)
0	(0)			(2)	3.21	0.97
1	(2)	3.60	0.55	(6)	3.99	0.59
2	(4)	3.99	0.59	(5)	4.14	0.49
3	(8)	3.86	0.52	(7)	4.33	0.58
4	(3)	4.25	0.72	(4)	3.81	0.77
5	(8)	4.30	0.51	(8)	4.74	0.58
6	97	4.25 (3.32–5.52)	0.60 (0.37–1.06)	102	4.46 (3.57–5.59)	0.64 (0.33–1.22)
7	52	4.33 (3.00–5.39)	0.56 (0.37–0.85)	60	4.40 (3.26–5.59)	0.59 (0.31–0.93)
8	103	4.33 (3.34–5.65)	0.61 (0.32–1.08)	100	4.46 (3.21–5.85)	0.61 (0.35–1.02)
9	74	4.35 (3.50–5.18)	0.60 (0.37–0.92)	68	4.56 (3.44–6.09)	0.68 (0.37–1.13)
10	118	4.32 (3.44–4.95)	0.60 (0.33–1.07)	131	4.38 (3.32–5.47)	0.68 (0.41–1.09)
11	68	4.32 (3.44–4.95)	0.59 (0.34–0.92)	65	4.35 (3.39–5.52)	0.69 (0.37–1.04)
12	120	4.43 (3.29–5.59)	0.71 (0.36–1.13)	128	4.17 (3.13–5.26)	0.77 (0.42–1.23)
13	88	4.12 (3.21–5.05)	0.65 (0.37–1.02)	70	4.33 (3.11–5.67)	0.79 (0.42–1.27)
14	91	4.07 (2.98–5.05)	0.65 (0.36–1.10)	80	4.17 (3.11–5.67)	0.75 (0.43–1.24)
15	138	3.99 (2.85–5.18)	0.66 (0.32–1.09)	150	4.30 (3.34–5.47)	0.72 (0.38–1.37)
16	200	4.07 (3.11–5.15)	0.64 (0.35–1.15)	192	4.20 (3.11–5.34)	0.67 (0.36–1.07)
17	126	4.04 (3.06–5.39)	0.74 (0.36–1.38)	83	4.30 (3.21–5.39)	0.70 (0.41–1.21)
18	(38)	4.22	0.74	(23)	4.33	0.75
19	(12)	4.14	0.68	(6)	4.20	0.75
20–29	97	4.62	0.92	129	4.61 (3.24–6.06)	0.81 (0.41–1.44)
30–39	178	4.96 (3.57–6.55)	1.21 (0.48–2.53)	291	4.83 (3.42–6.29)	0.86 (0.43–1.58)
40–49	167	5.36 (3.83–6.92)	1.42 (0.59–3.32)	188	6.23 (3.76–6.92)	1.08 (0.49–1.98)
50–59	70	5.35	1.61	80	5.58	1.15
60+	46	5.72	1.23	48	6.05	1.30